Second Edition

A First Course in Mathematical Modeling

Frank R. Giordano
Carroll College

Maurice D. Weir
U.S. Naval Postgraduate School

William P. Fox
U.S. Military Academy, West Point

Brooks/Cole Publishing Company

I(T)P® *An International Thomson Publishing Company*

Pacific Grove • Albany • Belmont • Bonn • Boston • Cincinnati • Detroit • Johannesburg • London
Madrid • Melbourne • Mexico City • New York • Paris • Singapore • Tokyo • Toronto • Washington

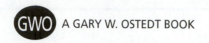 **GWO** A GARY W. OSTEDT BOOK

Publisher: *Gary W. Ostedt*
Marketing Team: *Jill Downey and Romy Taormina*
Editorial Associate: *Carol Benedict*
Production Editor: *Marlene Thom*
Manuscript Editor: *Harriet Screnkin*
Permissions Editor: *Carline Haga*
Interior Design: *Vernon T. Boes*

Cover Design: *Larry Molmud*
Art Editor: *Lisa Torri*
Interior Illustration: *Lori Heckelman*
Photo Editor: *Kathleen Olson*
Typesetting: *Integre Technical Publishing Co., Inc.*
Cover Printing: *Phoenix Color Corporation, Inc.*
Printing and Binding: *Quebecor Fairfield*

For more information, contact:

BROOKS/COLE PUBLISHING COMPANY
511 Forest Lodge Road
Pacific Grove, CA 93950
USA

International Thomson Publishing Europe
Berkshire House 168–173
High Holborn
London WC1V 7AA
England

Thomas Nelson Australia
102 Dodds Street
South Melbourne, 3205
Victoria, Australia

Nelson Canada
1120 Birchmount Road
Scarborough, Ontario
Canada M1K 5G4

International Thomson Editores
Seneca 53
Col. Polanco
11560 México, D.F., México

International Thomson Publishing GmbH
Königswinterer Strasse 418
53227 Bonn
Germany

International Thomson Publishing Asia
221 Henderson Road
#05–10 Henderson Building
Singapore 0315

International Thomson Publishing Japan
Hirakawacho Kyowa Building, 3F
2-2-1 Hirakawacho
Chiyoda-ku, Tokyo 102
Japan

Printed in the United States of America

10   9   8   7   6   5   4   3   2

Library of Congress Cataloging in Publication Data
Giordano, Frank R.
     A first course in mathematical modeling. — 2nd ed. / Frank R.
Giordano, Maurice D. Weir, William Fox.
        p.   cm.
     Includes index.
     ISBN 0-534-22248-X (hc)
     1. Mathematical models.   I. Weir, Maurice D.   II. Fox, William
P., [date]–   .   III. Title.
QA401.G55   1997
511'.8—dc21
                                                        97–76
                                                        CIP

Credits continue on p. 526.

Dedicated in loving memory of

RONALD SAMUEL GIORDANO, Ph.D., 1950–1996

PROFESSOR OF CHEMISTRY
WESTERN CONNECTICUT STATE UNIVERSITY
Educator, Scholar, Friend to All

The greatest good you can do for your students
  is to reveal to them their gifts,
Rather than showing off your own.
 DISRAELI

A teacher affects eternity,
He knows not where his influence stops.
 ADAMS

Contents

Chapter Ten MODELING WITH A DIFFERENTIAL EQUATION   345

Chapter Eleven MODELING WITH SYSTEMS OF DIFFERENTIAL EQUATIONS   381

Chapter Twelve PROBABILISTIC MODELING   420

Preface

The Mathematical Association of America's Committee on the Undergraduate Program in Mathematics (CUPM) has long recommended that "Students should have an opportunity to undertake 'real world' mathematical modeling projects, either as term projects in an operations research course, as independent study, or as an internship in industry."[1] That report goes on to add that a *modeling experience* should be included within the common core of all mathematical sciences majors. Further, this experience in modeling should start early: "... to begin the modeling experience as early as possible in the student's career and reinforce modeling over the entire period of study."[2]

To facilitate an early initiation of the modeling experience, the first edition of this text was designed to be taught concurrently or immediately after an introductory business or engineering calculus course. In this edition, we have added chapters treating discrete dynamical systems, linear programming and numerical search methods, and an introduction to probabilistic modeling. Additionally, we have expanded our introduction of simulation. The additional topics in discrete mathematics now make it possible to organize an entire course that does not require calculus. Chapters that require a concurrent introductory business or engineering calculus course are Chapter 8, *Continuous Optimization Models*; Chapter 10, *Modeling with a Differential Equation*; and Chapter 11, *Modeling with Systems of Differential Equations*. We have organized the text so that the course may begin in the first semester of freshman year. Chapter 1, *Graphs of Functions as Models*, requires only the notion of how the signs of the derivatives determine the shape of the graph of a function. The chapter can be skipped if desired. Chapter 5 uses calculus to derive the normal equations for the least-squares criterion of "best fit." The proof may be skipped if desired. Otherwise, the first seven chapters of the text do not require prior or concurrent experience in calculus. We now describe how our book is designed for a *first course* in modeling.

---

[1] Mathematical Association of America, Committee on the Undergraduate Program in Mathematics, *Recommendations for a General Mathematical Sciences Program* (Washington, DC: Mathematical Association of America, 1981), p. 13.
[2] *ibid.*, p. 77.

Goals and Orientation

The course is a bridge between the study of mathematics and the applications of mathematics to various fields, and it is a transition to the significant modeling experiences recommended by the CUPM. The course affords the student an early opportunity to see how the pieces of an applied problem fit together. The student investigates meaningful and practical problems chosen from common experiences encompassing many academic disciplines, including the mathematical sciences, operations research, engineering, and the management and life sciences.

This text provides an introduction to the entire modeling process. The student will have occasions to practice the following facets of modeling:

1. *Creative and Empirical Model Construction:* Given a real-world scenario, the student learns to identify a problem, make assumptions and collect data, propose a model, test the assumptions, refine the model as necessary, fit the model to data if appropriate, and analyze the underlying mathematical structure of the model to appraise the sensitivity of the conclusions when the assumptions are not precisely met.

2. *Model Analysis:* Given a model, the student learns to work backward to uncover the implicit underlying assumptions, assess critically how well those assumptions fit the scenario at hand, and estimate the sensitivity of the conclusions when the assumptions are not precisely met.

3. *Model Research:* The student investigates a specific area to gain a deeper understanding of some behavior and learns to use what has already been created or discovered.

We have designed the text to enhance a student's problem-solving capabilities. For purposes of discussion we identify the following steps of the problem-solving process:

1. Problem identification
2. Model construction or selection
3. Identification and collection of data
4. Model validation
5. Calculation of solutions to the model
6. Model implementation and maintenance

In many instances the undergraduate mathematical experience consists almost entirely of doing step 5: calculating solutions to models that are given. There is relatively little experience with "word problems," and what there is deals mainly with problems that are short (to accommodate a full syllabus) and often contrived. Such problems require the student to apply the mathematical technique currently being studied, from which solution to the model is calculated with great precision. For lack of experience, consequently, students often feel anxious when presented with a scenario for which the model is *not* given or for which there is no unique solution, and then are told to identify a problem and construct a model addressing the problem "reasonably well."

With this in mind, we feel that in an introductory modeling course students should spend a significant amount of time on the first several steps of the process just described—learning how to identify problems, construct or select models, and figure out what data need to be collected—progressing from relatively easy scenarios to more difficult ones. It is probably unreasonable to expect an average student to excel in a semester-long project on the first attempt. It takes time and experience to develop skill and confidence in the modeling process. We have found that involving students in the mathematical modeling process as early as possible, beginning with short projects, facilitates their progressive development and confidence in mathematics and modeling.

Many modeling texts present "type models" such as various inventory models for determining optimal inventory strategies. Students then learn to select an appropriate model for a particular situation. This approach has merit, and model selection is a valid step in the problem-solving process. However, undergraduate students often do not comprehend the assumptions behind a model, and only rarely do they take into consideration the appropriateness and sensitivity of those assumptions. Therefore, we emphasize *model construction* to promote student creativity and to demonstrate the artistic nature of model building, including the ideas of experimentation and simulation. Although we do discuss fitting data to chosen model types, our concentration is still on the entire model-building process, leaving the study of type models for more advanced courses.

Student Background and Course Content

Because our desire is to initiate the modeling experience as early as possible in the student's program, the only prerequisite for Chapters 8, 10, and 11 is a basic understanding of single-variable differential and integral calculus. Although some unfamiliar mathematical ideas are taught as part of the modeling process, the emphasis is on using mathematics already known by the students after completing high school. The modeling course will then motivate students to study the more advanced courses such as linear algebra, differential equations, optimization and linear programming, numerical analysis, probability, and statistics. The power and utility of these subjects are intimated throughout the text.

Although there are strong arguments to include such courses as advanced calculus, linear algebra, differential equations, probability, and statistics as prerequisites to an introductory modeling course, such a requirement necessitates postponing the course until the junior or senior undergraduate year, delaying the student's exposure to real-world applications. It also cuts off a number of student beneficiaries (namely, those non-mathematics majors who cannot satisfy all the prerequisites). Though our philosophy differs somewhat, this text still serves the more advanced student who has taken more mathematics courses. Certain sections of the text can be covered more rapidly by the advanced student, allowing more time for deeper extensions of the material as suggested by the projects for each chapter.

Further, the scenarios and problems in the text are not designed for the application of a particular mathematical technique. Instead, they demand thoughtful ingenuity in using fundamental concepts to find reasonable solutions to "open-ended" problems. Certain mathematical techniques (such as dimensional analysis, curve fitting, and Monte Carlo simulation) are presented because often they are not formally covered at the undergraduate level. Instructors should find great flexibility in adapting the text to meet the particular needs of students through the problem assignments and student projects. We have used this material to teach courses to both undergraduate and graduate students, and even as a basis for faculty seminars.

Organization of the Text

The organization of the text is best understood with the aid of Figure 1. The first five chapters are directed toward creative model construction and an overview of the entire modeling process. We begin with the construction of graphical models, which provides us with some concrete models to support our discussion of the modeling process in Chapter 2. This approach also provides a transition into model construction by first involving the student in model analysis. Next we classify models and analyze the modeling process. At this point students can really begin to analyze scenarios, identify problems, and determine the underlying assumptions and principal variables of interest in a problem. This work is preliminary to the models they will create beginning in Chapter 3. (The order of Chapters 1 and 2 may be reversed, although we have found the current order best for capturing student interest and reducing student anxiety.) In their first modeling experience, students are quite anxious about their "creative" abilities and how they are going to be evaluated. For these reasons we have found it advantageous to start them out on familiar ground by appealing to their understanding of graphs of functions and having them learn model analysis. The book blends mathematical modeling techniques with the more creative aspects of modeling for variety and confidence building, and gradually the transition is made to the more difficult creative aspects.

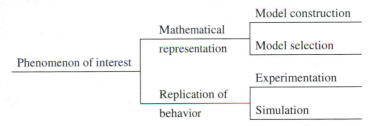

FIGURE 1   The organization of the text follows the above classification of the various models

In Chapter 3 we present the modeling of "change" using difference equations. Students are provided data from situations they have encountered in daily life. They are given the opportunity to construct and test models for various situations, beginning with scenarios that can be modeled exactly, before approximating more difficult scenarios. In Chapter 4 we present the concepts of proportionality and geometric similarity and use them to construct mathematical models for some of the previously identified scenarios. Students formulate tentative models or submodels and begin to learn how to test the appropriateness of the assumptions. In Chapter 5 model fitting is discussed, and in the process several optimization models are developed and are revisited in Chapters 8 and 9.

Chapters 6 and 7 are dedicated to empirical model construction. Chapter 6 begins with fitting simple one-term models to collected sets of data and progresses to more sophisticated interpolating models, including polynomial smoothing models and cubic splines. Simulation models are discussed in Chapter 7. An empirical model is fit to some collected data, and then Monte Carlo simulation is used to duplicate the behavior being investigated. The presentation motivates the eventual study of probability and statistics.

Chapters 8 and 9 are devoted to the study of optimization. At this point in the course students have encountered many situations that ask them to find the best solution. We begin Chapter 8 by allowing students to formulate optimization models of various types. We then classify the models by their mathematical structures. In Chapter 8 students get the opportunity to solve continuous optimization problems requiring only the application of elementary calculus. An introduction to constrained optimization problems is provided as well. In Chapter 9, linear programming and several numerical search methods are presented. The treatment of linear programming includes analytical as well as graphical methodologies. An introduction to the important topic of sensitivity analysis is provided. The chapter concludes with an introduction to numerical search methods including the dichotomous and golden section methods.

In Chapters 10 and 11 dynamic (time varying) scenarios are treated. These chapters build on the discrete analysis presented in Chapter 3 by now considering situations where time is varying continuously. We begin by modeling initial value problems in Chapter 10 and progress to interactive systems in Chapter 11, with students performing a graphical stability analysis. Students with a good background in differential equations can pursue analytical and numerical stability analyses as well, or they can investigate the use of difference equations or numerical solutions to differential equations by completing the projects.

Chapter 12 provides an introduction to probabilistic modeling. The topics of Markov processes, reliability, and linear regression are introduced, building on scenarios and analysis presented previously in the course. The text concludes with Chapter 13, which is devoted to dimensional analysis, a topic of great importance in the physical sciences and engineering because it represents a means of significantly reducing the experimental effort required when constructing models based on data collection. We also include an introduction to the construction of models of similitude.

The text is arranged in the order we prefer in teaching our modeling course. However, the order of presentation may be varied to fit the needs of a particular instructor or group of students. Figure 2 shows how the various chapters are interdependent or independent, allowing progression through the chapters without loss of continuity.

Student Projects

Student projects are an essential part of any modeling course. This text includes projects in creative and empirical model construction, model analysis, and model research. Thus we recommend a course consisting of a mixture of projects in all three

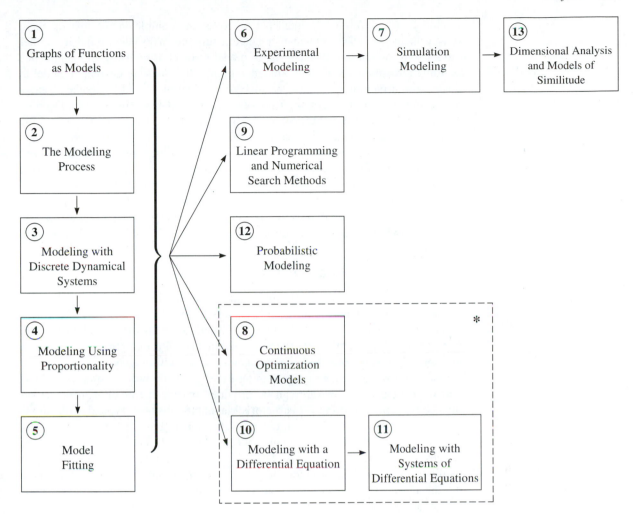

*Chapters 8, 10, and 11 require single-variable calculus as a corequisite.

FIGURE 2 Chapter organization and progression

facets of modeling. These projects are most instructive if they address scenarios that have no unique solution. Some projects should include *real* data that students are either given or can *readily* collect. A combination of individual and group projects can also be valuable. Individual projects are appropriate in those parts of the course in which the instructor wishes to emphasize the development of individual modeling skills. However, the inclusion of a group project early in the course gives students the exhilaration of a "brainstorming" session. A variety of projects is suggested in the text, such as constructing models for various scenarios, completing UMAP[3]

[3]UMAP modules are developed and distributed through COMAP, Inc., 57 Bedford Street, Suite 210, Lexington, MA 02173.

modules, or researching a model presented as an example in the text or class. It is valuable for each student to receive a mixture of projects requiring either model construction, model analysis, or model research for variety and confidence building throughout the course. Students might also choose to develop a model in a scenario of particular interest, or analyze a model presented in another course. We recommend six to eight short projects in a typical modeling course. Detailed suggestions on how student projects can be assigned and used are included in the Instructor's Manual that accompany this text.

In terms of the number of scenarios covered throughout the course, as well as the number of homework problems and projects assigned, we have found it better to pursue a few that are developed carefully and completely. Two or three good problems are about the maximum that an average student can handle in one week. We have provided many more problems and projects than can reasonably be assigned to allow for a wide selection covering many different application areas.

The Role of Computation

Although computing capability is not a requirement in using this book, computation does play an important role in several chapters, especially Chapters 3 and 9 where a spreadsheet is beneficial, and Chapter 7 where a means of generating random numbers is essential. We have found a combination of programmable calculators and microcomputers (or graphing calculators) to be advantageous throughout the course. Students who have programming experience can write computer code as part of a project, or software can be provided by the instructor as needed. We include some programs in the Instructor's Manual for this text. Typical applications for which students will find computers useful are in graphical displays of data, transforming data, least-squares curve fitting, divided difference tables and cubic splines, programming simulation models, linear programming and numerical search methods, and numerical solutions to differential equations. The use of computers has the added advantage of getting students to think early about numerical methods and strategies, and it provides insight into how "real-world" problems are tackled in business and industry. Students appreciate being provided with or developing software that can be taken with them after completion of the course.

Resource Materials

We have found material provided by the Consortium for Mathematics and Its Application (COMAP) to be outstanding and particularly well suited to the course we propose. COMAP was started from a National Science Foundation grant and has as its goal the production of instructional materials to introduce applications of mathematics into the undergraduate curriculum.

Individual modules for the undergraduate classroom, UMAP Modules, may be used in a variety of ways. First, they may be used as instructional material to support several lessons. In this mode a student completes the self-study module by working through its exercises (the detailed solutions provided with the module can be conveniently removed before it is issued). Another option is to put together a block of instruction using one or more UMAP modules suggested in the projects sections of the text. The modules also provide excellent sources for "model research," because they cover a wide variety of applications of mathematics in many fields. In this mode, a student is given an appropriate module to research and is asked to complete and report on the module. Finally, the modules are excellent resources for scenarios for which students can practice model construction. In this mode the instructor writes a scenario for a student project based on an application addressed in a particular module and uses the module as background material, perhaps having the student complete the module at a later date.

Projects with an interdisciplinary perspective (ILAPs) are being developed by a consortium led by West Point and distributed by COMAP under a grant (from 1996 to 2000) from the National Science Foundation. The projects are designed by interdisciplinary teams of faculty and include both individual and group work. Information on the availability of interdisciplinary projects can be obtained by writing COMAP at the address given previously, calling COMAP at 1-800-772-6627, or electronically: order@comap.com

The Mathematical Contest in Modeling

The first Mathematical Contest in Modeling (MCM) was held in 1985. Founded by Ben Fusaro, its purpose was to awaken interest in mathematical modeling. Each year two problems, one continuous and one discrete, are posed. On a designated weekend in February, teams may work from Friday morning until Monday afternoon and are free to consult any inanimate resource. Faculty advisors prepare the teams for the competition and ensure that the rules are complied with, but they are not consulted after the contest begins. Teams are encouraged to use any technology. Teams submit a complete solution with an abstract to the problem of their choice. Judges classify the papers into four categories: Outstanding, Meritorious, Honorable Mention, and Successful Participant. Outstanding teams are invited to present their papers at meetings of professional societies supporting the contest. Winning solutions are published in the Fall edition of the UMAP Journal. The contest problems for the years 1985–1996 are presented in Appendix A. A special edition of the UMAP journal, *MCM the First 10 Years*, contains winning solutions and hints for the faculty advisor, and is available from COMAP.

In 1996, 393 teams from 225 schools competed in the contest. The contest is sponsored by COMAP with funding support from the National Security Agency, the Society of Industrial and Applied Mathematics, the Institute for Operations Research and the Management Sciences, and the Mathematical Association of America. Additional information concerning the contest can be obtained by contacting COMAP.

Acknowledgments

It is always a pleasure to acknowledge individuals who have played a role in the development of a book. Several colleagues were especially helpful to us. We are particularly grateful to B.G. (retired) Jack M. Pollin and Dr. Carroll Wilde for stimulating our interest in teaching modeling and for support and guidance in our careers. We're indebted to many colleagues for reading the first edition manuscript and suggesting modifications and problems: Rickey Kolb, John Kenelly, Robert Schmidt, Stan Leja, Bard Mansager, and especially Steve Maddox and Jim McNulty. We thank the following individuals who reviewed preliminary versions of the manuscript for the first edition: Gilbert G. Walter, University of Wisconsin; Peter Salamon, San Diego State University; Peter A. Morris, Applied Decision Analysis, Inc.; Christopher Hee, Eastern Michigan University, Eugene Spiegel, University of Connecticut; David Ellis, San Francisco State University; David Sandell, U.S. Coast Guard Academy, Jeffrey Arthur, Oregon State University; Courtney Coleman, Harvey Mudd College; Don Snow, Brigham Young University.

We thank Marie Vanisko of Carroll College for her careful reading of the second edition and for her insightful suggestions. In addition, we thank the following reviewers for their helpful comments: Prem Bajaj, Wichita State University; Gary Grefsrud, Fort Lewis College; Kit Lumley, Columbus College; Peter E. Moore, Northern Kentucky University; and Robert Wheeler, Northern Illinois University–DeKalb.

We are indebted to a number of individuals who authored or co-authored UMAP materials that support the text: David Cameron, Brindell Horelick, Michael Jaye, Sinan Koont, Stan Leja, Michael Wells, and Carroll Wilde. In addition, we thank Solomon Garfunkel and the entire COMAP staff for their cooperation on this project. We acknowledge the National Science Foundation and the Mathematical Association of America for their support of modeling courses.

The production of any mathematics text is a complex process and we have been especially fortunate in having a superb and creative production staff at Brooks/Cole. In particular, we express our thanks to Craig Barth, our friend and editor for the first edition. For the second edition, we are especially grateful to Gary W. Ostedt, our editor; Carol Benedict, his assistant; Jill Downey, our marketing manager; Marlene Thom, our production editor; Vernon Boes, our designer; Lisa Torri, our art editor; and Kathleen Olson, our photo editor.

Finally, we are grateful to our wives—Judi Giordano, Gale Weir, and Wendy Fox—for their support and understanding. We additionally thank Gale for preliminary editing of the first edition and Judi for processing a large portion of the manuscript, and preliminary editing of both editions.

Frank R. Giordano
Maurice D. Weir
William P. Fox

Chapter One

GRAPHS OF FUNCTIONS AS MODELS

INTRODUCTION

Quite often we are interested in analyzing complex situations to predict qualitatively the effect of some course of action. One example is determining in a two-country nuclear arms race that is in a state of relative equilibrium what the effect will be on the number of nuclear missiles possessed by each side if one of the countries introduces mobile launching pads. In view of the many factors affecting the behavior of the parties involved, the assumptions of our model will necessarily be rather crude. Initially, we will be satisfied with a model that simply captures the general trend of the nuclear arms situation. Another example exists in the economics of the baseball sport industry. What is the effect of a baseball strike if it is successful at raising players' salaries? Will the salary increase actually reduce the demand of the fans for baseball? In cases like these, graphical models are useful in helping us understand the situation.

In mathematical modeling we often attempt to construct a mathematical function relating variables to serve as a model. In subsequent chapters we attempt to determine a more precise description of the function, but in this chapter we use the graph of the function to gain a qualitative understanding of the behavior under investigation. A graphical model has the important advantage of appealing to our visual intuition. It gives us a picture and a feel for what is happening that often eludes us in more symbolic analyses. Graphs are good for gaining qualitative information. On the other hand, graphical analysis does limit the number of variables we can study effectively and the precision we can attain. Nevertheless, these limitations are no disadvantage in cases where precision is restricted by the crudeness of the data or where the complexity of the situation confines our expectations of the model. Graphical analysis is also useful as a prelude to a more detailed analytical model, often providing clues on which factors should be considered more thoroughly in a subsequent analysis. We will use graphical models throughout the text to support other types of analyses.

Shape of a Curve

In a graphical model we are primarily interested in the general shapes of the curves representing functional relationships and the curve's intercepts with the coordinate axes. Recall from calculus that a **function** consists of a domain and a rule. The **domain** may be represented by one or several independent variables, and the **rule** assigns to each member of the domain one and only one value. For instance, the rule that assigns to every real number x the value $3x^2 + 1$ is a functional relationship of one independent variable whose graph is a parabola in the Cartesian plane. Likewise, the rule that assigns to every ordered pair of real numbers (x, y) the value $4 - x^2 - y^2$ is a functional relationship of two independent variables whose graph is a parabolic surface in Cartesian space. As the number of independent variables increases, a graphical analysis can become quite complex. To obtain visualizations in two-dimensional or three-dimensional space, it becomes necessary to fix the values of all but one or two of the independent variables, as if taking a snapshot for those particular values. Then new values can be assigned and another snapshot taken. We will discuss some of the ways the number of independent variables can be reduced in Chapters 4 and 13.

Let's review briefly some of the fundamental geometric ideas associated with the graph of a function of a single independent variable, say $y = f(x)$. Several concepts from calculus are useful in determining the shape of the graph of a function. The first derivative gives the slope of the line tangent to the graph at a point, so the sign of the first derivative indicates if the function is **increasing** (positive sign) or **decreasing** (negative sign). The magnitude of the first derivative gives the instantaneous rate of change of the function at the point. For example, the function depicted in Figure 1.1 has a first derivative that is positive and increasing at every point.

Because the first derivative itself is a function, the sign of its derivative (i.e., the sign of the second derivative) indicates whether the first derivative is increasing or decreasing. To interpret the second derivative geometrically, consider the function depicted in Figure 1.2. It has a second derivative that is everywhere positive because the first derivative is steadily increasing from negative to positive values. Notice that

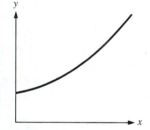

FIGURE 1.1 A graph showing y as an increasing function of x

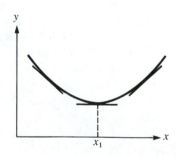

FIGURE 1.2 A steadily increasing first derivative corresponds to a positive second derivative

the tangent line at (x, y) turns continuously in the counterclockwise direction as x advances and that the graph lies everywhere above the tangent line. In this situation we say that the graph is **concave upward**. A graph is **concave downward** if the second derivative is negative. A graph that is concave upward is cupped to hold water; when it is concave downward, it would spill water. Note also in Figure 1.2 that the first derivative at the point x_1 is zero. Points where **relative maxima** and **minima** occur must have a first derivative equal to zero, provided the derivative exists. We will use these basic concepts from calculus to determine the general shapes of the curves in our graphical models.

1.1 AN ARMS RACE

You might ask, Why study the arms race? One reason is that almost all modern wars are preceded by unstable arms races. Strong evidence suggests that an unstable arms race between great powers, characterized by a sharp acceleration in military capability, is an early warning indicator of war. In a 1979 article, Michael Wallace of the University of British Columbia studied 99 international disputes during the 1816–1965 period.[1] He found that disputes preceded by an unstable arms race escalated to war 23 out of 28 times, whereas disputes *not* preceded by an arms race resulted in war only 3 out of 71 times. Wallace calculated an arms race index for the two nations involved in each dispute that correctly predicted war or no war in 91 out of the 99 cases studied. His findings do not mean that an arms race between the powers necessarily results in war nor that there is a causal link between arms races and conflict escalation. They do establish, however, that rapid competitive military growth is strongly associated with the propensity to war. Thus, by studying the arms race, we have the potential for predicting war. And if we can predict war, then there is hope that we can learn to avoid it.

There is another reason for studying the arms race. If the arms race can be approximated by a mathematical model, then it can be understood more concretely. You will see that the answers to such questions as, Will civil defense dampen the arms race? and Will the introduction of mobile missile launching pads help reduce the arms requirements? are not simply matters of political opinion. There is an objective reality to the arms race that the mathematical model intends to capture.

The former Soviet Union and the United States were engaged in a nuclear arms race during the Cold War. At that time political and military strategists asked how the United States should react to changes in numbers and sophistication of the Soviet nuclear arsenal. To answer the difficult question of How many weapons are enough? several factors had to be considered, including American objectives, Soviet objectives, and weapon technology. Former chairman of the Joint Chiefs of Staff

[1] Michael Wallace, "Arms Races and Escalation: Some New Evidence," in *Explaining War*, ed. J. David Singer (Beverly Hills, Calif.: Sage, 1979), pp. 240–252.

General Maxwell D. Taylor suggested the following objectives for the American strategic forces:

> The strategic forces, having the single capability of inflicting massive destruction, should have the single task of deterring the Soviet Union from resorting to any form of strategic warfare. To maximize their deterrent effectiveness they must be able to survive a massive first strike and still be able to destroy sufficient enemy targets to eliminate the Soviet Union as a viable government, society and economy, responsive to the national leaders who determine peace or war.[2]

Note especially that Taylor's strategy assumes the worst possible case: the Soviet's launching a preemptive first attack to destroy America's nuclear force.

How many weapons would be necessary to accomplish the objectives General Taylor suggests? After describing an appropriate system of Soviet targets (generally population and industrial centers), he goes on to say the following:

> The number of weapons we shall need will be those required to destroy the specific targets within this system of which few will be hardened silos calling for the accuracy and short flight time of ICBMs. As a safety factor, we should add extra weapons to compensate for losses that may be suffered in a first strike and for uncertainties in weapon performance. The total weapons requirement should be substantially less than the numbers available to us in our present arsenal.[3]

Thus a minimum number of missiles would be required to destroy specific enemy targets (generally population and industrial centers) chosen to inflict unacceptable damage on the enemy. Additional missiles would be required to compensate for losses incurred in the Soviet's presumed first strike. Implicitly, the number of such additional missiles depends on the size and effectiveness of the Soviet missile forces. Taylor concluded that meeting these objectives would allow for a reduction in America's nuclear arsenal.

In response to a question on expenditures for national defense, Admiral Hyman G. Rickover testified before a congressional committee as follows:

> For example, take the number of nuclear submarines, I'll hit right close to home. I see no reason why we have to have just as many as the Russians do. At a certain point you get where it's sufficient. What's the difference whether we have 100 nuclear submarines or 200? I don't see what difference it makes. You can sink everything on the ocean several times over with the number we have and so can they. That's the point I'm making.[4]

Again Admiral Rickover concluded that a reduction in arms would be possible.

[2]M. D. Taylor, "How to Avoid a New Arms Race," *The Monterey Peninsula Herald*, January 24, 1982, p. 3c.
[3]Ibid.
[4]Hyman C. Rickover, testimony before Joint Economic committee; *The New York Review*, March 18, 1982, p. 13.

We are going to develop a graphical model of the nuclear arms race based on the preceding remarks. The model will help answer the question, How many weapons is enough? Although the model applies to any kind of arms race, for purposes of discussion and illustration we focus on nuclear weapons delivered by long-range intercontinental ballistic missiles (ICBMs).

Developing the Graphical Model

Suppose two countries, Country X and Country Y, are engaged in a nuclear arms race and that *each* country adopts the following strategies.

Friendly strategy   To survive a massive first strike and inflict unacceptable damage on the enemy.

Enemy strategy   To conduct a massive first strike to destroy the friendly missile force.

That is, *each* country follows the friendly strategy when determining its own missile force and presumes the enemy strategy for the opposing country. Note especially that the friendly strategy implies targeting population and industrial centers, whereas the enemy strategy implies targeting missile sites. This was the policy of **nuclear deterrence** advocated during the Cold War.

First, let's define the variables x and y:

$$x = \textit{the number of missiles possessed by Country X}$$
$$y = \textit{the number of missiles possessed by Country Y}$$

Next, let $y = f(x)$ denote the function representing the minimum number of missiles required by Country Y to accomplish its strategies when Country X has x missiles. Similarly, let $x = g(y)$ represent the minimum number of missiles required by Country X to accomplish its objectives.

We begin by investigating the nature of the curve $y = f(x)$. Since a certain number of missiles y_0 is required by Country Y to destroy the selected population and industrial centers of Country X, y_0 is the intercept when $x = 0$. That is, Country Y considers that it needs y_0 missiles even if Country X has none (basically, a psychological defense in the sense that Y fears attack or invasion by X). As Country X increases its missile force, Country Y must add additional missiles because it assumes Country X is following the enemy strategy and targeting its missile force. Let's assume that the weapons technology is such that Country X can destroy no more than one of Country Y's missiles with each missile fired. Then the number of additional missiles Country Y needs for each missile added by Country X depends on the effectiveness of Country X's missiles. Convince yourself that the curve $y = f(x)$ must lie between the limiting lines shown in Figure 1.3. Line A, having slope 0, represents a state of absolute invulnerability of Country Y's missiles to any attack. At the other extreme, Line B, having slope 1, says that Country Y must add one new missile for each missile added by Country X.

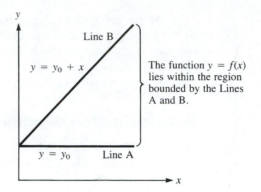

FIGURE 1.3 Bounding the function $y = f(x)$

To determine more precisely the shape of the graph of $y = f(x)$, we will analyze what happens for various cases relating the relative sizes of the two missile forces. To determine the cases, we subdivide the region between Lines A and B into smaller subregions defined by the lines $x = y, x = 2y, x = 3y, \ldots$, and so forth, as shown in Figure 1.4. We then approximate $y = f(x)$ in each of these subregions. Remember that when Country Y determines the number of missiles it needs to deter Country X for the graph of $y = f(x)$, Country Y is presumed to follow the friendly strategy whereas Country X follows the enemy strategy.

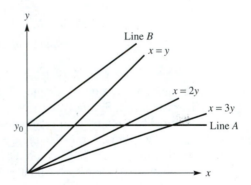

FIGURE 1.4 The subregions between Lines A and B

Case 1 $x < y$ In this situation, if Country X attacks it fires all of its x missiles at the same number of Country Y's missiles. Because the number $(y - x)$ of Country Y's missiles could not be attacked, at least that many would survive. Of the number x of Country Y's missiles that were fired on, a percentage s would survive, where $0 < s < 1$. Thus the total number of missiles surviving the attack is $y - x + sx$. Now Country Y must have y_0 missiles survive to inflict unacceptable damage on Country X. Hence,

$$y_0 = y - x + sx \quad \text{for} \quad 0 < s < 1$$

or solving for y,

$$y = y_0 + (1 - s)x \tag{1.1}$$

Equation (1.1) gives the minimum number of missiles Country Y must have to be confident that y_0 missiles will survive an attack by Country X.

Case 2 $y = x$ In this scenario, in firing all of its missiles Country X fires exactly one of its missiles at each of Country Y's missiles. Assuming the percentage s will survive the attack, the number $sx = sy$ survive, in which case Country Y needs

$$y = \frac{y_0}{s} \tag{1.2}$$

missiles to inflict unacceptable damage on Country X.

Case 3 $y < x < 2y$ Here in firing all its missiles, Country X targets each of Country Y's missiles once and a portion of them twice, as illustrated in Figure 1.5.

Country X:

Country Y:

FIGURE 1.5   An example of $y < x < 2y$

Convince yourself that $x - y$ of Country Y's missiles would be targeted twice and $y - (x - y) = 2y - x$ would be targeted once. Of those targeted once, a percentage $s(2y - x)$ will survive as before. Of those targeted twice, the percentage $s(x - y)$ will survive the first round. Of those that survive the first round, the percentage $s\left[s(x - y)\right] = s^2(x - y)$ will survive the second round. Hence Country Y must have

$$y_0 = s^2(x - y) + s(2y - x)$$

missiles survive, or, solving for y,

$$y = \frac{y_0 + x(s - s^2)}{2s - s^2} \tag{1.3}$$

is the minimum number of missiles required by Country Y.

Case 4 $x = 2y$ Here Country X will fire exactly two missiles at each of Country Y's missiles. If we reason as in Case 2, the number $s^2 y$ survive so that

$$y = \frac{y_0}{s^2} \tag{1.4}$$

is the minimum number of missiles required by Country Y.

Now let's combine all the cases into a single graph. For convenience we are going to assume that the discrete situation just discussed giving the minimum number of missiles can be represented by a continuous model (giving rise to fractions of missiles). First observe that Equations (1.1) and (1.3) both represent straight-line segments: the first segment for $x < y$ and the second segment for $y < x < 2y$. In Case 1, when $x < y$ we obtained the equation

$$y_0 = y - x + sx$$

As x approaches y, this last equation becomes (in the limit) $y_0 = sy$. In Case 3, when $y < x < 2y$ we obtained the equation

$$y_0 = s^2(x - y) + s(2y - x) \tag{1.5}$$

Again, as x approaches y, the equation becomes $y_0 = sy$. Thus the two line segments meet at $x = y$ with the common value $y = y_0/s$. Finally, as x approaches $2y$, Equation (1.5) becomes $y_0 = s^2y$.

These observations mean that the two line segments defined by (1.1) and (1.3) form a continuous curve meeting the lines $y = x$ and $2y = x$. Moreover, the slope $\frac{1-s}{2-s}$ for the line segment represented by (1.3) is less than the slope $1 - s$ of the line segment represented by (1.1), because $2 - s > 1$. Thus the curve is *piecewise linear with decreasing slopes.* The graphical model is depicted in Figure 1.6. Notice how the graph lies within the cone-shaped region between Lines A and B as discussed previously.

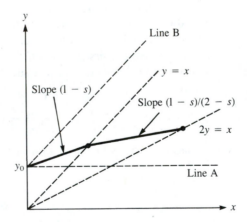

FIGURE 1.6   A graphical model relating the number of missiles for Country Y to the number of missiles for Country X when $0 \le x \le 2y$

We could continue to analyze additional cases, such as what happens when $2y < x < 3y$. Because we are interested only in qualitative information, however, let's see if we can determine the general shape of the curves more simply.

To simplify the analysis, let's replace our piecewise linear approximation by a single continuous smooth curve (one without corners) that passes through each of the points $(0, y_0)$, (x, x), $(x, \frac{x}{2})$, ... shown in Figure 1.6. (These are the points where the piecewise linear approximation crosses the y-axis and the lines $y = x$ and $2y = x$.) We want a curve given by a single equation rather than one represented by a different equation in each subregion. Generalizing from our analysis in Cases 2 and 4, one such curve is given by the following model:

$$y = \frac{y_0}{s^{x/y}} \quad \text{for} \quad 0 < s < 1 \tag{1.6}$$

An inspection of Equation (1.6) reveals that for every ratio x/y we can find y. Thus, the curve $y = f(x)$ crosses each line $x = y, x = 2y, \ldots, x = ny$, as illustrated in Figure 1.7, at the same points as did our piecewise linear approximation.

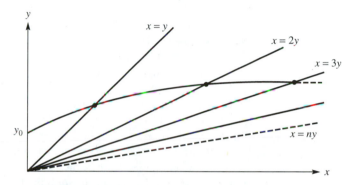

FIGURE 1.7 The curve $y = f(x)$ must cross every line $x = ny$

The situation for Country X is entirely symmetrical. (In determining its curve, Country X is assumed to have the friendly strategy to deter Country Y, and Country Y is presumed to have the enemy strategy.) Its minimal number of missiles is represented by a continuous curve $x = g(y)$ that crosses every line $y = x, y = 2x$, $\ldots, y = nx$. Thus the two curves must intersect.

The preceding discussion leads us to consider two idealized continuously differentiable curves such as those in Figure 1.8. Because the curve $y = f(x)$ represents the minimum number of missiles required by Country Y, the region above the curve represents missile levels satisfactory to Country Y. Likewise, the region to the right of the curve $x = g(y)$ represents missile levels satisfactory to Country X. Thus the darkest region in Figure 1.8 represents missile levels satisfactory to both countries.

The intersection point of the curves $y = f(x)$ and $x = g(y)$ represents the minimum level at which both sides are satisfied. To see that this is so, assume Country Y has y_0 missiles and observes that Country X has x_0 missiles. To meet its objectives, Country Y will have to add sufficient missiles to reach point 1 in Figure 1.8. In turn, Country X will have to add sufficient missiles to reach point 2 in Figure 1.8. This process will continue until both sides are satisfied simultaneously. Notice that any point in the darkest region will suffice to satisfy both countries, and there are many points in the darkest region that are likely to occur. The intersection

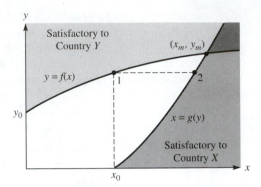

FIGURE 1.8   Regions of satisfaction to Country X and Country Y

point (x_m, y_m) in Figure 1.8 represents the minimum force levels required of both countries to meet their objectives.

We would like to know if the intersection point is unique. Note from Equation (1.6) that as the ratio x/y increases, y must increase; likewise, $x = g(y)$ increases. Because both curves are increasing, it is tempting to conclude that the intersection point is unique. Consider Figure 1.9, however. In the figure both curves are steadily increasing; the curve $y = f(x)$ crosses every line $x = ny$, and $x = g(y)$ crosses every line $y = nx$. Yet the curves have multiple intersection points. How can we ensure a unique intersection point? Notice that the slope of the curve $y = f(x)$ in Figure 1.9 is steadily decreasing until the point $x = x_1$, when it begins to increase. Thus the first derivative changes from a decreasing to an increasing function at $x = x_1$. That is, as x advances the tangent line changes from continuously turning in a clockwise direction to turning in a counterclockwise direction. In other words, the second derivative changes sign. If we can show such a sign change is impossible, then we can conclude that the intersection point is unique. In fact, we will show that the second derivative of $y = f(x)$ is always negative.

Taking the logarithm of Equation (1.6) yields

$$\ln y = \ln y_0 - \frac{x}{y} \ln s$$

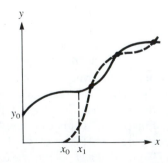

FIGURE 1.9   Steadily increasing curves with multiple intersection points

Multiplying both sides of the equation by y and simplifying gives

$$y \ln y - y \ln y_0 = -x \ln s$$

Differentiating implicitly with respect to x and simplifying yields

$$y'(1 + \ln y - \ln y_0) = -\ln s$$

or

$$y' = \frac{-\ln s}{1 + \ln y - \ln y_0}$$

Differentiating this last equation for the second derivative gives

$$y'' = \frac{-(-\ln s)\frac{1}{y}y'}{(1 + \ln y - \ln y_0)^2}$$

Next we determine the sign of y'. Rewrite y' as

$$y' = \frac{\ln s}{-1 + \ln \frac{y_0}{y}}$$

Because $0 < s < 1$, $\ln s$ is negative.

Now, for the cases we are considering, $y > y_0$, which implies that $\ln(y_0/y) < 0$. Thus $y' > 0$ everywhere, in which case $y'' < 0$ everywhere. Therefore, we can conclude that a unique intersection point does in fact exist. The model has the general shape shown in Figure 1.10.

Graphical Behavior of $y = f(x)$

The ways in which the graph of $y = y_0/s^{x/y}$ behaves depends on three factors: the constant y_0 (which is the minimal number of missiles required by Country Y after a preemptive first strike), the survivability percentage s (which is determined by the

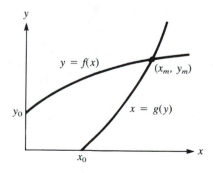

FIGURE 1.10   A graphical model of the nuclear arms race

technology and weapon effectiveness of Country X's missiles, as well as by how securely Country Y's missiles are protected), and the ratio x/y. If y_0 increases, then the curve $y = f(x)$ shifts upward and has a larger slope at each point than before (see Problem 5). If the survivability factor s increases, the curve rotates downward toward the horizontal Line A: $y = y_0$ and has a smaller slope at each point than before.

If the ratio x/y increases, then Country X can target Country Y's missiles more than once, requiring Country Y to need more of them. This results in an increase in the slope and upward rotation of the curve toward the Line B: $y = y_0 + x$.

If you have access to a computer graphing package, you can see these effects by plotting the graph $y = y_0/s^{x/y}$ for various values of the three factors.

Although the curve $x = g(y)$ for Country X displays similar behavior, we note that its constant x_0, survivability, and ratio factors are generally different from the values for Country Y. Let's consider several situations in which we use these ideas and the graphical model to analyze the effects on the intersection point for different political and military strategies likely to be entertained by the two countries.

Model Interpretation

Example 1   Suppose Country X decides to double its annual budget for civil defense. Presumably, Country Y will need more missiles to inflict an unacceptable level of damage on Country X's population centers. Thus y_0 increases. Because the effectiveness of Country X's weapons has not changed, the curve $y = f(x)$ shifts upward with increasing slope for every x. The curve $x = g(y)$ does not change because there is no change in Country Y's population centers nor in its weapons effectiveness. The net effect of the increased civil defense is shown in Figure 1.11. The dashed curve is the new position of the function $y = f(x)$ resulting from the civil defense of Country X. The point (x'_m, y'_m) is the new intersection point. Note that although the course of action seemed fairly passive, the net effect is to increase the minimum number of missiles required by both sides because $x'_m > x_m$ and $y'_m > y_m$.

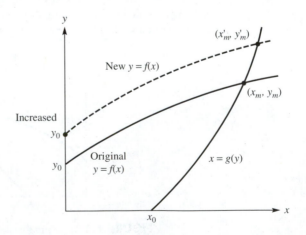

FIGURE 1.11   Country X increases its civil defense posture

Example 2 Next assume that Country X puts its missiles on mobile launching pads, which can be relocated during times of international crisis. There is no change in x_0 because Country X still requires the same number of missiles to inflict unacceptable damage on Country Y's population and industrial centers. Because Country X's missiles are less vulnerable than before, however, the curve $x = g(y)$ would flatten toward the y-axis, as shown by the dashed curve in Figure 1.12. The curve $y = f(x)$ does not change because Country Y requires the same number of missiles to inflict unacceptable damage on Country X as before and there is no change in the effectiveness of Country X's weapons. The net effect on the arms race of the mobile launching pads is depicted in Figure 1.12 and shows the new position of the equilibrium point (x_m', y_m'). Note that although this strategy is far less passive than the civil defense strategy, this alternative leads to a reduction in the minimum number of missiles required by both countries because $x_m' < x_m$ and $y_m' < y_m$.

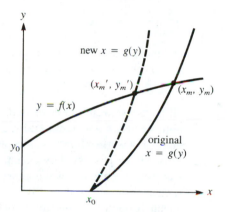

FIGURE 1.12 Country X uses mobile launching pads

Example 3 In this scenario suppose Country X and Country Y both use multiple warheads that can be targeted independently. Let's continue to count the numbers of missiles (not warheads) required by each country. Assume that each missile is armed with 16 smaller missiles, each possessing its own warhead. Because it still takes the same number of warheads to destroy the opponent's population and industrial centers, it is reasonable to expect the number of larger missiles, x_0 and y_0, to be reduced by the factor 16. Now consider the slope of the curve $y = f(x)$. Because the warheads can be targeted independently, an increase in x by 1 gives the capability of destroying 16 more of Country Y's missiles, so the ratio x/y increases. Thus $y = f(x)$ must rise more sharply than before to compensate for the increased destruction if it is to be able to meet its friendly strategy objective. A similar analysis applies to the ratio y/x for Country X and the curve $x = g(y)$. The new curves are represented in Figure 1.13. Because the reduction in values of the intercepts x_0 and y_0 coupled with the changes in the slopes of the curves give different effects on the new location of the intersection point (see Figure 1.13), it is difficult to tell from a graphical analysis whether the minimum number of missiles actually increases or decreases. This analysis demonstrates a limitation of graphical models. To determine the location of the equilibrium point (x_m', y_m') would require a more detailed analysis

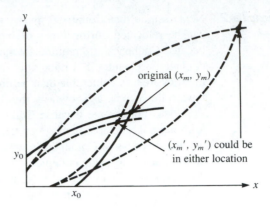

FIGURE 1.13   Both countries use multiple independently targeted warheads
on each missile

and more exact information concerning the weapon effectiveness and technological
capabilities of both countries (among other factors, such as military intelligence).

Example 4   Although we were unable to predict the effect of multiple warheads on the minimum
number of *missiles* required by each side in Example 3, we can analyze the total
number of warheads in this example's strategy. Let x and y now represent the *number
of warheads* possessed by Country X and Country Y, respectively. The number of
warheads needed by each country to inflict unacceptable damage on the opponent
remain at the levels x_0 and y_0, as before. An increase in x by 1 warhead, however,
enables Country X to destroy 16 of Country Y's warheads (rather than only 1 as in our
original assumption for the model), because they are all located on a single missile.
Thus Country Y will need more warheads, because the ratio x/y has increased and
$y = f(x)$ rises more steeply than before, as illustrated in Figure 1.14. The same

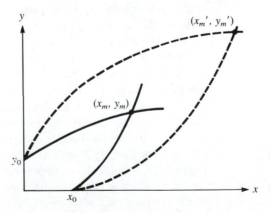

FIGURE 1.14   Multiple warheads on each missile increase the total number
of warheads required by each side

argument applies to Country X and the curve $x = g(y)$. Note that both countries require *more warheads* if multiple warheads are introduced on each missile because $x'_m > x_m$ and $y'_m > y_m$ in Figure 1.14.

1.1 PROBLEMS

1. Analyze the effect on the arms race of each of the following strategies:

   a. Country X increases the accuracy of its missiles by using a better guidance system.

   b. Country X increases the payload (destructive power) of its missiles without sacrificing accuracy.

   c. Country X is able to retarget its missiles in flight so that it can aim for missiles that previous warheads have failed to destroy.

   d. Country Y uses sea-launched ballistic missiles.

   e. Country Y adds long-range intercontinental bombers to its arsenal.

   f. Country X develops sophisticated jamming devices that dramatically increase the probability of neutralizing the guidance systems of Country Y's missiles.

2. Discuss the appropriateness of the assumptions used in developing the nuclear arms race model. What is the effect on the number of missiles if each country believes the other country is also following the friendly strategy? Is disarmament possible?

3. Develop a graphical model based on the assumptions that each side is following the enemy strategy. That is, each side desires first strike capability for destroying the missile force of the opposing side. What is the effect on the arms race if Country X now introduces antiballistic missiles?

4. Discuss how you might go about validating the nuclear arms race model. What data would you collect? Is it possible to obtain the data?

5. Use the polar coordinate substitution $x = r \cos \theta$ and $y = r \sin \theta$ in Equation (1.6) to show that a doubling of y_0 causes a doubling of r for every fixed θ. Show that if y_0 increases, then $y = f(x)$ shifts upward with an increasing slope.

1.1 PROJECTS

For Projects 1 through 4, complete the requirements in the referenced UMAP module, and prepare a short summary for classroom discussion.

1. "The Distribution of Resources," by Harry M. Schey, UMAP 60, 61, 62 (one module). The author investigates a graphical model that can be used to measure the distribution of resources. It presents an excellent review of the geometric interpretation of the derivative as applied to the economics of the distribution of

a resource. It also discusses numerical calculation of the derivative and definite integral.

2. "Nuclear Deterrence," by Harvey A. Smith, UMAP 327. The stability ability of the arms race is analyzed assuming objectives similar to those suggested by General Taylor. Analytic models are developed using probabilistic arguments. An understanding of elementary probability is required.

3. "The Geometry of the Arms Race," by Steven J. Brams, Morton D. Davis, and Philip D. Straffin, Jr., UMAP 311. In this module the possibilities of both parties disarming is analyzed introducing elementary game theory. Interesting conclusions are based on Country *X*'s ability to detect Country *Y*'s intentions and vice versa.

4. "The Richardson Arms Race Model," by Dina A. Zinnes, John V. Gillespie, and G. S. Tahim, UMAP 308. A model is constructed based on the classical assumptions of Lewis Fry Richardson. Difference equations are introduced.

1.1 Further Reading

SAATY, Thomas L. *Mathematical Models of Arms Control and Disarmament*. New York: Wiley, 1968.
SCHRODT, Philip A. "Predicting Wars With the Richardson Arms-Race Model." BYTE 7, no. 7 (July 1982): 108–134.
WALLACE, Michael D. "Arms Races and Escalation; Some New Evidence." In *Explaining War*, edited by J. David Singer, pp. 240–252. Beverly Hills, CA: Sage, 1979.

1.2   MANAGING RESOURCES: MAJOR LEAGUE BASEBALL

The Controversy

Over the last several decades baseball owners and players have debated issues concerning wages and the rights of players to act as free agents (the right to have an agent solicit among all owners for the highest bid for a player's services) in the marketplace. Players argue that owners impose restrictions that prevent them from being paid their market value. Owners argue that the players are receiving their market value while they, the owners, are making only a reasonable business profit. The owners further argue that because there are no excess profits, any increase in player salaries will have to be passed on to the fans. They also maintain that the demand of the fans for baseball will decrease if ticket prices are raised to cover increases in player salaries. Finally, the owners insist that the increase in the price charged to fans coupled with the decrease in demand will result in *less revenue* being generated, not more. Thus the owners would be forced to hire fewer quality players and pay them less or to sell the franchise. The end result predicted by the owners is that increasing salaries is bad for the *long-term interests* of the baseball industry (players, fans, and owners).

These differing views have resulted in baseball strikes that affect not only the owners and the players but also the fans. During the 1994–1995 strike, for example, for the first time in the history of baseball, sports fans were denied the World Series. There is now piqued interest in professional sports and how the interactions of owners and players affect the consumer, the fan. Certainly, the situation is complex. Here we will study the baseball market by constructing a graphical model.

First we build a graphical model of the baseball industry. Then we will use the model to analyze qualitatively what would happen if the owners tried to pass a salary increase for the quality players on to the fans in the form of increased prices, much like a tax being added to a product. We would like to know who pays for the salary increase as well as the short-term and long-term effects on the industry.

Baseball Industry and the Theory of the Firm

If we consider the fans to be the consumers and a baseball franchise to be a firm, then the baseball industry follows many of the economic principles of the theory of the firm. The product the franchises produce is entertainment by a team of players. In particular, baseball has many characteristics of a **monopsony** (single buyer). There exists a rigid set of rules governing contracts between players and the teams, free agency, arbitration, and services to the team. Consequently, once a player signs a contract in organized baseball, he is no longer eligible to sell his services in any manner he chooses. Under these rules, the players believe they receive less than they would if the labor market for them were *perfectly competitive*.

Let's construct a graphical model to address the following questions qualitatively:

1. Are the players underpaid for their services?
2. If the players do receive an increase in salary, can the owners pass the increase on to the fans? If not, who pays the lion's share of the increase, the owners or the fans?
3. If the fans pay the major share of the increase and there is a fundamental change in consumer demand as the result of a strike, what is the long-term effect on the baseball industry?

In the analysis to follow, we are concerned with gaining a qualitative understanding of the principal factors involved in the baseball dispute. A graphical analysis will help us, especially because precise data would be difficult to obtain. We begin by graphically analyzing general economic principles in a perfect competition. In ensuing sections we interpret the results of the graphical model as they apply to our baseball situation as the imperfect competition of a monopsony.

Economics of Perfect Competition

A baseball franchise creates a team that provides entertainment for the consumer, the fan. To form this team, a franchise has a complex organization, including a minor league farm system to produce major league players, scouts to find talent, coaches, and a management structure. Additionally, the franchise owns or rents a large infrastructure consisting of stadiums, training areas, equipment, and the like.

But what does the fan pay for? To begin to answer this question, let's suppose the fan comes to see the quality players. The more *quality players* a team has, the more often fans will come to the stadium to watch them play. Although distinct quality players differ in their ability to attract fans, let's assume that each quality player attracts a constant number of fans. Revenue is generated when the fan pays a fee for baseball entertainment (attending a game, parking, concessions, TV advertisements, and so on). If we assume that the price of the entertainment is constant, then the total revenue generated by the team is a constant times the number q of quality players the team possesses, as indicated in Figure 1.15.

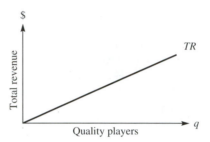

FIGURE 1.15 Total revenue is a constant times the number of quality players

We assume there is a limited number of quality players and the owners, players, and fans know who they are. A serious question facing the owners is how many quality players to procure to maximize their own profits. If we assume the industry is large enough that the number of quality players belonging to a single franchise does not affect the revenue generated by a single quality player, then the owner need only consider the difference between the **revenue generated** by the additional quality player being considered and the owner's **total cost** in fielding the additional player.

What is the total cost in fielding an additional quality player? Individual teams encounter **fixed costs** that are independent of the number of players procured. These costs include purchasing and maintaining stadiums, equipment, insurance, administration, and transportation. The **variable costs** depend on the number of baseball players in the entire franchise, including the minor leagues. Variable costs include salaries, fringe benefits, food, and uniforms. (See Figure 1.16.)

When the fixed costs are divided by the number of quality players, the share apportioned to each player is obtained. This per unit share is relatively high when the entire fixed costs of the franchise are borne by having only a few quality players. As a franchise hires more quality players, however, the per unit share of the fixed costs diminishes. (See Figure 1.17.) But when a team hires more quality players, eventually the talents of the quality players become duplicate talents that are already procured and the cost of fielding the major league team increases dramatically, straining the capabilities of the team to make a profit.

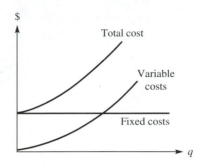

FIGURE 1.16   Fixed costs and variable costs as a function of the number q of quality players

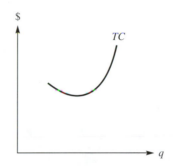

FIGURE 1.17   The per unit share of the total cost apportioned to each player

Because the per unit costs tend to be relatively high when the number of quality players hired is either very low or very high, we intuitively expect the existence of a level q^* that yields a maximum profit over the ranges being considered. This idea is illustrated in Figure 1.18 where we plot total profit as a function of the number q of quality players hired by a franchise. Our goal now is to characterize q^* mathematically.

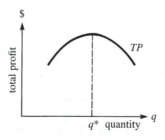

FIGURE 1.18   Total profit is maximized at q^*

At a given level q, total profit $TP(q)$ is the difference between total revenue $TR(q)$ and total cost $TC(q)$. That is,

$$TP(q) = TR(q) - TC(q)$$

A necessary condition for a relative maximum to exist is that the derivative of *TP* with respect to *q* must be zero (0):

$$TP' = TR' - TC' = 0$$

or, at the level q^* of maximum profit,

$$TR'(q^*) = TC'(q^*) \tag{1.7}$$

Thus, at q^* it is necessary that the slope of the total revenue curve equal the slope of the total cost curve. This condition is depicted in Figure 1.19.

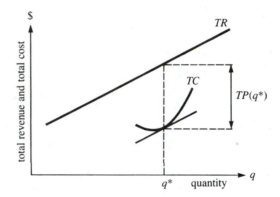

FIGURE 1.19 At q^* the slopes of the total cost and total revenue curves are equal

Let's interpret economically the meaning of the derivatives *TR'* and *TC'*. From the definition of the derivative,

$$TR'(q) \approx \frac{TR(q + \Delta q) - TR(q)}{\Delta q}$$

for Δq small (where the symbol $\approx$ means "is approximately equal to"). Thus if $\Delta q = 1$ unit (i.e., player), you can see that $TR'(q)$ approximates $TR(q + 1) - TR(q)$, which is the revenue generated by the next unit, or the *marginal revenue MR* of the $q + $ 1st unit, or quality player. Because total revenue is the price per unit times the number of units, it follows that the marginal revenue of the $q + $ 1st unit is the revenue generated by that unit less the revenue lost on the previous units resulting from price reductions (see Problem 4). In the problem at hand, we are assuming the revenue generated by a quality player is constant so there are no price reductions resulting from an additional unit on the marketplace. Thus $MR(q)$ is simply the (constant) price p of the entertainment generated by a quality player. Similarly, $TC'(q)$ represents the *marginal cost MC* of the $q + $ 1st unit—that is, the *extra* cost in changing output to include one additional player. If Equation (1.7) is interpreted in these new terms, a necessary condition for maximum profit to occur at q^* is that marginal revenue equals marginal cost:

$$MR(q^*) = MC(q^*) \tag{1.8}$$

For the critical point defined by Equation (1.8) to be a relative maximum, it is sufficient that the second derivative TP'' be negative. Because $TP' = MR - MC$, we have

$$TP''(q^*) = MR'(q^*) - MC'(q^*) < 0$$

or

$$MR'(q^*) < MC'(q^*) \tag{1.9}$$

In words, at the level q^* of maximum profit the slope of the marginal revenue curve is less than the slope of the marginal cost curve. The results of (1.8) and (1.9) together imply that the marginal revenue and the marginal cost curves intersect at q^* with the marginal cost curve rising more rapidly. Intuitively, it is profitable for an owner to continue to procure quality players until the cost of the next quality player exceeds the additional revenue he would generate. These results are illustrated in Figure 1.20.

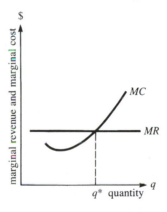

FIGURE 1.20   At q^*, $MR = MC$ and $MR' < MC'$

Interpreting the Graphical Model

Now let's interpret the graphical models represented by Figure 1.20 in terms of the baseball industry. The *MR* curve represents the revenue generated by the next quality player. This revenue is the price of the entertainment provided by the player in the form of costs of tickets, concessions, TV advertisements, and so forth. The curve is drawn horizontally because in the large competitive baseball industry, the number of quality players a single franchise produces seldom influences the market value of a quality player (interpreted as the revenue generated by the entertainment provided by a single quality player), so there is no loss in revenue generated by previous players resulting from any reductions in market value as more quality players are fielded. Thus the price of the entertainment is a constant for each quality player. The implication for the baseball industry is that an owner attempting to maximize profit will continue to procure quality players until the costs of the next quality player exceed the additional revenue he will generate. Verify that this situation is suggested by the graphical model in Figure 1.20.

Imperfect Competition

Suppose the total revenue generated by a team is a function only of the number of quality players a team possesses. An owner (called a **monopsonist**—a single buyer) who desires to maximize profit will purchase quality players as long as the extra revenues derived from the additional player are at least as large as the extra costs for this input. In other words, do not pay a quality player more than he will bring to the owner in total revenues minus his salary and share of the total cost of fielding the team. As we saw in a perfect competition, marginal revenue will equal marginal cost at equilibrium.

   In the preceding analysis, market forces help sellers and buyers find a price and quantity that is satisfactory to both. Assume the baseball industry is at an equilibrium state with respect to supply and demand at the time the players sought higher salaries through free agency, salary arbitration, and so forth. The players justify their actions by arguing that the owners can absorb the salary increases from their profits and quality players attract additional revenue. From the owners' perspective, if equilibrium has been achieved, additional entertainment cannot be sold profitably unless the price is reduced (theory of the firm) or new fans are discovered (thereby increasing the revenue generated by each quality player). Further, if the owners are making only a reasonable profit (as they argue), then they have to generate additional revenues to maintain the same profit. If they cannot generate new fans, then perhaps they can pass the cost of the salary increase on to the fans as a surcharge above the current price (much like a tax added at the counter to goods purchased to generate revenues for states and counties). If the price of baseball entertainment is raised, what is the effect on demand in the short term and the long term? Is additional revenue actually generated? Can the owners really pass the surcharge on to the fan? Who pays the cost of the salary increase: the fan or the owner? Do owners and fans share the cost in some manner? What is the effect of the strike on the player in the long term? In the next section we build a graphical model to answer these questions.

1.2 PROBLEMS

1. Justify mathematically and interpret economically the graphical model for the theory of the firm in Figure 1.21. What are the major assumptions suggested by the graphical model?

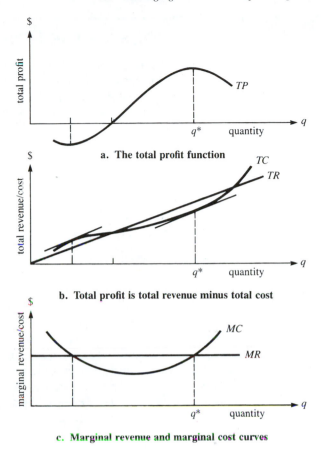

a. The total profit function

b. Total profit is total revenue minus total cost

c. Marginal revenue and marginal cost curves

FIGURE 1.21 A graphical model for the theory of the firm

2. Show that for total profit to reach a relative minimum, $MR = MC$ and $MC' < MR'$.

3. Suppose a large competitive market is the concession industry at the ballpark and the firm within the industry is a concession stand. How well does the model in Figure 1.21 reflect the reality of the situation? How would you adjust the graphical model to make improvements?

4. Verify the result that the marginal revenue of the $q + 1$st unit equals the price of that unit minus the loss in revenue on previous units resulting from price reductions.

5. In 1996 Shaquille O'Neill received a $120 million contract with the Los Angeles Lakers. Very shortly thereafter, the Lakers announced that the price of a ticket would increase by $12.50. Can the increased price of Shaquille's contract be passed on to the fans? Explain your position.

1.3 THE EFFECTS OF THE STRIKE ON BASEBALL

The players' perceptions of not being paid their full worth coupled with the owners' inability to convince the players that there are no excess profits led to a strike. One result of the long strike was that the owners gave in to an increase in salaries for the players. If the owners did not have excess profits as they claimed, then a key question for them is this:

Can the owners pass a salary increase per quality player on to the fans?

Let's construct a graphical model to find out.

Who Pays the Salary Increase: The Fan or the Owner?

Assume for a given team that an owner was maximizing total profit before the salary increase; that is, given a market revenue generated by a quality player *MR*, the owner can support owning q^* quality players, as suggested by Figure 1.20. Assume further that the salary increase for players caused a shift in the marginal cost curve as shown in Figure 1.22. Because the owner must pay each player the salary increase, the marginal cost to the owner of each player increases by the amount of the increased salary. Geometrically, this means the marginal cost curve shifts upward by the amount of the salary increase. Assume for the moment that the entire industry is able to increase its revenue by adding on the amount of the salary increase to the revenues generated by the quality players. Basically, this means the number of fans generated by each quality player remains constant, but the price each fan pays for baseball entertainment (in the form of ticket prices, advertising, and so on) increases to cover the additional salaries of the quality players. Under this assumption the marginal revenue curve also shifts upward by the amount of the salary increase, as depicted in Figure 1.22. Note from the figure that the optimal quantity is still q^* players. Hence the model predicts *no change* in the quantity of quality players as a result of the salary increase. Rather, the owners will acquire the same number of players but charge a higher price to fans for the entertainment.

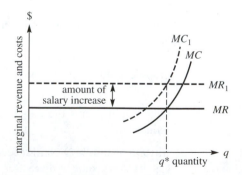

FIGURE 1.22 Both the marginal revenue and marginal cost curves shift upward by the amount of the salary increase, leaving the optimal number of quality players to procure at the same level q^*

Note, too, that it's the *fan who pays* the full amount of the salary increase in the form of price increases for tickets, TV, souvenirs, concessions, and so forth. (In the case of television, advertisers will pay more and try to pass on those costs to consumers by charging more for the products advertised.) The situation is what many people predicted: The fans would continue to support baseball regardless of increased prices.

The shortcoming with the model in Figure 1.22 is that it does not reveal whether the *entire* baseball industry can in fact continue to maintain the same quantity of players at their new salaries with the higher prices fans must pay. The total number of fans who purchase baseball entertainment must at least remain constant to generate the additional revenue promised by increasing the cost of the entertainment. To find out whether the number of fans at least remains constant, we need to construct a model for the entire industry.

Baseball Industry

For each owner in the industry, consider the intersection of the owners' various marginal revenue curves with each marginal cost curve. (Remember that each horizontal *MR* curve corresponds to the revenue generated by a single quality player.) This situation is depicted for one owner in Figure 1.23a.

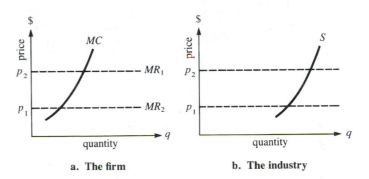

a. The firm   **b. The industry**

FIGURE 1.23   The industry's supply curve is obtained by summing the number of quality players each owner would field at each price for the entertainment provided by the player

For each price of the entertainment, sum over all owners in the baseball industry the number of quality players each owner would optimally acquire. This summing yields a curve for the entire baseball industry. Because the curve represents the number of quality players the baseball industry would supply at various price levels for the entertainment they provide, it is called a **supply curve** (an example is shown in Figure 1.23b). Qualitatively, as the market price of baseball entertainment increases, the baseball industry is willing to acquire greater numbers of expensive players.

Next, consider aggregate consumer demand for the product at various price levels *charged* for the entertainment (by the owners.) From a fan's point of view, the quantity of entertainment demanded is a function of what they have to pay for it. It is traditional, however, to plot this market price as a function of quantity (see Figure 1.24a). Conceptually, for each price level, individual consumer demands could be summed as in the procedure for obtaining the baseball industry's supply curve. This summation is depicted graphically in Figure 1.24b. Qualitatively, as the price increases, we expect the aggregate demand for baseball entertainment to decrease as consumers begin to go to a fewer number of games and substitute alternative forms of entertainment (see Figure 1.24b).

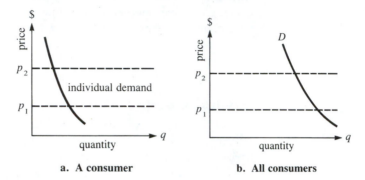

a. A consumer   **b. All consumers**

FIGURE 1.24   The baseball industry's demand curve D represents the aggregate demand for baseball entertainment at various price levels and is obtained by summing individual consumer demands at those levels; notice that we plot price versus quantity for demand curves rather than vice versa

Finally, consider the baseball industry's supply and demand curves together. Suppose the two curves intersect at a unique point (q^*, p^*) as depicted in Figure 1.25. If the industry supplies q^* and charges p^* (supply curve), then the consumers are willing to buy the amount q^* at the price p^* (demand curve). Thus there is **equilibrium** in the sense that no excess supply exists at that price, and both owners and fans are satisfied.

Obviously an industry does not know the precise demand curve for its product. Therefore, it is important to determine what occurs if the industry supplies other

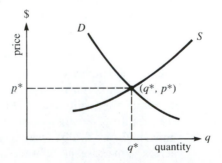

FIGURE 1.25   The intersection of the supply and demand curves gives a market price and a market quantity that satisfy both consumers and suppliers alike

than q^*. For example, suppose the industry supplies an amount q_1 greater than q^* (see Figure 1.26a). Then the consumers are willing to buy an amount as large as q_1 if the price is as low as p_1, forcing a reduction in price. If the market price drops to p_1, however, industry is willing to supply only q_2 units and would cut quality players. Then at q_2 the unsatisfied demand would drive the price back up to p_2. Convince yourself that this process converges to (q^*, p^*) in Figure 1.26a, where the supply curve is steeper than the demand curve. There are forces that actually drive supply and demand to the equilibrium point.

On the other hand, consider Figure 1.26b in which the supply curve is more horizontal than the demand curve. In this case, the equilibrium point (q^*, p^*) will not be achieved by the iterative process just described. Instead, there are likely to be wild fluctuations in the amount supplied and the price charged as the owners and players search for an equilibrium point (does this sound like the aftermath of the strike?). Convince yourself from Figure 1.26b that the equilibrium point is difficult to achieve when the supply curve is not as steep as the demand curve.

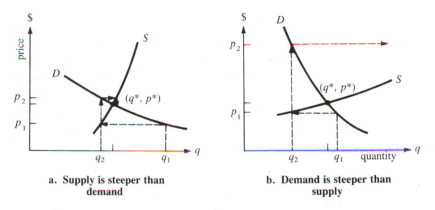

a. **Supply is steeper than demand**

b. **Demand is steeper than supply**

FIGURE 1.26   The ease with which the equilibrium point of supply and demand is achieved depends on the relative slopes of the supply and demand curves

The Player Ultimately Pays

Now consider the effect of a salary increase on the supply and demand curves. Suppose a particular franchise is in the equilibrium position (q^*, p^*) when a salary increase is added for each quality player. Because the owner has to pay the salary increase to each quality player, each marginal cost curve shifts upward by the amount of the salary increase (see Figure 1.22). These individual shifts cause the aggregate supply curve to shift upward by the amount of the salary increase as well. This phenomenon is depicted in Figure 1.27. If there is no reason for a shift in the demand curve, the intersection of the demand curve with the new supply curve shifts upward toward the left to a new equilibrium point (q_1, p_1), indicating an increase in the price of the entertainment. Furthermore, notice from Figure 1.27 that the increase in price from p^* to p_1 is less than the salary increase. Thus the model predicts that the *fans and the owners share the cost of the salary increase*. Study Figure 1.27 carefully and convince yourself that the portion of the salary increase the fans pay and the reduction in baseball entertainment purchased depend on the slopes of the supply and demand curves at the time the salary increase is imposed. The flatter the demand curve, the more owners have to pay for the increase.

Let's summarize. If the quality players are given a salary increase while the owners have no excess profit, it is unlikely that the owners can pass the entire increase on to the fans. The flatter the demand curve, the greater is the reduction in demand for the entertainment purchased and the more owners will have to pay for the salary increases. The result is a new equilibrium with *fewer quality players q_1* being demanded by the fans and being supplied by the owners. There is likely to be turmoil as fans and owners search for the new equilibrium.

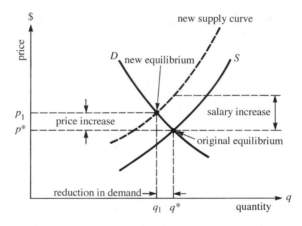

FIGURE 1.27 A salary increase for each quality player causes a decrease in the quantity of quality players procured by the owner coupled with an increase in the price of the entertainment

The Future of Baseball: Alternative Strategies

Today, the owners are planning to reconfigure baseball economics. They are convinced that the high players' salaries are driving them into financial ruin. Alternative strategies currently being considered include:

1. Share revenues among club owners so richer clubs subsidize poorer clubs.
2. Curtail free agency, which means teams could not bid against one another for a particular player's services. Rather, the player would be assigned to a single team with whom he must negotiate a salary.
3. Abolish salary arbitration, which permits an arbiter to settle salary disputes between owners and players.

4. Establish team salary caps as an upper limit a franchise is allowed to spend on the salaries of players.
5. Establish a tax on baseball to offset increases in player salaries.

In the problem set we ask you to analyze graphically each of the strategies in terms of the graphical models we have developed.

1.3 PROBLEMS

1. Show that when the demand curve is very steep, a salary increase added to each player will primarily fall on the consumer (fan). If the demand curve is more nearly horizontal, show that the salary increase is paid more by the owners. Which do you think is the case for baseball today? What if the supply curve is very steep? What if the supply curve is nearly horizontal? Interpret these supplies in terms of baseball players.

2. Consider the baseball industry. Discuss the conditions for which the demand curve will be steep near the equilibrium. Are there any conditions in which the curve will be flat?

3. Criticize the following quote: "The effect of a tax on a commodity might be seen at first sight to be an advance in price to the consumer. But an advance in price will diminish the demand. And a reduced demand will send the price down again. It is not certain, therefore, after all, that the tax will really raise prices."[5]

4. What if a tax is imposed on baseball to cover common costs shared by all owners? Discuss the effects of the tax on the equilibrium point of the supply and demand curves. What happens to the new price and the new quantity?

5. Criticize the graphical model of the baseball industry. Name some major factors that have been neglected. Which of the underlying assumptions are not satisfied by the strike or settlement? Did the graphical model provide you with any valuable insights? How could you adjust the model?

6. Analyze each of the following strategies for reconfiguring the economics of baseball:

   a. Share revenues among club owners.

   b. Curtail free agency.

   c. Abolish salary arbitration.

   d. Establish team salary caps.

   e. Establish a tax on baseball (to offset increases in player salaries).

[5]H. D. Henderson, *Supply and Demand* (Chicago: University of Chicago Press, 1958), p. 22.

1.3 PROJECTS

For Projects 1 through 4, complete the requirements in the referenced UMAP module and prepare a short summary for classroom discussion.

1. "Differentiation, Curve Sketching, and Cost Functions," by Christopher H. Nevison, UMAP 376. In this module costs and revenue for a firm are discussed using elementary calculus. The author discusses several of the economic ideas presented in this chapter.

2. "Price Discrimination and Consumer Surplus: An Application of Calculus to Economics," by Christopher H. Nevison, UMAP 294. The topics in the title are analyzed in a competitive market, and two-tier price discrimination is discussed. The module discusses several of the economic ideas presented in this chapter.

3. "Economic Equilibrium: Simple Linear Models," by Philip M. Tuchinsky, UMAP 208. In this module linear supply and demand functions are constructed, and the equilibrium market position is analyzed for an industry producing one product. The result is then extended to *n* products. The author concludes by briefly considering nonlinear and discontinuous functions.

4. "I Will If You Will... A Critical Mass Model," by Jo Anne S. Growney, UMAP 539. A graphical model is presented to treat the problem of individual behavior in a group when the individual makes a choice dependent on his or her perception of the behavior of fellow group members. The model can provide insight into paradoxical situations in which members of a group prefer one type of behavior but actually engage in the opposite behavior (like not cheating versus cheating in class).

1.3 Further Reading

ASIMAKOPULOS, A. *An Introduction to Economic Theory: Microeconomics.* New York: Oxford University Press, 1978.

COHEN, Klaman J. and Richard M. Cyert. *Theory of the Firm.* Englewood Cliffs, NJ: Prentice-Hall, 1975.

HENDERSON, Hubert D. *Supply and Demand.* Chicago: University of Chicago Press, 1958.

MANSFIELD, Edwin. *Microeconomics: Theory and Applications.* 2nd Edition. New York: Norton, 1975.

THOMPSON, Arthur A., Jr. *Economics of the Firm: Theory and Practice.* Englewood Cliffs, NJ: Prentice-Hall, 1973.

Chapter Two

THE MODELING PROCESS

INTRODUCTION

In Chapter 1 we presented graphical models representing nuclear deterrence and a sports strike. Now we examine more closely the process of mathematical modeling.

To gain an understanding of the processes involved in mathematical modeling, consider the two worlds depicted in Figure 2.1. Suppose we want to understand some behavior or phenomenon in the real world. We may wish to make predictions about that behavior in the future and analyze the effects various situations have on it. In the baseball strike, for instance, we were interested in predicting the effects of higher players' salaries on the long-term health of the baseball industry. As another example, when studying the populations of two interacting species, we may wish to know if the species can coexist within their environment, or if one species will eventually dominate and drive the other to extinction. Or in the management of a fishery, it may be important to determine the optimal sustainable yield of a harvest and the sensitivity of the species to population fluctuation caused by harvesting.

How can we construct and use models in the mathematical world to help us better understand real-world systems? Before discussing how we link the two worlds together, let's consider what we mean by a real-world system and why we would be interested in constructing a mathematical model for a system in the first place.

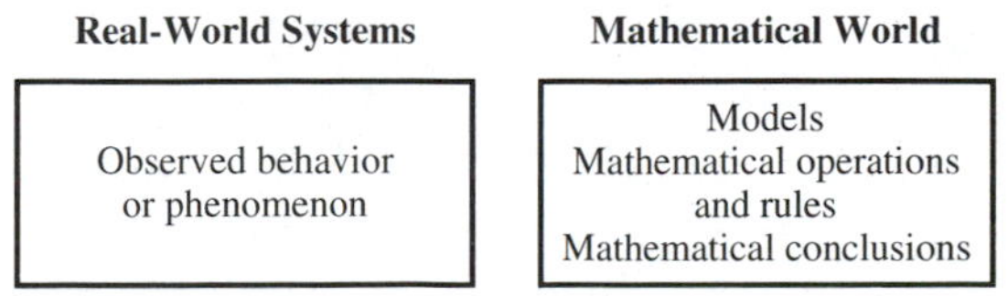

Real-World Systems	Mathematical World
Observed behavior or phenomenon	Models Mathematical operations and rules Mathematical conclusions

FIGURE 2.1   The real and mathematical worlds

system   A **system** is an assemblage of objects joined in some regular interaction or interdependence. The modeler is interested in understanding how a particular system works, what causes changes in the system, and the sensitivity of the system to certain changes. He or she is also interested in predicting what changes might occur and when they occur. How might such information be obtained?

For instance, suppose the goal is to draw conclusions about an observed phenomenon in the real world. One procedure would be to conduct some real-world behavior trials or experiments and observe their effect on the real-world behavior. This is depicted on the left side of Figure 2.2. Although such a procedure might minimize the loss in fidelity incurred by a less direct approach, there are many situations in which we would not want to follow such a course of action. For instance, there may be prohibitive costs for conducting even a single experiment such as determining the level of concentration at which a drug proves fatal or studying the radiation effects of a failure in a nuclear power plant near a major population area. Or we may not be willing to accept even a single experimental failure, such as when investigating different designs for a heat shield for a manned spacecraft. Moreover, it may not even be possible to produce a trial, as in the case of investigating some specific change in the composition of the ionosphere and its corresponding effect on the polar ice cap. Further, we may be interested in generalizing the conclusions beyond the specific conditions set by one trial (such as a cloudy day in New York with temperature 82°F, wind 15–20 miles per hour, humidity 42%, and so on). Finally, even though we succeed in predicting the real-world behavior under some very specific conditions, we have not necessarily *explained* why the particular behavior occurred. (Although the ability to predict and explain are often closely related, the ability to predict a behavior does not necessarily imply an understanding of it. In Chapter 6 we will study techniques specifically designed to help us make predictions even though we cannot explain satisfactorily all aspects of the behavior.) The preceding discussion underscores the need to develop indirect methods for studying real-world systems.

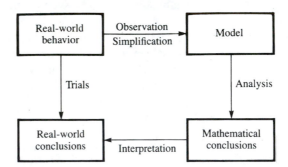

FIGURE 2.2 Reaching conclusions about the behavior of real-world systems

An examination of Figure 2.2 suggests an alternative way of reaching conclusions about the real world. First, we make specific observations about the behavior being studied and identify the factors that seem to be involved. Usually we cannot consider, or even identify, all the factors involved in the behavior, so we make simplifying assumptions that eliminate some factors. For instance, we may choose to neglect the humidity in New York City, at least initially, when studying radioactive effects from the failure of a nuclear power plant. Next, we conjecture tentative relationships among the factors we have selected, thereby creating a rough model of the behavior. Having constructed a model, we then apply appropriate mathematical analysis leading to conclusions about the model. Note that these conclusions pertain only to the model, not to the actual real-world system under investigation. Because we made some simplifications in constructing the model and the observations on which the model is based invariably contain errors and limitations, we must carefully account for these anomalies before drawing any inferences about the real-world behavior. In summary, we have the following rough modeling procedure:

1. Through observation, identify the primary factors involved in the real-world behavior, possibly making simplifications.
2. Conjecture tentative relationships among the factors.
3. Apply mathematical analysis to the resultant model.
4. Interpret mathematical conclusions in terms of the real-world problem.

Figure 2.3 portrays the entire modeling process as a closed system. Given some real-world system, we gather sufficient data to formulate a model. Next we analyze the model and reach mathematical conclusions about it. Then we interpret the model and make predictions or offer explanations. Finally, we test our conclusions about the real-world system against new observations and data. We may then find we need to go back and refine the model to improve its predictive or descriptive capabilities. Or perhaps we will discover that the model really doesn't fit the real world accurately, so we must formulate a new model. We will study the various components of this modeling process in detail throughout the book.

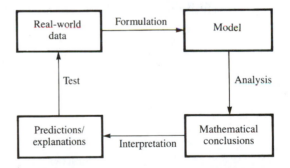

FIGURE 2.3 The modeling process as a closed system

2.1 ## MATHEMATICAL MODELS

Until now we have been intentionally vague about the nature of the model itself. The models used to represent or approximate a real-world system can be different in both appearance and purpose. For instance, one kind of model is a miniature replication of a real-world object of interest, such as a model spacecraft that might be used to study certain design features under experimental conditions. Another kind of model is a mathematical model. In this book our main concern is with the latter type.

mathematical model   For our purposes we define a **mathematical model** as a mathematical construct designed to study a particular real-world system or phenomenon. We include graphical, symbolic, simulation, and experimental constructs. An example of a graphical model is that of the arms race presented in Chapter 1. A symbolic model can be a formula or equations describing how the underlying factors of the model are interrelated, such as Newton's second law, which states that force = mass × acceleration. Still another kind of mathematical model is a simulation model, such as a computer program that randomly generates 1000 integers from 1 to 6 as representative of 1000 tosses of a six-sided die. Another example is using a scaled-down prototype to simulate the drag force on a proposed design for a submarine.

Mathematical models can be differentiated further. There are existing mathematical models that can be identified with a particular real-world phenomenon and used to study it. Then there are those mathematical models that we construct specifically to study a special phenomenon. Figure 2.4 depicts this differentiation between models. Starting from some real-world phenomenon, we can represent it mathematically by constructing a new model or selecting an existing model. On the other hand, we can replicate the phenomenon experimentally or with some kind of simulation.

When it comes to the question of constructing a mathematical model, a variety of conditions can cause us to abandon hope of achieving any success. The mathematics involved may be so complex and intractable that there is little hope of analyzing or solving the model, thereby defeating its utility.

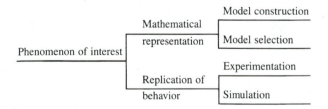

FIGURE 2.4   The nature of the model

This complexity can occur when attempting to use a model given by a system of partial differential equations or a system of nonlinear algebraic equations, for instance. Or the problem may be so large (in terms of the number of factors involved) that it is impossible to capture all the necessary information in a single mathematical model. Predicting the global effects of the interactions of population, use of resources, and pollution is an example of such an impossible situation. In such cases we may attempt to replicate the behavior *directly* by conducting various experimental trials. Then we collect data from these trials and analyze the data in some way, possibly using statistical techniques or curve-fitting procedures. From the analysis, we can reach certain conclusions.

In other cases, we may attempt to replicate the behavior *indirectly.* We might use an analog device such as an electrical current to model a mechanical system. We might use a scaled-down model such as a scaled model of a jet aircraft in a wind tunnel. Or we might attempt to replicate a behavior on a digital computer—for instance, simulating the global effects of the interactions of population, use of resources, and pollution or simulating the operation of an elevator system during morning rush hour.

The distinction between the various model types as depicted in Figure 2.4 is made solely for ease of discussion. For example, the distinction between experiments and simulations is based on whether the observations are obtained directly (experiments) or indirectly (simulations). In practical models this distinction is not nearly so sharp; one master model may use several models as submodels, including selections from existing models, simulations, and experiments. Nevertheless, it is informative to contrast these types of models and compare their various capabilities for portraying the real world.

To that end, consider the following properties of a model:

Fidelity   The preciseness of a model's representation of reality

Costs   The total cost of the modeling process

Flexibility   The ability to change and control conditions affecting the model as required data are gathered

It is useful to know the degree to which a given model possesses each of these characteristics. However, because specific models vary greatly, even within the classes identified in Figure 2.4, the best we can hope for is a comparison in the relative performance between the classes of models for each of the characteristics. The comparisons are depicted in Figure 2.5, where the ordinate axis denotes the degree of effectiveness of each class.

Let's summarize the results shown in Figure 2.5. First, consider the characteristic of fidelity. We would expect observations made directly in the real world

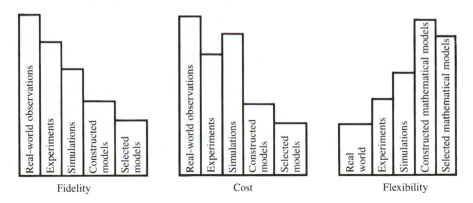

FIGURE 2.5   Comparisons among the model types

to demonstrate the greatest fidelity, even though some testing bias and measurement error may be present. We would expect experimental models to show the next greatest fidelity, because behavior is being observed directly in a more controlled environment, such as a laboratory. Because simulations are a step further from the real world and introduce indirect observations (such as constructing and analyzing a scaled model), simulations suffer from further loss in fidelity. Whenever a mathematical model is constructed, real-world conditions are simplified, resulting in more loss of fidelity. Finally, any selected model is based on additional simplifications that are not even tailored to the specific problem, and these simplifications imply still further loss in fidelity.

Next, consider cost. Generally we would expect any selected mathematical model to be the least expensive. Constructed mathematical models bear an additional cost of tailoring the simplifications to the phenomenon being studied. Experiments are usually expensive to set up and operate. Likewise, simulations use indirect devices that are often expensive to develop, and simulations commonly involve large amounts of computer space, time, and maintenance.

Finally, consider the characteristic of flexibility. Constructed mathematical models are generally the most flexible because different assumptions and conditions can be chosen relatively easily. Selected models are less flexible because they are developed under specific assumptions; nevertheless, specific conditions often can be varied over wide ranges. Simulations usually entail the development of some other indirect device to alter assumptions and conditions appreciably. Experiments are even less flexible, because some factors are difficult to control beyond specific ranges. Observations of real-world behavior have little flexibility because the observer is limited to the specific conditions that pertain at the time of the observation. Moreover, other conditions might be highly improbable, or impossible, to create. It is important to understand that our discussion is only qualitative in nature and that there are many exceptions to these generalizations.

Construction of Models

In the preceding discussion we viewed modeling as a process and considered briefly the form of the model itself. Now let's focus attention on the construction of mathematical models. We begin by presenting an outline of a procedure that is helpful in constructing models. In the next section we illustrate the various steps in the procedure by discussing several real-world examples.

Step 1. Identify the problem.   What is it you would like to do or find out? Typically this is a difficult step because people often have great difficulty in deciding what must be done. In real-life situations no one simply hands us a mathematical problem to solve. Usually we have to sort through large amounts of data and identify some particular aspect of the situation we wish to study. Moreover, we must be sufficiently precise (ultimately) in the formulation of the problem to allow for translation into mathematical symbology of the verbal statements describing it. This translation is accomplished through the next steps. It is important to realize that the answer to the question posed might not lead directly to a usable problem identification.

Step 2. Make assumptions.   Generally we cannot hope to capture in a usable mathematical model all the factors influencing the problem that has been identified. The task is simplified by reducing the number of factors under consideration. Then relationships among the remaining variables must be determined. Again, the complexity of the problem can be reduced by assuming relatively simple relationships. Thus the assumptions fall into two main activities:

a. Classify the variables. What things influence the behavior you identified in Step 1? List these things as variables. The variables the model seeks to explain are the dependent variables and there may be several of these. The remaining variables are the independent variables. Each variable is classified as dependent, independent, or neither. For example, in the arms race model the strength of the friendly force is a dependent variable that varies according to the number of enemy weapons, the enemy and friendly weapon technologies, and so forth. These latter variables are independent variables from the point of view of the friendly country.

   You may choose to neglect some of the independent variables for either of two reasons. First, the effect of the variable may be relatively small compared with other factors involved in the behavior. For example, in the arms race model we neglected such factors as the state of the economy and political considerations to reach a qualitative assessment of the behavior. We may also neglect a factor that affects the various alternatives in about the same way, even though it may have an important influence on the behavior under investigation. For example, consider the problem of determining the optimal shape for a lecture hall where readability of a chalkboard or overhead projection is a dominant criterion. Lighting is certainly a crucial factor, but it would affect all possible shapes in about the same way. We can simplify the analysis considerably by neglecting such a variable, possibly incorporating it later in a separate, more refined model.

b. Determine interrelationships among the variables selected for study. Before we can hypothesize a relationship among the variables, we generally must make some additional simplifications. The problem may be sufficiently complex so we cannot see a relationship among all the variables initially. In such cases it may be

submodel

possible to study **submodels**. That is, we study one or more of the independent variables separately. Eventually we will connect the submodels together. For instance, in the arms race model we hypothesized that the growth of the friendly arsenal depends on its current size, the size of the enemy arsenal, and the respective weapon technologies. We then proposed a submodel for the weapon technologies by assuming the weapons could not be retargeted in flight. This assumption yielded a relatively simple submodel to incorporate into the master model. In later chapters we will study various techniques, such as proportionality, that will aid in hypothesizing relationships among the variables.

   Step 3. Solve or interpret the model.   Now put together all the submodels to see what the model is telling us. For example, the assumptions made in the arms race led to a graphical model that could be interpreted for various values of the independent variables and changes in the assumptions. In some cases the model may consist of mathematical equations or inequalities that must be solved to find the information we are seeking. Often a problem statement requires a best or *optimal solution* to the model. Models of this type are discussed in Chapters 8 and 9.

Although it was possible to construct a master model in the arms race model, often we will find we are not quite ready to complete this step. Or we may end up with a model so unwieldy we cannot solve or interpret it. In such situations we might return to Step 2 and make additional simplifying assumptions. Sometimes we will even want to return to Step 1 to redefine the problem. This point will be amplified in the following discussion.

Step 4. Verify the model.   Before we use the model, we must test it out. There are several questions we should ask before designing these tests and collecting data—a process that can be expensive and time consuming. First, does the model answer the problem identified in Step 1, or did it stray from the key issue as we constructed the model? Second, is the model usable in a practical sense; that is, can we really gather the data necessary to operate the model? Third, does the model make common sense? In the arms race model, for instance, did we make a mathematical error in Step 3 or a faulty assumption in Step 2?

Once the commonsense tests are passed, we will want to test many models using actual data obtained from empirical observations. We need to be careful to design the test in such a way as to include observations over the *same range* of values of the various independent variables we expect to encounter when actually using the model. The assumptions made in Step 2 may be reasonable over a restricted range of the independent variables but very poor outside of those values. For instance, a frequently used interpretation of Newton's second law states that the net force acting on a body is equal to the mass of the body times its acceleration. This law is a reasonable model until the speed of the object approaches the speed of light.

Be careful about the conclusions you draw from any tests. Just as we cannot prove a theorem simply by demonstrating many cases in which the theorem does hold, likewise, we cannot extrapolate broad generalizations from the particular evidence we gather about our model. A model does not become a law just because it is verified repeatedly in some specific instances. Rather, we *corroborate the reasonableness* of the model through the data we collect.

Step 5. Implement the model.   Of course our model is of no use just sitting in a filing cabinet. We will want to explain it in terms that the decision makers and users can understand if it is ever to be of use to anyone. Further, unless the model is placed in a user-friendly mode, it will quickly fall into disuse. Expensive computer programs sometimes suffer such a demise. Often the inclusion of an additional step to facilitate the collection and input of the data necessary to operate the model determines its success or failure.

Step 6. Maintain the model.   Remember that the model is derived from the specific problem identified in Step 1 and from the assumptions made in Step 2. Has the original problem changed in any way, or have some previously neglected factors become important? Does one of the submodels need to be adjusted? For instance, should we now consider the possibility of disarmament in the arms race model? Does the current economic situation dictate a cap on military expenditures? Is there a change in weapon technology that permits the retargeting of missiles in flight? What happens now that the Cold War has ended?

We summarize the steps for constructing mathematical models in Figure 2.6. We should not be too enamored with our work. Like any model, our procedure is an approximation process and therefore has its limitations. For example, the procedure seems to consist of discrete steps leading nicely to a usable result, but that's rarely the case in practice. Before offering an alternative procedure that emphasizes the iterative nature of the modeling process, let's discuss the advantages of the methodology depicted in Figure 2.6.

Step 1. Identify the problem.
Step 2. Make assumptions.
    a. Identify and classify the variables.
    b. Determine interrelationships between the variables and submodels.
Step 3. Solve the model.
Step 4. Verify the model.
    a. Does it address the problem?
    b. Does it make common sense?
    c. Test it with real-world data.
Step 5. Implement the model.
Step 6. Maintain the model.

FIGURE 2.6 Construction of a mathematical model

The process shown in Figure 2.6 provides a methodology for progressively focusing on those aspects of the problem we wish to study. Furthermore, it demonstrates a curious blend of creativity with the scientific method used in the modeling process. The first two steps are more artistic or original in nature. They involve abstracting the essential features of the problem under study, neglecting any factors judged to be unimportant, and postulating relationships precise enough to help answer the questions posed by the problem yet simple enough to permit the completion of the remaining steps. Whereas these steps admittedly involve a degree of craftsmanship, we will learn some scientific techniques we can apply to appraise the importance of a particular variable and the preciseness of an assumed relationship. Nevertheless, when generating numbers in Steps 3 and 4, remember that the process has been largely inexact and intuitive.

scientific method

Let's contrast the modeling process presented in Figure 2.6 with the scientific method. One version of the **scientific method** is as follows:

Step 1. Make some general observations of a phenomenon.
Step 2. Formulate a hypothesis about the phenomenon.
Step 3. Develop a method to test that hypothesis.
Step 4. Gather data to use in the test.
Step 5. Test the hypothesis using the data.
Step 6. Confirm or deny the hypothesis.

By design, the mathematical modeling process and scientific method have similarities. For instance, both processes involve making assumptions or hypotheses, gathering real-world data, and testing or verification using that data. These similarities should not be surprising; though recognizing that part of the modeling process is an art, we do attempt to be scientific and objective whenever possible.

There are also subtle differences between the two processes. One difference lies in the primary goal of the two processes. In the modeling process, assumptions are made in selecting which variables to include or neglect and postulating the interrelationships among the included variables. The goal in the modeling process is to *hypothesize a model*, and as with the scientific method, evidence is gathered to corroborate that model. Unlike the scientific method, however, the objective is not to *confirm* or *deny* the model (we already know it is not precisely correct because of the simplifying assumptions we have made), but rather to test its *reasonableness*. We may decide that the model is quite satisfactory and useful and elect to accept it. Or we may decide that the model needs to be refined or simplified. In extreme cases we may even need to redefine the problem, in a sense rejecting the model altogether. We will see in subsequent chapters that this decision process really constitutes the heart of mathematical modeling.

Figure 2.7 amplifies these ideas in viewing the modeling process and graphically displays its iterative nature. The figure shows how we begin by examining some system and identifying the particular behavior we wish to predict or explain. Next we identify the variables and simplifying assumptions and generate a model. We then attempt to validate the model with appropriate tests. If the results of the tests are satisfactory, we can use the model for its intended purpose. If the results are not satisfactory, there are several possibilities to pursue. We may decide the model

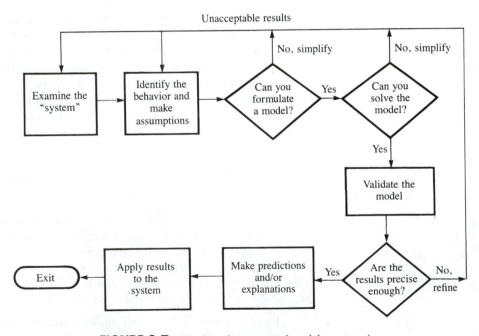

FIGURE 2.7   The iterative nature of model construction

needs to be refined by incorporating additional variables or by restructuring a particular submodel. In some cases the test results may be so unsatisfactory that the original problem must be redefined because it turns out to be entirely too ambitious.

The process depicted in Figure 2.7 not only emphasizes the iterative nature of model construction but also introduces the trade-offs between model simplification and model refinement. We generally start with a simple model, progress through the modeling process, and then refine the model as the results of our validation procedures dictate. If we cannot come up with a model or solve the one we have, we must *simplify*. We simplify a model by treating some variables as constants, by neglecting or aggregating some variables, by assuming simple relationships (such as proportionality) in any submodels, or by restricting further the problem under investigation. On the other hand, if our results are not precise enough, we must *refine* the model. Refinement of a model is generally achieved in the opposite way: We introduce additional variables, assume more sophisticated relationships among the variables, or expand the scope of the problem. By trading off between simplification and refinement, we determine the generality, realism, and precision of our model. This trading-off process cannot be overemphasized and constitutes the art of modeling.

robust
fragile

sensitivity

We complete this section by introducing several terms that are useful in describing models. A model is said to be **robust** when its conclusions do not depend on the precise satisfaction of the assumptions. A model is **fragile** if its conclusions do depend on the precise satisfaction of some set of conditions. The term **sensitivity** refers to the degree of change in a model's conclusions as some condition on which they depend is varied; the greater the change, the more sensitive is the model to that condition. For example, the arms race model is both fragile and sensitive.

2.1 PROBLEMS

In Problems 1 through 8, the scenarios are vaguely stated. From these vague scenarios, identify a problem you would like to study. Which variables affect the behavior you have identified in the problem identification? Which variables are the most important? Remember, there are really no right answers.

1. The population growth of a single species.

2. A retail store intends to construct a new parking lot. How should the lot be illuminated?

3. A farmer wants to maximize the yield of a certain crop of food grown on his land. Has the farmer identified the correct problem? Discuss alternative objectives.

4. How would you design a lecture hall for a large class?

5. An object is to be dropped from a great height. When and how hard will it hit the ground?

6. How should a manufacturer of some product decide how many units of that product should be manufactured each year and how much to charge for each unit?

7. The United States Food and Drug Administration is interested in knowing whether a new drug is effective in the control of a certain disease in the population.

8. How fast can a skier ski down a mountain slope?

For the scenarios presented in Problems 9 through 12, identify a problem worth studying and list the variables that affect the behavior you have identified. Which variables would be neglected completely? Which might be considered as constants initially? Identify any submodels you would want to study in detail and any data you would want collected.

9. A botanist is interested in studying the shapes of leaves and the forces that mold them. She clips some leaves from the bottom of a white oak tree and finds the leaves to be rather broad, not very deeply indented. When she goes to the top of the tree, she gets deeply indented leaves with hardly any broad expanse of blade.

10. Animals of different size work differently. The small ones talk in squeaky voices, their hearts beat faster, and they breathe more often than large animals. The skeleton of a larger animal is more robustly built than that of a small animal. The ratio of the diameter to the length in a larger animal is greater than it is in a smaller one. So there are regular distortions in the proportions of animals as the size increases from small to large.

11. A physicist is interested in studying properties of light. She wants to understand the path of a ray of light as it travels through the air into a smooth lake, particularly at the interface of the two different media.

12. A company with a fleet of trucks faces rising maintenance costs as the age and mileage of the trucks increase.

13. People are fixated by speed. Which computer systems offer the greatest speed?

14. How can we improve our ability to sign up for the right classes each term?

15. How should we save a portion of our earnings?

2.2 ILLUSTRATIVE EXAMPLES

We now demonstrate the modeling process presented in the previous section with several illustrative examples. Special emphasis is placed on identifying the problem and important variables. In subsequent chapters we will carry out the entire model-building process with these problems and use the techniques being studied there to suggest models we can test.

Example 1   Vehicular Stopping Distance

Scenario   Consider the following general rule often given in driver education classes:

> Allow one car length for every 10 miles of speed under normal driving conditions, but more distance in adverse weather or road conditions. One way to accomplish this is to use the Two-Second Rule for measuring the correct following distance no matter what your speed. To obtain that distance, watch the vehicle ahead of you pass some definite point on the highway, like a tar strip or overpass shadow. Then count to yourself "one thousand and one, one thousand and two"; that's two seconds. If you reach the mark before you finish saying those words, then you are following too close behind.

The preceding rule is implemented easily enough, but how good is it?

Problem identification   Our ultimate goal is to test this rule and suggest another rule if it fails. However, the statement of the problem—How good is the rule?— is rather vague. We need to be more specific and spell out a problem, or ask a question, whose solution or answer will help us accomplish our goal while at the same time permit a more exact mathematical analysis. Consider the following problem statement: *Predict a vehicle's total stopping distance as a function of its speed.*

Assumptions   We begin our analysis with a rather obvious model for total stopping distance:

$$\text{total stopping distance} = \text{reaction distance} + \text{braking distance}$$

By *reaction distance*, we mean the distance the vehicle travels from the instant the driver perceives a need to stop to the instant when the brakes are actually applied. *Braking distance* is the distance required for the brakes to bring the vehicle to a complete stop.

First let's develop a submodel for reaction distance. The reaction distance is a function of many variables, and we start by listing just two of them:

$$\text{reaction distance} = f(\text{response time, speed})$$

We could continue developing the submodel in as much detail as we like. For instance, response time is influenced by individual driving factors as well as by the vehicle operating system. System time is the time from which the driver touches the brake pedal until the brakes are mechanically applied. For modern cars we would probably neglect the influence of the system because it is quite small in comparison with the human factors. The portion of the response time determined by the driver depends on many things, such as reflexes, alertness, and visibility. Because we are developing only a general rule, we could just incorporate average values and conditions for the latter variables. Once all the variables deemed important to the submodel have been identified, we can begin to determine interrelationships among them. We will suggest a submodel for reaction distance in Chapter 4.

Next consider the braking distance. The weight and speed of the vehicle are certainly important factors to be taken into account. The efficiency of the brakes, type and condition of the tires, road surface, and weather conditions are other legitimate factors. As before, we would most likely assume average values and conditions for these latter factors. Thus our initial submodel gives braking distance as a function of vehicular weight and speed:

$$\text{braking distance} = h(\text{weight, speed})$$

In Chapter 4 we will also suggest and analyze a submodel for braking distance.

Finally let's discuss briefly the last three steps in the modeling process for this problem. We would want to test our model against real-world data. Do the predictions afforded by the model agree with real driving situations? If not, we would want to assess some of our assumptions and perhaps restructure one (or both) of our submodels. If the model does predict real driving situations accurately, then does the rule stated in the opening discussion agree with the model? The answer gives an objective basis for answering How good is the rule? Whatever rule we come up with (to implement the model), it must be easy to understand and easy to use if it is going to be effective. In this example, maintenance of the model does not seem to be a particular issue. Nevertheless, we would want to be sensitive to the effects on the model of changes such as power brakes or disc brakes, a fundamental change in tire design, and so on.

Example 2   *Automobile Gasoline Mileage*

Scenario   During periods of concern when oil shortages and embargoes create an energy crisis, there is always interest in how fuel economy varies with vehicular speed. We suspect that at very low speeds, when driving in low gears, automobiles convert power relatively inefficiently, and at very high speeds drag forces on the vehicle increase rapidly. It seems reasonable, then, to expect that automobiles have one or more speeds that yield optimum fuel mileage (the most miles per gallon of fuel). If this is so, fuel mileage would decrease beyond that optimum speed, but it would be beneficial to know just how this decrease takes place. Moreover, is the decrease significant? Consider the following excerpt from a newspaper article (when a national 55-mph speed limit existed):

> Observe the 55-mile-an-hour national highway speed limit. For every five miles an hour over 50, there is a loss of one mile to the gallon. Insisting that drivers stay at the 55-mile-an-hour mark has cut fuel consumption 12 percent for Ryder Truck Lines of Jacksonville, Florida—a savings of 631,000 gallons of fuel a year. The most fuel-efficient range for driving generally is considered to be between 35 and 45 miles an hour.[1]

[1] "Boost Fuel Economy," *Monterey Peninsula Herald*, May 16, 1982.

Note especially the suggestion that there is a loss of 1 mile to the gallon for every 5 mph over 50 mph. How good is this general rule?

Problem identification   *What is the relationship between the speed of a vehicle and its fuel mileage?* By answering that question, we can assess the accuracy of this rule.

Assumptions   Let's consider the factors influencing fuel mileage. First, there are propulsion forces that drive the vehicle forward. These forces depend on the power available from the type of fuel being burned, the engine's efficiency in converting that potential power, gear ratios, air temperature, and many other factors, including vehicular velocity. Next there are drag forces that tend to retard the vehicle's forward motion. The drag forces include frictional effects that depend on the vehicle's weight, type and condition of the tires, and condition of the road surface. Air resistance is another drag force and depends on the vehicular speed, vehicular surface area and shape, the wind, and air density. Another factor influencing fuel mileage relates to the driving habits of the driver. Does he or she drive at constant speeds or constantly accelerate? Does he or she drive on level or mountainous terrain? Thus fuel mileage is a function of several factors, summarized in the following equation.

$$\text{fuel mileage} = f(\text{propulsion forces, drag forces, driving habits, and so on})$$

It is clear that the answer to the original problem will be quite detailed considering all the possible combinations of car types, drivers, and road conditions. Because such a study is far too ambitious to be undertaken here, we restrict the problem we are willing to address.

Restricted problem identification   *For a particular driver, driving his or her car on a given day on a level highway at constant highway speeds near the optimum speed for fuel economy, provide a qualitative explanation of how fuel economy varies with small increases in speed.*

Under this restricted problem, environmental conditions such as air temperature, air density, and road conditions can be considered as constant. Because we have specified that the driver is driving his or her car, we have fixed the tire conditions, shape and surface of the vehicle, and fuel type. By restricting the highway driving speeds to be near the optimal speed, we obtain the simplifying assumptions of constant engine efficiency and constant gear ratio over small changes in vehicular velocity. Restricting a problem as originally posed is a powerful technique for obtaining a manageable model. We will model the fuel economy problem under the stated restricted conditions in Chapter 4.

Example 3   The Assembly Line

Scenario   On one hand, there is increasing concern by both individual laborers and labor unions for both wages and job quality. On the other hand, corporations are feeling the competition from abroad and need to keep their costs down. To increase domestic productivity, more assembly line operations have evolved. A question management faces is how to assign employees to the various jobs on the assembly line, realizing that the jobs are often quite tedious and repetitive and demand various worker skill levels.

Problem identification   The classical approach to making employee job assignments is to ensure maximum company profits. Typically, people are assigned to machines in such a way that the profit margin on the product being produced is maximized. In situations of fixed product demand, this approach amounts to minimizing the costs of production.

   The difficulty with the classical approach is that it focuses on only one aspect of the problem—namely, short-term profits. In many situations the results of such assignment procedures include job dissatisfaction, excessive absenteeism, poor workmanship and product quality, and diminished worker efficiency and overall productivity. The net effect often is increasing production costs coupled with reduced consumer demand. Thus, whereas management attempts to maximize short-term profits, long-term profits suffer.

   Although the maximization of profit is certainly a primary consideration, other factors, such as product quality and job satisfaction, must be taken into account. The precise reconciliation of the various factors into a problem statement depends on the particular situation. Let's assume that the firm has a contract and has the plant capacity to produce a fixed number of items per month. We then define the problem as follows: *Minimize production cost while meeting specified levels of demand, quality control, and job satisfaction.*

Assumptions   Under this problem identification, production cost is a function of individual wages, each individual's productivity on each job, the expected number of defects each individual makes on each job per unit time, and the reduction in both quality and productivity as a function of the time an individual spends at a particular job assignment.

   Let's ask whether we can obtain the needed data. Certainly the hourly wage presents no problem. In many instances we can build a history of an individual's productivity and workmanship by measuring the hourly output on each machine as well as the number of defects produced. The reduction in efficiency incurred by assigning an individual the same job for a prolonged period may be difficult to measure. We hope that, after observing and interviewing several employees, we can establish guidelines to prevent excessive diminution in productivity while simultaneously maintaining job satisfaction. For example, we may determine that an individual should not be assigned to a particular tedious job for more than 2 hours each day.

The advantage of this approach is that it forces the modeler to consider many more aspects of the problem. *The mathematical modeler must guard against the tendency to ignore factors just because they are difficult to quantify.* In the preceding problem there may also be salutary effects from involving management in the data collection process. A lesson to be learned from this example is that a careful modeler considers all pertinent factors and then attempts to determine the sensitivity of the model to the various assumptions made.

Example 4   *Morning Rush Hour*

The next model stresses the importance of data collection.

Scenario   You have been hired as a consultant by the manager of several skyscrapers used for offices in New York City. The clientele are particularly concerned with the poor elevator service they receive during the morning rush hour. They complain of long waits in the lobby and excessive stops by the elevator to discharge a few passengers. The manager is unwilling to construct additional elevators, being convinced the problem can be resolved through better scheduling.

Problem identification   A number of formulations of the elevator problem are worth considering. Many people are sensitive to the amount of time they spend waiting in queues, which suggests scheduling the elevators in such a way to minimize the *average waiting time* of a customer. Even then we need to be careful: Drastic reduction of the service provided to floors used by relatively few customers could reduce the average waiting time yet significantly increase the *longest waiting time*. A similar argument applies to the amount of time spent in an elevator. If the vast majority of the people in an elevator require delivery to the top floor, the average travel time could be reduced by going to the top floor first and then delivering the remaining customers on the way back down. However, the longest delivery time might increase considerably. Moreover, consider the psychological effect on the customer who wants to go only to the second floor.

In this type of problem, it is important to be sensitive to people's perceptions about the service being provided. How do the manager and customers measure a successful operation? It may well be based on the *length of the longest queue* or on the *number of complaints* received from dissatisfied customers. The point is that each of these criteria leads to a distinct mathematical model with generally different optimal solutions. Unless we consider such possibilities, we may succeed in formulating and solving a difficult mathematical problem yet fail to improve the manager's situation.

Although any of the problem formulations might be appropriate in a given situation, let's consider the total delivery time for our problem identification: *Minimize the average total elevator delivery time of a morning's rush hour customers as measured by the difference between individual arrival time at the lobby and arrival time at the floor of destination.*

This formulation takes into account both the time spent in the lobby and the time spent in the elevator. A weighted combination of these two times might be more appropriate. Whereas we have identified a particular problem to be solved here, the alternative problems mentioned earlier should make us sensitive to other criteria for formulating a problem, such as minimizing the longest total delivery time or the longest queue.

Assumptions   Factors influencing the total delivery time include the building layout (location of offices, warehouses, factories, and so on) and the distribution of starting times of the various agencies by the number of customers, number of floors, and commuter schedules. To determine interrelationships among the variables, we need to gather additional information about them. We may determine that different agencies have distinct starting times, so the present poor service is caused by a few early arrivals or a few stragglers. In such cases we could use express elevators for high-density floors to give them priority during their peak arrival periods. In other cases we might find that the arrival times are largely driven by commuter schedules, and we would want to investigate the predictability of the appropriate schedules. It may be determined that the distribution of arrivals and destinations approximate well-studied distribution functions. Then we could use properties of those distributions to estimate the effects of various schemes for routing the elevators on the average total delivery time. Taking another approach, we may find it desirable to simulate the arrivals via a computer and then test various schemes accordingly. Among the schemes worth considering are the following:

1. Assigning elevators to the even and odd floors.
2. Splitting the building into two or more groups of contiguous floors and assigning a different set of elevators to each group of floors.
3. Reserving express elevators for high-density floors during peak periods.
4. Using express elevators for specified floors and local elevators in between those floors.

Model verification   How can we verify the solution to our model once it is obtained? One way is to sample customers during the rush hour and measure their total delivery time. Before and after implementing the solution to our model, we might be interested in gathering statistics such as average and longest time spent in the lobby, average and longest time in the elevator, and length of the longest queue. The manager's clients may, however, object to the harassment of such an approach. Then we would be forced to use more easily obtained measures such as queue length and running time versus stopped time of the elevators. A word of caution is in order here. When using indirect measures such as those just mentioned, we may have a tendency to experiment with the solution to improve the indirect measure. But the improvement of an indirect measure may not result in a more satisfied customer. For example, we might work to minimize the longest queue, if that is the measure of effectiveness. As a result, customers may find themselves spending less time in lobbies but more time in elevators and may still feel dissatisfied with the service.

2.2 PROBLEMS

1. Consider the assumptions in the example on vehicular stopping distance. Name some variables other than the two listed in the text that influence the reaction distance. Once a model is constructed for this problem, how would you go about validating it? What kind of data would you collect? Is it possible to obtain the data? How would you use the model, once it is validated to your satisfaction, to test the rule stated in the scenario?

2. In the automobile gasoline mileage example, discuss the difficulties associated with constructing a model for the original problem: What is the relationship between the speed of a vehicle and its fuel mileage?

3. Consider a new company that is just getting started in producing a single product in a competitive market situation. Discuss some of the short-term and long-term goals the company might have as it enters into business. How do these goals affect employee job assignments? Would the company necessarily decide to maximize profits in the short run? Why?

4. Discuss the differences between using a model to predict, versus to explain, a real-world system. Think of some situations in which you would like to explain a system; likewise, imagine others in which you would want to predict a system. Relate this question to the examples presented in the text.

5. Suppose the manager of the skyscraper in the elevator problem asks you (the consultant), How long does it take to get to the top? What questions would you begin to ask to formulate a problem for purposes of constructing a model?

   a. If the elevator is an express to the top, how might a graph of its speed versus time appear? Assume a constant acceleration and deceleration of the elevator of α ft per sec per sec.

   b. What is the total distance traveled during both acceleration and deceleration?

   c. If D is the distance in feet to the top, answer the manager's question.

2.2 PROJECTS

1. Consider the taste of brewed coffee. What are some of the variables affecting taste? Which variables might be neglected initially? Suppose you hold all variables fixed except water temperature. Most coffeepots use boiled water in some manner to extract the flavor from the ground coffee. Do you think boiled water is optimal for producing the best flavor? How would you test this submodel? What data would you collect and how would you gather it?

2. A transportation company is considering transporting people between skyscrapers in New York City via helicopter. You are hired as a consultant to determine the number of helicopters needed. Identify an appropriate problem precisely. Use the model-building process to identify the data you would like to have to

determine the relationships among the variables you select. You may want to redefine your problem as you proceed.

3. Consider wine making. Suggest some objectives a commercial producer might have. Consider taste as a submodel. What are some of the variables affecting taste? Which variables might be neglected initially? How would you relate the remaining variables? What data would be useful to determine the relationships?

4. Should a couple buy or rent a home? As the cost of a mortgage rises, intuitively, it would seem that there is a point at which it no longer pays to buy a house. What variables determine the total cost of a mortgage?

5. Consider the operation of a medical office. Records have to be kept on individual patients, and accounting procedures are a daily task. Should the office buy or lease a small computer system? Suggest objectives that might be considered. What variables would you consider? How would you relate the variables? What data would you like to have to determine the relationships among the variables you select? Why might solutions to this problem differ from office to office?

6. When should a person replace his or her vehicle? What factors should affect the decision? Which variables might be neglected initially? Identify the data you would like to have to determine the relationships among the variables you select.

7. How far can a person long jump? In the 1968 Olympic Games in Mexico City, Bob Beamon of the United States increased the record by a remarkable 10%, a record that stood through the 1996 Olympics. List the variables that affect the length of the jump. Do you think the low air density of Mexico City accounts for the 10% difference?

8. Is college a financially sound investment? Income is forfeited for 4 years, and the cost of college is extremely high. What factors determine the total cost of a college education? How would you determine the circumstances necessary for the investment to be profitable?

Chapter Three

MODELING WITH DISCRETE DYNAMICAL SYSTEMS[1]

INTRODUCTION

In mathematical modeling we are often interested in building models to explain behavior or make predictions. In this chapter we direct our attention to modeling *change*. A powerful paradigm to use is

$$\textit{future value} = \textit{present value} + \textit{change}$$

Often, we wish to predict the future on what we know now, in the present, and add in the change that has been carefully observed. In such cases, we actually begin by studying the change itself according to the formula

$$\textit{change} = \textit{future value} - \textit{present value}$$

difference equation

differential equation

interactive systems

By collecting data over a period of time and plotting that data, we often can discern patterns to model that capture the trend of the change. If the behavior is taking place over *discrete time periods*, the preceding construct leads to a **difference equation**, which we study in this chapter. If the behavior is taking place *continuously* with respect to time, the construct leads to a **differential equation**, which we begin studying in Chapter 10. Both are powerful methodologies for studying change to explain and predict behavior.

We begin our study of *discrete change* in Section 3.1 by examining behavior that can be modeled *exactly* by difference equations. In Section 3.2 we use proportionality to *approximate change* we have observed. In Section 3.3 we construct *numerical solutions* to the difference equations we have built to determine the types of *long-term behaviors* they predict. In Section 3.4 we model **interactive systems**, such as ecological systems involving predators and prey, using systems of difference

[1]Part of this chapter is adapted with permission from Chapter 1, *Principles and Practices of Mathematics*, COMAP, written by Frank Giordano, Chris Arney, and Shelley Gordon.

dynamical system

equations. Again we examine the long-term behaviors of these systems by examining numerical solutions. We begin by modeling the arms race as a **dynamical system**.

The Arms Race Revisited

In the arms race model of Section 1.1, we assumed each country follows a deterrent strategy that requires it to have a given number of weapons to deter the enemy (inflict unacceptable damage) even if the enemy has no weapons. Under this strategy, as the enemy adds weapons, the friendly force increases its arms inventory by some percentage of the number of attacking weapons, which depends on how effective the friendly force perceives the enemy's weapons to be. For example, suppose Country Y feels it needs 120 weapons to deter the enemy. Further, for every two weapons possessed by Country X, Country Y feels it needs to add one additional weapon (to ensure 120 weapons remain after a strike by Country X). Thus the number of weapons needed by Country Y (y weapons) as a function of the number of weapons it thinks Country X has (x weapons) is

$$y = 120 + \frac{1}{2}x$$

Now suppose Country X is following a similar strategy, feeling it needs 60 weapons even if Country Y has no weapons. Further, for every three weapons it thinks Country Y possesses, Country X feels it must add one weapon. Thus the number x of weapons needed by Country X as a function of the number y of weapons it thinks Country Y has is

$$x = 60 + \frac{1}{3}y$$

How does the arms race proceed?

Modeling the Dynamics of the Arms Race

Suppose initially that Countries Y and X do not think the other side has arms. Then they build 120 weapons and 60 weapons, respectively. Now assume each has perfect intelligence; that is, each knows the other has built weapons. In the next stage Country Y increases its inventory to 150 weapons:

$$y = 120 + \frac{1}{2}(60) = 150 \text{ weapons}$$

Similarly, Country X notes that Country Y had 120 weapons during the previous stage and increases its inventory to 100 weapons:

$$x = 60 + \frac{1}{3}(120) = 100 \text{ weapons}$$

The arms race would proceed *dynamically*; that is, in successive stages. At each stage a country adjusts its inventory based on the strength of the enemy during the *previous* stage. In stage $n = 2$, Country Y realizes that Country X now has 100 weapons and reacts by increasing its inventory to $y = 120 + \frac{1}{2}(100) = 170$. Similarly, Country X increases its inventory to $x = 60 + \frac{1}{3}(150) = 110$. If we let n represent the stage of the arms race, convince yourself that the data in the following table represent the growth of the arms race under the assumptions we have just made.

Stage n	0	1	2	3	4
Country Y	120	150	170	175	175
Country X	60	100	110	117	118

Note that the growth in the arms race appears to be diminishing. The number of weapons needed by Country Y appears to be headed for about 180, whereas the number for Country X appears headed for about 120 weapons. (See Figures 3.1 and 3.2.) Does this model actually predict an equilibrium value as we suggested in the model developed in Section 1.1? Is the equilibrium position *stable* in the sense that small changes in the number of weapons *initially* possessed by either side has little change on the final outcome? Is the outcome sensitive to changes in the coefficients of the model? In this chapter we learn how to model dynamical systems such as this arms race and to use our model for predictions. Finally, we determine how *sensitive* the predictions are to the assumptions we have made.

Before we begin, let's introduce some notation and formulate the preceding model in a common format we will use throughout this chapter.

$$\text{Let } n = \text{stage (years, decades, fiscal periods, etc.)}$$

$$x_n = \text{number of weapons possessed by } X \text{ in stage } n$$

$$y_n = \text{number of weapons possessed by } Y \text{ in stage } n$$

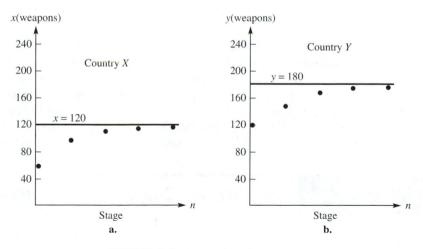

FIGURE 3.1   Dynamics of an arms race

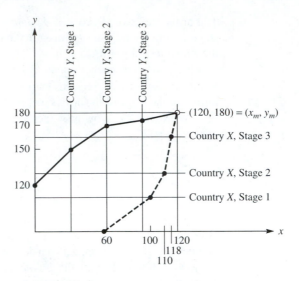

FIGURE 3.2 Arms race curves for each country

Then our assumptions imply that at stage $n + 1$

$$y_{n+1} = y_0 + \frac{1}{2} x_n$$

$$x_{n+1} = x_0 + \frac{1}{3} y_n$$

with

$$x_0 = 60$$

and

$$y_0 = 120$$

initial values   The values x_0 and y_0 are called **initial values**. Along with the coefficients
parameters $\frac{1}{2}$ and $\frac{1}{3}$, they are **parameters** we ultimately would like to vary to determine the
sensitivity of the predictions. In the preceding format, it is clear that the number of
weapons needed by either side in stage $n + 1$ depends on the number of missiles
in the previous stage. Thus we are modeling the *change* in the system under study.
Often in mathematical modeling we are interested in capturing the change in a
system so we can predict the future, so let's learn how to model dynamical systems.

3.1   MODELING CHANGE WITH DIFFERENCE EQUATIONS

In this section we build mathematical models to describe change in an observed
behavior. When we observe change, we are often interested in understanding why
the change occurs in the way it does, perhaps to analyze the effects of different

conditions or to predict what will happen in the future. A mathematical model helps us better understand a behavior while allowing us to experiment mathematically with different conditions affecting it. Let's begin with an example.

Example 1   A Savings Certificate

Consider the value of a savings certificate initially worth $1,000 that accumulates interest paid each month at 1% per month. The following *sequence* of numbers represents the value of the certificate in consecutive months:

$$A = \{1000, 1010, 1020.10, 1030.30, \ldots\}$$

Definition 1   For a sequence of numbers $A = \{a_1, a_2, a_3, \ldots\}$ the **first differences** are defined as follows:

$$\Delta a_1 = a_2 - a_1$$
$$\Delta a_2 = a_3 - a_2$$
$$\Delta a_3 = a_4 - a_3, \ldots$$

In general, the **nth first difference** is defined as

$$\Delta a_n = a_{n+1} - a_n$$

Note from Figure 3.3 that the first difference represents the rise or fall; that is, the *change* in the graph of the sequence during one time period. For example, several of the first differences for the sequence representing the value of the savings certificate

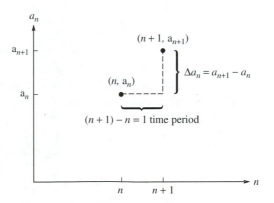

FIGURE 3.3   The first difference of a sequence is the rise in the graph

are as follows:

$$\Delta a_1 = a_2 - a_1 = 1010 - 1000 = 10$$
$$\Delta a_2 = 1020.10 - 1010 = 10.10$$
$$\Delta a_3 = 1030.30 - 1020.10 = 10.20$$

Note that the first differences represent the *change in the sequence* during the period, or the *interest earned* in the case of the savings certificate example.

The first difference is useful for modeling change taking place in discrete intervals. In the current example, we know that the change in the value of the certificate from one period to the next is merely the interest paid during that period. If n is the number of months and a_n the value of the certificate after n months, then the change in each period is represented by the nth difference

$$\Delta a_n = a_{n+1} - a_n = 0.01a_n$$

This expression can be rewritten as the following difference equation:

$$a_{n+1} = a_n + 0.01a_n$$

We also know the initial deposit (initial value) that then gives the **dynamical system model**

$$a_{n+1} = 1.01a_n, \quad n = 0, 1, 2, 3, \ldots \tag{3.1}$$
$$a_0 = 1000$$

where a_n represents the amount accrued after n months. Note that since n represents the nonnegative integers $\{0, 1, 2, 3, \ldots\}$, Equation (3.1) represents an *infinite set* of algebraic equations, called a dynamical system. Also note that dynamical systems describe the *change* from one period to the next. The difference equation formula provides a method for computing the next term knowing the immediately previous term in the sequence, but it does not allow us to compute a specific term directly (say the savings after 100 periods).

Because it is change we often observe, we can construct a difference equation by representing or approximating the change from one period to the next. To modify our example, if we were to withdraw $50 from the account each month, the change during a period would be the interest earned during that period minus the monthly withdrawal, or

$$\Delta a_{n+1} = a_{n+1} - a_n = 0.01a_n - 50$$

In most examples, mathematically describing the change is not going to be as precise a procedure as illustrated here. Often it is necessary to *plot the change, observe a pattern*, and then *describe the change* in mathematical terms. That is, we

will be trying to find

$$change = \Delta a_n = some\ function$$

The change may be a function of previous terms in the sequence (as was the case with no monthly withdrawals), or it may also involve some external terms (such as the amount of money withdrawn in the current example or an expression involving the period n). Thus, in constructing models representing change we will be **modeling change in discrete intervals**, where

$$change = \Delta a_n = a_{n+1} - a_n = f\ (terms\ in\ the\ sequence,\ external\ terms)$$

Modeling change in this way becomes the art of determining or approximating a function f that represents the change.

Let's consider a second example in which a difference equation exactly models a behavior being followed in the real world.

Example 2   *Mortgaging a Home*

Six years ago your parents purchased a home by financing $80,000 for 20 years, paying monthly payments of $880.87 with a monthly interest of 1%. They have made 72 payments and wish to know how much they owe on the mortgage, which they are considering paying off with an inheritance they received. Or they could be considering refinancing the mortgage with several interest rate options, depending on the length of the payback period. The change in the amount owed each period increases by the amount of interest and decreases by the amount of the payment:

$$\Delta b_n = b_{n+1} - b_n = 0.01b_n - 880.87$$

Solving for b_{n+1} and incorporating the initial condition gives the dynamical system model

$$b_{n+1} = b_n + 0.01b_n - 880.87$$
$$b_0 = 80000$$

where b_n represents the amount owed after n months. Thus,

$$b_1 = 80000 + 0.01(80000) - 880.87 = 79919.13$$
$$b_2 = 79919.13 + 0.01(79919.13) - 880.87 = 79837.45$$

yielding the sequence

$$B = \{80000, 79919.13, 79837.45, \ldots\}$$

The sequence is graphed in Figure 3.4.

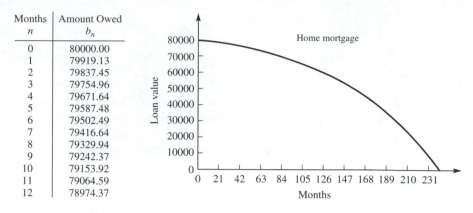

| Months | Amount Owed |
n	b_n
0	80000.00
1	79919.13
2	79837.45
3	79754.96
4	79671.64
5	79587.48
6	79502.49
7	79416.64
8	79329.94
9	79242.37
10	79153.92
11	79064.59
12	78974.37

FIGURE 3.4   Numerical solution and graph for Example 2

In the problem set we will see other behaviors that follow procedures in the world that can be modeled exactly by difference equations. In the next section, we will use difference equations to approximate change that has been observed. After collecting data for the change and discerning patterns of the behavior, we will use the concept of proportionality to test and fit models that we propose. But first, let's summarize the important definitions and practice using difference equations to model exactly real-world behavior.

Definition 2   A **sequence** is a function whose domain is the set of all nonnegative integers and whose range is a subset of the real numbers.

Definition 3   A **dynamical system** is a relationship among terms in a sequence.

Definition 4   A **numerical solution** is a table of values satisfying the dynamical system.

3.1 PROBLEMS

Sequences

1. Write out the first five terms of the following sequences.

   a. $a_{n+1} = 3a_n,\ a_0 = 1$

   b. $a_{n+1} = 2a_n + 6,\ a_0 = 0$

   c. $a_{n+1} = 2a_n(a_n + 3),\ a_0 = 4$

   d. $a_{n+1} = a_n^2,\ a_0 = 1$

Difference Equations

2. By examining the following sequences, write a difference equation to represent the change during the nth interval as a function of the previous term in the sequence.

a. $\{2, 4, 6, 8, 10, \ldots\}$
b. $\{2, 4, 16, 256, \ldots\}$
c. $\{1, 2, 5, 11, 23, \ldots\}$
d. $\{1, 8, 29, 65, \ldots\}$

Dynamical Systems

3. By substituting $n = 0, 1, 2, 3$ write out the first four algebraic equations represented by the following dynamical systems.

a. $a_{n+1} = 3a_n, \; a_0 = 1$
b. $a_{n+1} = 2a_n + 6, \; a_0 = 0$
c. $a_{n+1} = 2a_n(a_n + 3), \; a_0 = 4$
d. $a_{n+1} = a_n^2, \; a_0 = 1$

Modeling Change Exactly

For Problems 4–7, formulate a dynamical system that models change exactly for the following situations.

4. You currently have $5,000 in a savings account that pays 0.5% interest each month. You add another $200 each month.

5. You owe $500 on a credit card that charges 1.5% interest each month. You can pay $50 each month and make no new charges.

6. Your parents are considering a 30-year $100,000 mortgage that charges 0.5% interest each month. Formulate a model in terms of a monthly payment p that allows the mortgage (loan) to be paid off after 360 payments. *Hint:* If a_n represents the amount owed after n months, what are a_0 and a_{360}?

7. Your grandparents have an annuity. The value of the annuity increases each month as 1% interest on the previous month's balance is deposited. Your grandparents withdraw $1,000 each month for living expenses. Presently, they have $50,000 in the annuity. Model the annuity with a dynamical system. Will the annuity run out of money? When? *Hint:* What value will a_n have when the annuity is depleted?

8. Name several behaviors you think can be modeled by dynamical systems.

3.1 PROJECTS

1. You wish to buy a new car upon graduation from college. You narrow your choices to a Saturn, Cavalier, and Tercel. Each company offers you its prime deal:

Saturn	$11,990	$1,000 down	3.5% interest for up to 60 months
Cavalier	$11,550	$1,500 down	4.5% interest for up to 60 months
Tercel	$10,900	$500 down	6.5% interest for up to 48 months

You are able to spend at most $475 a month on a car payment. Use a dynamical system to determine which car to buy.

3.2 APPROXIMATING CHANGE WITH DIFFERENCE EQUATIONS

In most examples, describing the change mathematically will not be as precise a procedure as in the cases of the savings certificate and mortgage examples. Typically, we must plot the change, observe a pattern, and then approximate the change in mathematical terms. In this section we approximate some observed change to complete the expression

$$change = \Delta a_n = some\ function\ f$$

Modeling change is then the art of determining or approximating the function f, which represents the change. We begin by distinguishing between change that takes place continuously and change that occurs in discrete time intervals.

Discrete Versus Continuous Change

discrete
continuous

When constructing models involving change, an important distinction is that some change takes place in **discrete** time intervals (such as the depositing of interest in an account) and other change happens **continuously** (such as the change in the temperature of a cold can of soda on a warm day). Difference equations represent change in the case of discrete time intervals. Later we will see the relationship between discrete change and continuous change (for which calculus was developed). For now, in the several models that follow, we approximate a continuous change by examining data taken at discrete time intervals.

Approximating Change

Few models represent exactly the real world. Generally, mathematical models simply *approximate* real-world behavior. That is, some *simplification* is required to represent a real-world behavior with a mathematical construct. For example, suppose we want to represent the spotted owl population in a habitat to predict the effect of changes in environmental policy (such as intense logging operations). The spotted owl population depends on many variables, including the birth rate, death rate, availability of resources, competition for resources, predators, and natural disasters.

In a mathematical construct we cannot hope to capture precisely every detail. Eventually we will build models considering each of the variables listed, but first let's construct some simple models.

Example 1 Growth of a Yeast Culture

Consider the data in Figure 3.5 that was collected from an experiment measuring the growth of a yeast culture. The graph represents the assumption that the change in population is proportional to the current size of the population. That is, $\Delta p_n =$

Time in hours n	Observed yeast biomass p_n	Change in biomass $p_{n+1} - p_n$
0	9.6	
1	18.3	8.7
2	29.0	10.7
3	47.2	18.2
4	71.1	23.9
5	119.1	48.0
6	174.6	55.5
7	257.3	82.7

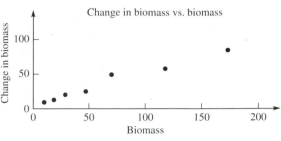

FIGURE 3.5   Growth of a yeast culture versus time in hours; data from R. Pearl, "The Growth of Population," *Quart. Rev. Biol.* 2(1927): 532–548.

$p_{n+1} - p_n = kp_n$, where p_n represents the size of the biomass after n hours (or other time measurement) and k is a positive constant.

Although the graph of the data does not appear to lie precisely along a straight line passing exactly through the origin, it can be *approximated* by such a straight line. By placing a ruler over the data to approximate a straight line through the origin, we can estimate the slope of the line to be about 0.6. Using the estimate $k = 0.6$ for the slope of the line, we might hypothesize the proportionality model

$$\Delta p_n = p_{n+1} - p_n = 0.6 p_n$$

yielding the prediction $p_{n+1} = 1.6 p_n$. Note that the model predicts a population that increases forever.

Model Refinement: Modeling Births, Deaths, and Resources

If both births and deaths during a period are proportional to the population, then the change in population itself should be proportional to the population, as was illustrated. However, certain resources (food, for instance) can support only a maximum population level rather than one that increases indefinitely. As these maximum levels are approached, growth should slow. The data in Figure 3.6 show what actually happens to the yeast culture growing in a restricted area as time increases beyond the eight observations given in Figure 3.5.

Notice from the third column of the worksheet in Figure 3.6 that the change in population per hour becomes smaller as the resources become more limited or constrained. From the graph of population versus time, the population appears to be **carrying capacity** approaching a limiting value or **carrying capacity**. Suppose that based on our graph we estimate the carrying capacity to be 665. (Actually, the graph doesn't precisely tell us the correct number is 665 and not 664 or 666, for example.) Nevertheless, as p_n approaches 665, the change does slow considerably. Because $665 - p_n$ does get

Time in hours n	Yeast biomass p_n	Change/ hour $p_{n+1} - p_n$
0	9.6	
1	18.3	8.7
2	29.0	10.7
3	47.2	18.2
4	71.1	23.9
5	119.1	48.0
6	174.6	55.5
7	257.3	82.7
8	350.7	93.4
9	441.0	90.3
10	513.3	72.3
11	559.7	46.4
12	594.8	35.1
13	629.4	34.6
14	640.8	11.4
15	651.1	10.3
16	655.9	4.8
17	659.6	3.7
18	661.8	2.2

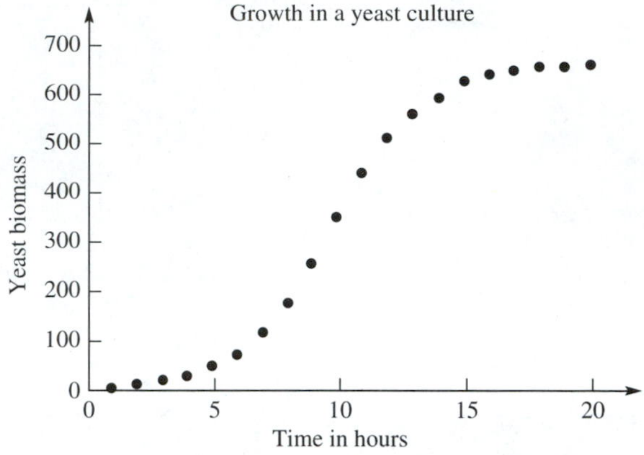

FIGURE 3.6 Yeast biomass approaches a limiting population level

smaller as p_n approaches 665, consider the model

$$\Delta p_n = p_{n+1} - p_n = k(665 - p_n)p_n$$

which causes Δp_n to become increasingly small as p_n approaches 665. Let's check this hypothesized model against data.

To test the model, plot $(p_{n+1} - p_n)$ versus $(665 - p_n)p_n$ to see if there is a reasonable proportionality. Then find the proportionality constant k.

Examining Figure 3.7, we see that the plot reasonably approximates a straight line projected through the origin. If we accept the proportionality argument, we can estimate the slope of the line approximating the data to be $k \approx 0.00082$, which gives the model

$$p_{n+1} - p_n = 0.00082(665 - p_n)p_n$$

Solve the Model Numerically and Verify the Results

Solving for p_{n+1} gives

$$p_{n+1} = p_n + 0.00082(665 - p_n)p_n$$

Observe that the right-hand side of the last equation is a quadratic in p_n. Such **nonlinear** dynamical systems are classified as **nonlinear** and generally cannot be solved for analytical solutions. That is, we usually cannot find a formula expressing p_n in terms of n. However, if we are given that $p_0 = 9.6$, we can substitute in the expression to

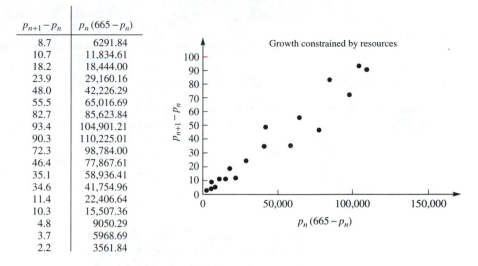

$p_{n+1}-p_n$	$p_n(665-p_n)$
8.7	6291.84
10.7	11,834.61
18.2	18,444.00
23.9	29,160.16
48.0	42,226.29
55.5	65,016.69
82.7	85,623.84
93.4	104,901.21
90.3	110,225.01
72.3	98,784.00
46.4	77,867.61
35.1	58,936.41
34.6	41,754.96
11.4	22,406.64
10.3	15,507.36
4.8	9050.29
3.7	5968.69
2.2	3561.84

FIGURE 3.7 Testing the constrained growth model

compute p_1:

$$p_1 = p_0 + 0.00082(665 - p_0)p_0 = 9.6 + 0.00082(665 - 9.6)9.6 = 14.76$$

iterating
numerical solution
model predictions

In a similar manner, we can substitute $p_1 = 14.76$ to compute $p_2 = 28.00$. **Iterating** in this way we can compute a table of values to provide a **numerical solution** to the model. This numerical solution of **model predictions** is presented in Figure 3.8. The predictions and observations are plotted together versus time on

Time in hours	Observations	Predictions
0	9.6	9.60
1	18.3	14.76
2	29.0	28.00
3	47.2	44.12
4	71.1	71.11
5	119.1	105.73
6	174.6	172.41
7	257.3	244.81
8	350.7	343.32
9	441.0	441.08
10	513.3	522.00
11	559.7	577.15
12	594.8	608.03
13	629.4	629.04
14	640.8	647.77
15	651.1	653.52
16	655.9	658.52
17	659.6	660.79
18	661.8	662.52

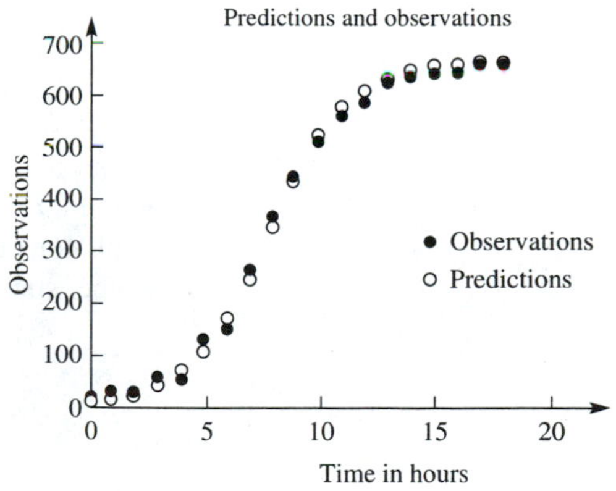

FIGURE 3.8 Model predictions and observations

the same graph. Note that the model captures very well the *trend* of the observed data.

We next construct several models that are solved in subsequent sections.

Example 2 **Spotted Owl Population**

Here we build four models of the spotted owl population, each incorporating different assumptions. The first is a simple model that assumes an abundance of resources. The second adds the assumption that resources are restricted. In the third and fourth models we assume the existence of another species living in the ecosystem. More specifically, in the third model we assume the two species compete against each other for the scarce resources, and in the fourth model we assume the two species have a predator–prey relationship.

Unconstrained growth Assume the population changes by only births and deaths. Also assume that during each time period the number of births is a percentage of the current population bp_n for b a positive constant. Similarly, assume during each period that a percentage of the current population dies; say dp_n for d a positive constant. Neglecting all other variables, the change in population is the births minus the deaths,

$$\Delta p_n = p_{n+1} - p_n = bp_n - dp_n = kp_n$$

where $k = b - d$ represents the growth constant (because for most populations k is a positive number).

Constrained growth Suppose the habitat can only support an owl population of size M, where M represents the carrying capacity of the environment. That is, if there are more than M spotted owls, the growth rate becomes negative. Also assume the growth rate slows as p nears M. One model capturing these assumptions is

$$\Delta p_n = p_{n+1} - p_n = k(M - p_n)p_n$$

for k a positive constant.

Competing species Now suppose a second species lives in the habitat. Let's denote the population of the competing species after n periods as c_n. Suppose in the absence of the other species, each individual species exhibits unconstrained growth:

$$\Delta c_n = c_{n+1} - c_n = k_1 c_n$$

and

$$\Delta p_n = p_{n+1} - p_n = k_2 p_n$$

where k_1 and k_2 represent the constant (positive) growth rates. The effect of the presence of a second species is to diminish the growth rate of the other species, and

vice versa. Whereas there are many ways to model the mutually detrimental interaction of the two species, we will assume that this decrease is roughly proportional
to the number of possible interactions between the two species. So one submodel
is to assume that the decrease is proportional to the product of c_n and p_n. These
considerations are modeled by the equations

$$\Delta c_n = c_{n+1} - c_n = k_1 c_n - k_3 c_n p_n$$

and

$$\Delta p_n = p_{n+1} - p_n = k_2 p_n - k_4 c_n p_n$$

The positive constants k_3 and k_4 represent the *relative intensities* of the competitive
interactions.

Predator–prey species Next let's assume that the spotted owl's primary food
source is a single prey, say mice. We denote the size of the mouse population
after n periods as m_n. In the absence of the predatory spotted owl, the mouse
population prospers. We can model the detrimental effect on the mouse growth rate
by the spotted owl (predator) population similar to how we modeled detrimental
competition:

$$\Delta m_n = m_{n+1} - m_n = k_1 m_n - k_2 p_n m_n$$

where k_1 and k_2 are positive constants. On the other hand, if the single food source of
the spotted owl is the mouse, then in the absence of mice the spotted owl population
would diminish to zero, say at a rate proportional to p_n (or $-k_3 p_n$ for k_3 a positive
constant). The presence of more mice increases the growth rate of owls, and we
assume the increase in the growth rate of the predatory owl is proportional to the
product of p_n and m_n. One such model is

$$\Delta p_n = p_{n+1} - p_n = -k_3 p_n + k_4 p_n m_n$$

Summarizing our predator–prey model we have the system

$$\Delta m_n = m_{n+1} - m_n = k_1 m_n - k_2 p_n m_n$$
$$\Delta p_n = p_{n+1} - p_n = -k_3 p_n + k_4 p_n m_n$$

Notice the similarities and differences between the predator–prey model and the
competing species model presented earlier.

Example 3 Spread of a Contagious Disease

Suppose there are 400 students in a college dormitory and that one or more students
has a severe case of the flu. Let i_n represent the number of infected students after
n time periods. Assume some interaction between those infected and those not
infected is required to pass on the disease. If all are susceptible to the disease,

$(400 - i_n)$ represents those susceptible but not yet infected. If those infected remain contagious, we can model the change in those infected as a proportionality to the product of those infected by those susceptible but not yet infected, or

$$\Delta i_n = i_{n+1} - i_n = ki_n(400 - i_n)$$

In this model the product $i_n(400 - i_n)$ represents possible interactions between those infected and those not infected. A fraction k would now become infected.

There are many refinements to this model. For example, we can consider that a segment of the population is not susceptible to the disease, that the infection period is limited, or that infected students are removed from the dorm to prevent interaction with the uninfected. More sophisticated models treat the infected and susceptible populations separately, as we did in the predator–prey model.

Example 4   Cooling of a Heated Object

Now we consider a physical example in which the behavior is taking place continuously. Suppose a cold can of soda is taken from a refrigerator and placed in a warm classroom and we measure the temperature periodically. Suppose the temperature of the soda is initially 40°F and the room temperature is 72°F. Temperature is a measure of energy per unit volume. Because the volume of soda is small relative to the volume of the room, we would expect the room temperature to remain constant. Further, we may want to assume the entire can of soda is the same temperature neglecting variation within the can. We might expect the change in temperature per time period to be greater when the difference in temperatures between the soda and room is large, and the change in temperature per unit time to be less when the difference in temperatures is small. Letting t_n represent the temperature of the soda after n time periods and k a positive constant of proportionality, we propose

$$\Delta t_n = t_{n+1} - t_n = k(72 - t_n)$$

Again, many refinements are possible for this model. Although we have assumed k is constant, it actually depends on the shape and conductivity properties of the container, the time period between the measurements, and so on. Also, the temperature of the environment may not be constant in many instances and it may be necessary to take into account that the temperature is not uniform throughout the object. The temperature of the object may vary in one dimension (as in the case of a thin wire), in two dimensions (such as for a flat plate), or in three dimensions (as in the case of a space capsule reentering the Earth's atmosphere).

We have presented only a glimpse of the power of difference equations to model change in the world about us. In the next section, we build numerical solutions to some of these models and observe the patterns they exhibit. Having observed certain patterns for various types of difference equations, we will then classify them by their mathematical structures. This will assist us in determining the long-term behavior of the dynamical systems under study.

3.2 PROBLEMS

Approximating Change Using Proportionality

1. Digoxin is used in the treatment of heart disease. In the following table, y represents the amount of digoxin in the bloodstream and t represents the time in days after taking a single dose. The initial dosage is 0.5 mg.

t	0	1	2	3	4	5	6	7	8
y	0.500	0.345	0.238	0.164	0.113	0.078	0.054	0.037	0.026

   a. Formulate a model using a difference equation for the *change* in concentration per day being proportional to the amount of digoxin present. Test your assumption by plotting the change versus the amount present at the beginning of the period.

   b. Assume that after the initial dose of 0.5 mg, each day a maintenance dose of 0.1 mg is taken. Formulate the refined model.

   c. Build a table of values or numerical solution for part b for 15 days.

2. A certain drug is effective in treating a disease if the concentration remains above 100 mg/L. The initial concentration is 640 mg/L. It is known from laboratory experiments that the drug decays at the rate of 20% of the amount present each hour.

   a. Formulate a model representing the concentration at each hour.

   b. Build a table of values and determine when the concentration reaches 100 mg/L.

3. A humanoid skull is discovered near the remains of an ancient campfire. Archaeologists are convinced the skull is the same age as the original campfire. It is determined from laboratory testing that only 1% of the original amount of carbon-14 remains in the burned wood taken from the campfire. It is known that carbon-14 decays at a rate proportional to the amount remaining and that carbon-14 decays 50% over 5700 years. Formulate a model for carbon-14 dating.

4. Sociologists recognize a phenomenon called *social diffusion*, which is the spreading of a piece of information, a technological innovation, or a cultural fad among a population. The members of the population can be divided into two classes: those who have the information and those who do not. In a fixed population whose size is known, it is reasonable to assume that the rate of diffusion is proportional to the number who have the information times the number yet to receive it. If a_n denotes the number of people who have the information in a population of N people after n days, formulate a dynamical system to approximate the change in the number of people in the population who have the information.

5. Consider the spreading of a highly communicable disease on an isolated island with population size N. A portion of the population travels abroad and returns to the island infected with the disease. Formulate a dynamical system to

approximate the change in the number of people in the population who have the disease.

6. Assume we are considering the survival of whales and that if the number of whales falls below a minimum survival level *m* the species will become extinct. Assume also that the population is limited by the carrying capacity *M* of the environment. That is, if the whale population is above *M*, it will experience a decline because the environment cannot sustain that large a population level. In the following model, a_n represents the whale population after *n* years; discuss the model:

$$a_{n+1} = k(M - a_n)(a_n - m)$$

7. The following data were obtained for the growth of a sheep population introduced into a new environment on the island of Tasmania.[2]

Year	1814	1824	1834	1844	1854	1864
Population	125	275	830	1200	1750	1650

Plot the data. Is there a trend? Plot the change in population versus years elapsed after 1814. Formulate a discrete dynamical system that reasonably approximates the change you have observed.

8. The following data represent the U.S. population from 1790 to 1990.

Year	Population
1790	3,929,000
1800	5,308,000
1810	7,240,000
1820	9,638,000
1830	12,866,000
1840	17,069,000
1850	23,192,000
1860	31,443,000
1870	38,558,000
1880	50,156,000
1890	62,948,000
1900	75,995,000
1910	91,972,000
1920	105,711,000
1930	122,755,000
1940	131,669,000
1950	150,697,000
1960	179,323,000
1970	203,212,000
1980	226,505,000
1990	248,710,000

[2]Adapted from J. Davidson, "On the Growth of the Sheep Population in Tasmania," *Trans. Roy. Soc. S. Australia* 62(1938): 342–346.

Find a dynamical system model that fits the data fairly well. Test your model by plotting the predictions of the model against the data.

9. The gross national product (GNP) represents the sum of consumption purchases of goods and services, government purchases of goods and services, and gross private investment (which is the increase in inventories plus buildings constructed and equipment acquired). Assume that the GNP is increasing at the rate of 3% per year and that the national debt is increasing at a rate proportional to the GNP. Let a_n represent the GNP after n years and b_n represent the national debt after n years. Construct a dynamical system to model the change in the GNP and national debt.

10. In 1868, the accidental introduction into the United States of the cottony-cushion insect (*Icerya purchasi*) from Australia threatened to destroy the American citrus industry. To counteract this situation, a natural Australian predator, a ladybird beetle (*Novius cardinalis*), was imported. The beetles kept the insects down to a relatively low level. When DDT (an insecticide) was discovered to kill scale insects, farmers applied it in the hopes of reducing even further the scale insect population. However, DDT turned out to be fatal to the beetle as well, and the overall effect of using the insecticide was to increase the numbers of the scale insect. Modify the predator–prey model to reflect a predator–prey system where farmers apply (on a regular basis) an insecticide that destroys both the insect predator and the insect prey at a rate proportional to the numbers present.

11. Consider two species whose survival depends on their mutual cooperation. An example would be a species of bee that feeds primarily on the nectar of one plant species and simultaneously pollinates that plant. Letting a_n and b_n represent the bee and plant population levels after n days, we have the model

$$a_{n+1} = a_n - k_1 a_n + k_2 a_n b_n$$
$$b_{n+1} = b_n - k_3 b_n + k_4 a_n b_n$$

where the k_i are positive constants.

a. Discuss the meaning of each k_i in terms of mutual cooperation.

b. What assumptions are implicitly being made about the growth of each species in the absence of cooperation?

12. Place a cold can of soda in a room. Measure the temperature of the room and periodically measure the temperature of the soda. Formulate a model to predict the change in the temperature of the soda. Estimate any constants of proportionality from your data. What are some of the sources of error in your model?

3.2 PROJECTS

1. Complete the UMAP module "The Diffusion of Innovation in Family Planning," by Kathryn N. Harmon, UMAP 303. This module gives an interesting

application of finite difference equations to study the process through which public policies are diffused to understand how national governments might adopt family planning policies.

2. Complete the UMAP module "Difference Equations With Applications," by Donald R. Sherbert, UMAP 322. This module presents a good introduction to solving first- and second-order linear difference equations, including the method of undetermined coefficients for nonhomogeneous equations. Applications to problems in population and economic modeling are included.

3.2 Further Reading

FRAUENTHAL, James C. *Introduction to Population Modeling*. Lexington, MA: COMAP, 1979.

HUTCHINSON, G. Evelyn. *An Introduction to Population Ecology*. New Haven, CT: Yale University Press, 1978.

LEVINS, R. "The Strategy of Model Building in Population Biology." *American Scientist* 54 (1966): 421–431.

LOTKA, A. J. *Elements of Mathematical Biology*. New York: Dover, 1956.

ODUM, E. P. *Fundamentals of Ecology*. Philadelphia: Saunders, 1971.

PEARL, R., & L. J. Reed. "On the Rate of Growth of the Population of the United States Since 1790." *Proceedings of the National Academy of Science* 6 (1920): 275–288.

3.3 NUMERICAL SOLUTIONS

In this section we build numerical solutions to dynamical systems by starting with an initial value and iterating a sufficient number of subsequent values to determine the patterns involved. In some cases we will see that the behavior predicted by dynamical systems can often be characterized by the mathematical structure of the system. In other cases, we will see wild variations in the behavior caused by only small changes in the initial values of the dynamical system. We will also examine dynamical systems for which small changes in the proportionality constants cause wildly different predictions.

Long-Term Behavior

As seen in the previous section, dynamical systems often represent real-world behavior we are trying to understand. Often we want to predict future behavior and gain insight into how to influence the behavior, if necessary. Consequently, we have great interest in the predictions of the model. Does the sequence in the dynamical system grow without bound (see Figure 3.9a)? (The answer would be of great interest if the model were of global warming.) Does the sequence become small without bound (i.e., negative with an absolute value that grows large as illustrated in Figure 3.9b)? Does the behavior approach a limiting value (Figure 3.9c)? If we were modeling the motion of an automobile suspension system dampened by springs and shock absorbers, we would want a design that dampened external disturbances.

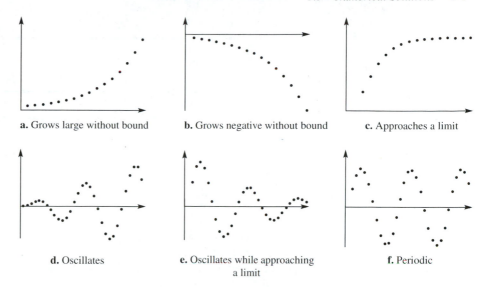

a. Grows large without bound   **b.** Grows negative without bound   **c.** Approaches a limit

d. Oscillates   **e.** Oscillates while approaching a limit   **f.** Periodic

FIGURE 3.9   Long-term behavior

Does the motion oscillate (Figure 3.9d)? Oscillations, if too large or frequent, can cause great damage to structures such as bridges, buildings, and automotive systems. Does the sequence oscillate as it approaches a limit (Figure 3.9e)? We might expect this to happen to the vertical displacements in a properly designed automobile after hitting a pothole. If the model is of global warming, it would be reassuring to know that the temperature will return to an equilibrium value after soaring past it. Is the motion periodic, such as in a swinging pendulum, the seasonal variations in weather, or locations of heavenly bodies (Figure 3.9f)? These are but a few of the interesting possibilities summarized in Figure 3.9.

Linear Dynamical Systems $a_{n+1} = ra_n$, for r Constant

We begin our numerical explorations by looking at equations having the form $a_{n+1} = ra_n$, where r can be any positive or negative constant. We illustrate a process we can use to explore other dynamical systems in the exercises. Our goal is to gain insight into the nature of the long-term behavior for all possible values of r and any initial value. That is, what is the long-term behavior (as described in Figure 3.9) and how does it vary with the parameter r? Also, what are the critical values of r for which the nature of the long-term behavior changes, as suggested in Figure 3.10?

Values of r	$r = 0$	$r = 1$	$\lvert r \rvert > 1$	$\lvert r \rvert < 1$	$r < 0$
Long-term behavior	?	?	?	?	?

FIGURE 3.10   Long-term behavior of $a_{n+1} = ra_n$ for key values of r

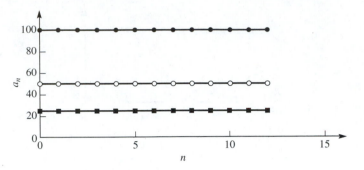

FIGURE 3.11 Every solution of $a_{n+1} = a_n$ is a constant solution

Let's consider some significant values of r. If $r = 0$, then all the values of the sequence are zero, so there is no need for further analysis. If $r = 1$, then the sequence becomes $a_{n+1} = a_n$. This is a rather interesting case in the sense that, no matter where the sequence starts, it stays there forever (as illustrated in Figure 3.11 where the sequence is enumerated and graphed for $a_0 = 50$, as well as graphed for other starting values). Values for which a dynamical system remains constantly **equilibrium values**   at that value, once reached, are called **equilibrium values** of the system. We will define that term more precisely later, but for now note that in Figure 3.11 any starting value would be an equilibrium value.

We encountered the form $a_{n+1} = ra_n$, r some constant, in some of our modeling applications. If a_n represents the amount in a savings account n months after investing \$1,000 at 1% monthly interest, then the sequence can be iterated and graphed, as in Figure 3.12. From the graph and an analysis of the model, we expect that if $r > 1$ then the sequence grows without bound.

n	a_n
0	1000.00
1	1010.00
2	1020.10
3	1030.30
4	1040.60
5	1051.01
6	1061.52
7	1072.14
8	1082.86
9	1093.69
10	1104.62
11	1115.67
12	1126.83
13	1138.09
14	1149.47
15	1160.97
16	1172.58
17	1184.30

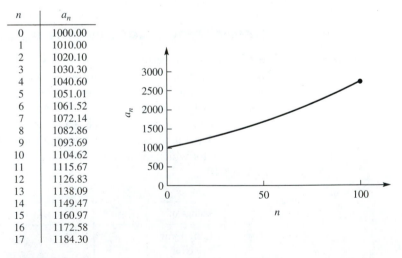

FIGURE 3.12 For $r > 1$ and constant, the solutions to $a_{n+1} = ra_n$ grow without bound

Now, what happens if r is negative? If we replace 1.01 with -1.01 in the preceding example, we obtain the graph in Figure 3.13. Note the oscillation between both positive and negative values. Because the negative sign causes the next term in the sequence to be of opposite sign from the previous term, we conclude that, in general, negative values of r cause oscillations in the linear form $a_{n+1} = ra_n$.

What happens if $|r| < 1$? We know what happens if $r = 0$, if $r > 1$, if $r < -1$, and if r is negative in general, so let's next consider positive fractions with an example. Suppose digoxin decays in the bloodstream such that each day one-half of the concentration of digoxin remains from the previous day. If we start with 0.6 mg and a_n represents the amount after n days, we can represent the behavior with the following model (a numerical solution appears in Figure 3.14):

$$a_{n+1} = 0.5a_n, \quad n = 0, 1, 2, 3, \ldots$$
$$a_0 = 0.6$$

Note that the behavior approaches the value 0 (which is an *equilibrium value* for this system because if we start with no digoxin in the bloodstream, we will never have any). The value 0 is a *limit* for the system, meaning that the difference between the value 0 and the value for a_n (starting with $a_0 = 0.6$) can be made as small as we please once n is large enough. The knowledge of equilibrium values and limits aids our understanding of the behavior of dynamical systems immensely.

Growth and decay problems of the form investigated in this section are important in modeling. For example, carbon-14 is contained in all living organisms and begins to decay after death at such a predictable rate that the amount remaining is used to date archaeological findings, rare paintings, documents, and the like. Up to this point we have made observations for a few values of r and only a few starting values. But from those examples, can you guess the nature of the behavior for all

n	a_n
0	1000
1	-1010.00
2	1020.10
3	-1030.30
4	1040.60
5	-1051.01
6	1061.52
7	-1072.14
8	1082.86
9	-1093.69
10	1104.62
11	-1115.67
12	1126.83
13	-1138.09
14	1149.47
15	-1160.97
16	1172.58

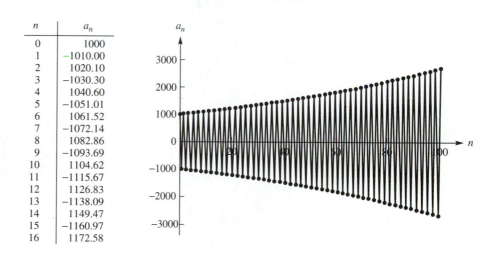

FIGURE 3.13 Negative value of r causes oscillation

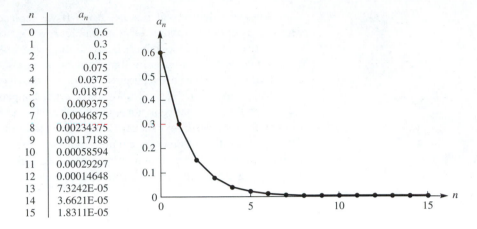

n	a_n
0	0.6
1	0.3
2	0.15
3	0.075
4	0.0375
5	0.01875
6	0.009375
7	0.0046875
8	0.00234375
9	0.00117188
10	0.00058594
11	0.00029297
12	0.00014648
13	7.3242E-05
14	3.6621E-05
15	1.8311E-05

FIGURE 3.14 A positive fractional value of r causes decay

values of r and all possible starting values? Can you complete Figure 3.10 now? Try it and see based on our examples.

Let's summarize our observations so far:

$r = 0$	Constant solution and equilibrium value at 0
$r = 1$	All initial values are constant solutions
$r < 0$	Oscillation
$\lvert r \rvert < 1$	Decay to limiting value of 0
$\lvert r \rvert > 1$	Growth without bound

Finding Equilibrium Values

Consider a dynamical system of the form $a_{n+1} = ra_n$. For an equilibrium value to occur, $a_{n+1} = a_n$. Let's call the equilibrium value a. Substituting $a_{n+1} = a_n = a$ and solving gives

$$a = \frac{0}{1 - r}, \quad r \neq 1$$

Thus if $r \neq 1$, the number 0 is an equilibrium value (as we saw in Figure 3.14). If $r = 1$, then every initial value results in a constant solution (as we saw in Figure 3.11). Hence every value is an equilibrium value. Summarizing our observations, we have

THEOREM 3.1 For the dynamical system $a_{n+1} = ra_n$:

If $r \neq 1$, an equilibrium value exists at $a = 0$.

If $r = 1$, every number is an equilibrium value.

Dynamical Systems of the Form $a_{n+1} = ra_n + b$, where r and b Are Constants

Now let's add a constant b to the dynamical system we just studied. Again we'd like to classify the nature of the long-term behavior for all possible cases. We consider three examples to gain some insight into the behavior.

Example 1 — Prescription for Digoxin

Consider again the digoxin problem. Recall that digoxin is used in the treatment of heart patients. The objective of the problem is to consider the decay of digoxin in the bloodstream in order to prescribe a dosage that keeps the concentration between acceptable (safe and effective) levels. Suppose we prescribe a daily drug dosage of 0.1 mg. This results in the dynamical system

$$a_{n+1} = 0.5a_n + 0.1$$

Now let's consider three starting values, or initial doses:

$$A: a_0 = 0.1$$
$$B: a_0 = 0.2$$
$$C: a_0 = 0.3$$

In Figure 3.15 we compute the numerical solutions for each case.

Note that the value 0.2 is an equilibrium value, because once reached the system remains at 0.2 forever. Further, if we start below the equilibrium (as in Case A) or above the equilibrium (as in Case C), apparently we approach the equilibrium value as a *limit*. In the exercises you are asked to compute solutions for starting values even closer to 0.2, lending evidence that 0.2 is a *stable* equilibrium value.

	A	B	C	
n	a_n	a_n	a_n	
0	0.1	0.2	0.3	
1	0.15	0.2	0.25	
2	0.175	0.2	0.225	
3	0.1875	0.2	0.2125	
4	0.19375	0.2	0.20625	
5	0.196875	0.2	0.203125	
6	0.1984375	0.2	0.2015625	
7	0.19921875	0.2	0.20078125	
8	0.19960938	0.2	0.20039063	
9	0.19980469	0.2	0.20019531	
10	0.19990234	0.2	0.20009766	
11	0.19995117	0.2	0.20004883	
12	0.19997559	0.2	0.20002441	
13	0.19998779	0.2	0.20001221	
14	0.1999939	0.2	0.2000061	
15	0.19999695	0.2	0.20000305	

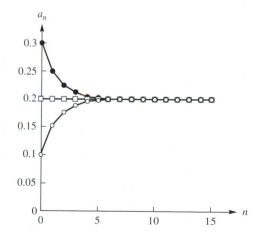

FIGURE 3.15 Three initial digoxin doses

When prescribing digoxin, the concentration level must stay above an effective level for a period of time without exceeding a safe level. In the exercises you are asked to find initial and subsequent doses that are both *safe* and *effective*.

Example 2   An Investment Annuity

Return now to the bank account problem and consider an *annuity*. Annuities are often planned for retirement purposes. They are basically savings accounts that pay interest on the amount present and allow the investor to withdraw a fixed amount each month until the account is depleted. An interesting issue (posed in the exercises) is to know how much we must save each month to build an annuity allowing for withdrawals beginning at a certain age with a specified amount for a desired number of years before depletion of the account. For now, consider 1% as the monthly interest rate and a monthly withdrawal of $1,000. This gives the dynamical system

$$a_{n+1} = 1.01a_n - 1000$$

Now suppose we made the following initial investments:

$$A: a_0 = 90,000$$
$$B: a_0 = 100,000$$
$$C: a_0 = 110,000$$

The numerical solutions for each case are graphed in Figure 3.16.

Notice that the value 100,000 is an equilibrium value because once reached, the system remains there for all subsequent values. If we start above that equilibrium, however, there is growth without bound. (Try plotting with $a_0 = \$100,000.01$.) On

	A	B	C		
n	a_n	a_n	a_n		
0	90000	100000	110000		
1	89900	100000	110100		
2	89799	100000	110201		
3	89697	100000	110303		
4	89594	100000	110406		
5	89490	100000	110510		
6	89385	100000	110615		
7	89279	100000	110721		
8	89171	100000	110829		
9	89063	100000	110937		
10	88954	100000	111046		
11	88843	100000	111157		
12	88732	100000	111268		
13	88619	100000	111381		
14	88505	100000	111495		
15	88390	100000	111610		

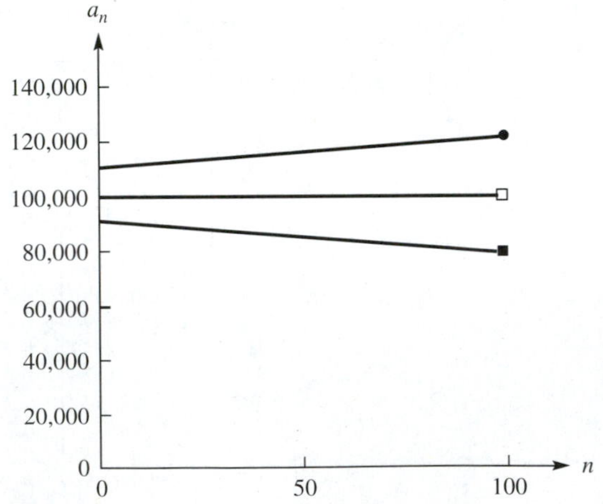

FIGURE 3.16   An annuity with three initial investments

the other hand, if we start below $100,000, the savings are used up at an increasing rate. (Try $999,999.99.) Note how drastically different the long-term behaviors are even though the starting values differ by only $0.02. In this situation we say that the equilibrium value 100,000 is *unstable*: If we start close to the value (even within a penny) we do not remain close. Look at the dramatic differences displayed by Figures 3.15 and 3.16. Both systems show equilibrium values, but the first is stable and the second is unstable.

We've considered here the cases where $|r| < 1$ and $|r| > 1$. Let's see what happens if $r = 1$.

Example 3   A Checking Account

Most students can't keep enough cash in their checking account to earn any interest. Suppose you have an account that pays no interest and that each month you pay only your dorm rent of $300, giving the dynamical system

$$a_{n+1} = a_n - 300$$

The result is probably obvious from the application, but when compared with Figures 3.15 and 3.16 the graph is revealing. In Figure 3.17 we plot the numerical solution for a starting value of 3000. Do you see how drastically the graph of the solution differs from the previous examples? Can you find an equilibrium as in Examples 1 and 2?

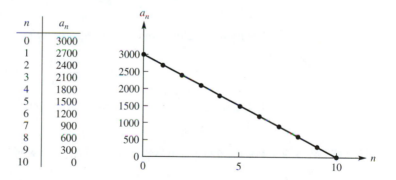

n	a_n
0	3000
1	2700
2	2400
3	2100
4	1800
5	1500
6	1200
7	900
8	600
9	300
10	0

FIGURE 3.17   A checking account to pay dorm costs

Now let's collect our observations thus far, classifying the three examples and their long-term behavior according to the value of the constant r.

Dynamical System $a_{n+1} = ra_n + b, b \neq 0$

Value of r	Long-term behavior observed
$|r| < 1$	Stable equilibrium
$|r| > 1$	Unstable equilibrium
$r = 1$	Graph is a line with no equilibrium

Finding Equilibrium Values

Determining if equilibrium values exist and classifying them as stable or unstable immensely assists in analyzing the long-term behavior of a dynamical system. Consider again Examples 1 and 2. In Example 1, how did we know that a starting value of 0.2 would result in a constant solution or equilibrium value? Similarly, how do we know the answer to the same question for an investment of $100,000 in Example 2? For a dynamical system of the form

$$a_{n+1} = ra_n + b \qquad (3.2)$$

denote by a the equilibrium value, if one exists. From the definition of an equilibrium value, if we start at a we must remain there for all n; that is, $a_{n+1} = a_n = a$ for all n. Substituting a for a_{n+1} and a_n in Equation (3.2) yields

$$a = ra + b$$

and solving for a we find,

$$a = \frac{b}{1 - r}, \quad \text{if } r \neq 1$$

Thus, for Example 1, the equilibrium value is

$$a = \frac{0.1}{1 - 0.5} = 0.2$$

For Example 2, we compute the equilibrium as

$$a = \frac{-1000}{1 - 1.01} = 100,000$$

In Example 3, $r = 1$ and no equilibrium exists. The next theorem summarizes our observations.

THEOREM 3.2   For the dynamical system $a_{n+1} = ra_n + b$ where $b \neq 0$:

If $r \neq 1$, an equilibrium exists at $a = \dfrac{b}{1 - r}$.

If $r = 1$, no equilibrium value exists.

Nonlinear Systems

An important advantage of discrete systems is that numerical solutions can be constructed for any dynamical system when given an initial value. We have seen thus far that long-term behavior can be sensitive to the starting value and to the values of the parameter r. Recall the model for the yeast biomass from Section 3.2:

$$p_{n+1} = p_n + 0.00082(665 - p_n)p_n$$

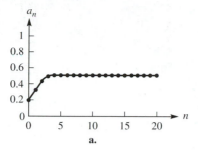
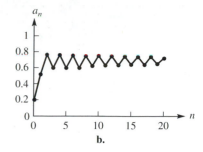
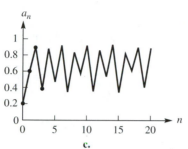
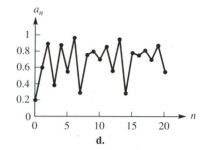

FIGURE 3.18 Numerical solutions to the equation $a_{n+1} = r(1 - a_n)a_n$ for the values of r: (a) $r = 2$, (b) $r = 3$, (c) $r = 3.6$, (d) $r = 3.7$

For simplicity, consider the dynamical system with the same form:

$$a_{n+1} = r(1 - a_n)a_n$$

In Figure 3.18 we plot a numerical solution for the four values of r indicated in the caption beginning with $a_0 = 0.2$.

Note how remarkably different the behavior is in each of the four cases. In Figure 3.18a, we see that for $r = 2$ the behavior approaches a limit. In Figure 3.18b, we see that for $r = 3$ the behavior oscillates between the values of about 0.6 and 0.7. We call such behavior periodic with a two-cycle. In Figure 3.18c, the motion repeats itself every four iterations, or is periodic with a four-cycle. Although it is not evident in Figure 3.18d, the value of r is sufficiently large that there is no periodicity and the motion never repeats itself. Thus there is no pattern and it is impossible to predict long-term behavior from the model. We call such behavior *chaotic*. Chaotic systems demonstrate sensitivity to the constant parameters of the system. A given chaotic system can be quite sensitive to initial conditions.

3.3 PROBLEMS

1. For the following problems, find an equilibrium value if one exists. Classify the equilibrium value as stable or unstable.

   a. $a_{n+1} = 1.1a_n$

   b. $a_{n+1} = 0.9a_n$

    c. $a_{n+1} = -0.9a_n$

    d. $a_{n+1} = a_n$

    e. $a_{n+1} = -1.2a_n + 50$

    f. $a_{n+1} = 1.2a_n - 50$

    g. $a_{n+1} = 0.8a_n + 100$

    h. $a_{n+1} = 0.8a_n - 100$

    i. $a_{n+1} = -0.8a_n + 100$

    j. $a_{n+1} = a_n - 100$

    k. $a_{n+1} = a_n + 100$

2. Build a numerical solution for the following *initial value problems*. Plot your data to observe patterns in the solution. Is there an equilibrium solution? Is it stable or unstable?

    a. $a_{n+1} = -1.2a + 50, a_0 = 1000$

    b. $a_{n+1} = 0.8a_n - 100, a_0 = 500$

    c. $a_{n+1} = 0.8a_n - 100, a_0 = -500$

    d. $a_{n+1} = -0.8a_n + 100, a_0 = 1000$

    e. $a_{n+1} = a_n - 100, a_0 = 1000$

3. You currently have $5,000 in a savings account that pays 0.5% interest each month. You add another $200 each month. Build a numerical solution to determine when the account reaches $20,000.

4. You owe $500 on a credit card that charges 1.5% interest each month. You can pay $50 each month, and you make no new charges. What is the equilibrium value? What does the equilibrium value mean in terms of the credit card? Build a numerical solution. When will the account be paid off? How much is the last payment?

5. Your parents are considering a 30-year $100,000 mortgage that charges 0.5% interest each month. Formulate a model in terms of a monthly payment p that allows the mortgage (loan) to be paid off after 360 payments. *Hint:* If a_n represents the amount owed after n months, what are a_0 and a_{360}? Experiment by building numerical solutions to find a value of p that works.

6. Your parents are considering a 30-year mortgage that charges 0.5% interest each month. Formulate a model in terms of a monthly payment p that allows the mortgage (loan) to be paid off after 360 payments. Your parents can afford a monthly payment of $1,500. Experiment to determine the maximum amount of money they can borrow. *Hint:* If a_n represents the amount owed after n months, what are a_0 and a_{360}?

7. Your grandparents have an annuity. The value of the annuity increases each month as 1% interest on the previous month's balance is deposited. Your grandparents withdraw $1,000 each month for living expenses. Presently, they have $50,000 in the annuity. Model the annuity with a dynamical system. Find the equilibrium value. What does the equilibrium value represent for this problem? Build a numerical solution to determine when the annuity is depleted.

8. *Continuation of Problem 1, Section 3.2:*

   a. Find the equilibrium value of the digoxin model. What is the significance of the equilibrium value?

   b. Experiment with different initial and maintenance doses. Find a combination that is convenient, with time between doses and amount to be taken considered as measures of convenience. (Rounded amounts are considered to be more convenient.)

9. *Continuation of Problem 2, Section 3.2:* Experiment with different initial and maintenance doses. Find a combination that is convenient, with time between doses and amount to be taken considered as measures of convenience.

10. *Continuation of Problem 3, Section 3.2:* Determine the age of the humanoid skull found near the remains of an ancient campfire.

3.3 PROJECTS

1. You plan to invest part of your paycheck after graduation to finance your children's education. You want to have enough in the account to draw $1,000 a month, every month for 8 years beginning 20 years from now. The account pays 0.5% interest each month.

   a. How much money will you need 20 years from now to accomplish the financial objective? Assume you stop investing when your first child begins college—a safe assumption.

   b. How much must you deposit each month over the next 20 years?

2. Consider the following dynamical system:

$$a_{n+1} = ra_n(1 - a_n)$$
$$a_0 = 100$$

   For the following values of r, build a numerical solution and plot the results: $r = 0.75, 1.5, 2.75, 3, 3.25, 3.45, 3.55, 3.57, 3.59, 3.65$. For each value of r, observe the pattern of your numerical solution. Does there appear to be an equilibrium value? Would you be able to make predictions from the model? Experiment with different values for a_0. Are the results sensitive to the starting value?

3. Assume we are considering the survival of whales and that if the number of whales falls below a minimum survival level m the species will become extinct. Assume also that the population is limited by the carrying capacity M of the environment. That is, if the whale population is above M, it will experience a decline because the environment cannot sustain that large a population level. In the following model, a_n represents the whale population after n years; discuss the model:

$$a_{n+1} = k(M - a_n)(a_n - m)$$

Now experiment with different values for m, M, and k. Try several starting values. What does your model predict?

4. Complete the modules "The Growth of Partisan Support I: Model and Estimation" (UMAP 304) and "The Growth of Partisan Support II" (UMAP 305), by Carol Weitzel Kohfeld. UMAP 304 presents a simple model of political mobilization, refined to include the interaction between supporters of a particular party and recruitable nonsupporters. UMAP 305 investigates the mathematical properties of the first-order quadratic-difference equation model. The model is tested using data from three U.S. counties.

3.3 Further Reading

TUCHINSKY, Philip M. *Man in Competition with the Spruce Budworm*, UMAP Expository Monograph. The population of tiny caterpillars periodically explodes in the evergreen forests of Eastern Canada and Maine. They devour the trees' needles and cause great damage to forests that are central to the economy of the region. The province of New Brunswick is using mathematical models of the budworm/forest interaction in an effort to plan for and control the damage. The monograph surveys the ecological situation and examines the computer simulation and models currently in use.

3.4   SYSTEMS OF DIFFERENCE EQUATIONS

In this section we consider systems of difference equations. For selected starting values we build numerical solutions to get an indication of the long-term behavior of the system. As seen in Section 3.3, equilibrium values are values of the dependent variable(s) for which there is no change in the system once the equilibrium values are obtained. For the systems considered in this section, we will find the equilibrium values. Then we explore starting values in the vicinity of the equilibrium values. If we start close to an equilibrium value, we want to know if the system will

a. Remain close

b. Approach the equilibrium value

c. Not remain close

What happens near these values gives great insight concerning the long-term behavior of the system. Does the system demonstrate periodic behavior? Are there oscillations? Does the long-term behavior described by the numerical solution appear to be sensitive to

a. The initial conditions

b. Small changes in the constants of proportionality used to model the behavior under study

Although our approach in this section is numerical, we will revisit the scenarios described in this section in Chapter 11 when we treat systems of differential equations. Our goal now is to model the behavior with difference equations and explore numerically the behavior predicted by the model.

Example 1   A Car Rental Company

Consider a car rental company with distributorships in Orlando and Tampa. The company specializes in catering to travel agents who want to arrange tourist activities in both Orlando and Tampa. Consequently, a traveler will rent a car in one city and drop the car off in the second city. Travelers may begin their itinerary in either city. The company is trying to determine how much to charge for this drop-off convenience. Because cars are dropped off in both cities, will a sufficient number of cars end up in each city to satisfy the demand for cars in that city? If not, how many cars must the company transport from Orlando to Tampa or from Tampa to Orlando? The answers to these questions will help the company figure out its expected costs.

Suppose we analyzed the historical records and determined that 60% of the cars rented in Orlando are returned to Orlando, whereas the other 40% end up in Tampa. Of the cars rented from the Tampa office, 70% are returned to Tampa, whereas 30% end up in Orlando. When analyzing problems such as this, we may find a diagram such as the one in Figure 3.19 helpful.

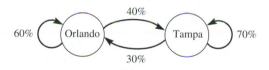

FIGURE 3.19   Car rental offices in Orlando and Tampa

Dynamical systems model   Let's develop a model of the system. Let n represent the number of business days. Now define

$$O_n = \text{the number of cars in Orlando at the end of day } n$$
$$T_n = \text{the number of cars in Tampa at the end of day } n$$

Thus the historical records reveal the system

$$O_{n+1} = .6O_n + .3T_n$$
$$T_{n+1} = .4O_n + .7T_n$$

Equilibrium values   The equilibrium values for the system are those values of O_n and T_n for which no change in the system takes place. Let's call the equilibrium values, if they exist, O and T, respectively. Then $O = O_{n+1} = O_n$ and $T = T_{n+1} = T_n$ simultaneously. Substitution in our model yields the following requirements for the equilibrium values:

$$O = 0.6O + 0.3T$$
$$T = 0.4O + 0.7T$$

This system is satisfied whenever $O = \frac{3}{4}T$. For example, if the company owns 7000 cars and starts with 3000 in Orlando and 4000 in Tampa, then our model predicts that

$$O_1 = 0.6(3000) + 0.3(4000) = 3000$$
$$T_1 = 0.4(3000) + 0.7(4000) = 4000$$

Thus this system remains at $(O, T) = (3000, 4000)$ if we start there.

Next let's explore what happens if we start at values other than the equilibrium values. Let's iterate the *system* for the following four initial conditions:

Four Starting Values for the Car Rental Problem

	Orlando	Tampa
Case 1	7000	0
Case 2	5000	2000
Case 3	2000	5000
Case 4	0	7000

A numerical solution, or table of values, for each of the starting values is graphed in Figure 3.20.

n	Orlando	Tampa
0	7000	0
1	4200	2800
2	3360	3640
3	3108	3892
4	3032.4	3967.6
5	3009.72	3990.28
6	3002.916	3997.084
7	3000.875	3999.125

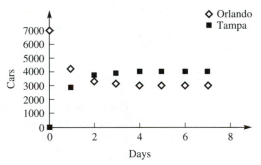

a. Case 1

n	Orlando	Tampa
0	5000	2000
1	3600	3400
2	3180	3820
3	3054	3946
4	3016.2	3983.8
5	3004.86	3995.14
6	3001.458	3998.542
7	3000.437	3999.563

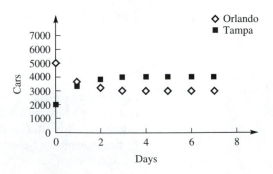

b. Case 2

FIGURE 3.20 The rental car problem (*continues*)

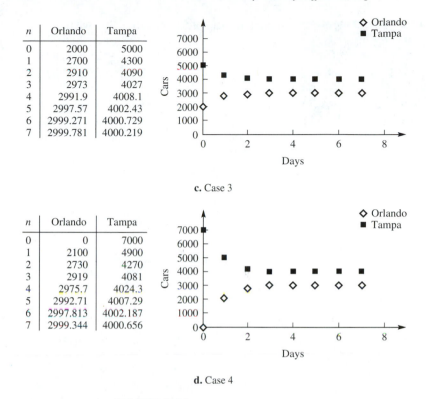

n	Orlando	Tampa
0	2000	5000
1	2700	4300
2	2910	4090
3	2973	4027
4	2991.9	4008.1
5	2997.57	4002.43
6	2999.271	4000.729
7	2999.781	4000.219

c. Case 3

n	Orlando	Tampa
0	0	7000
1	2100	4900
2	2730	4270
3	2919	4081
4	2975.7	4024.3
5	2992.71	4007.29
6	2997.813	4002.187
7	2999.344	4000.656

d. Case 4

FIGURE 3.20 The rental car problem

Sensitivity to initial conditions and long-term behavior   Note that within a week the system is very close to the equilibrium value $(3000, 4000)$ even if initially no cars were at one of the sites. Our results suggest that the equilibrium value is stable and insensitive to the starting values. Based on these explorations, we would be inclined to predict that the system will approach the equilibrium value where $\frac{3}{7}$ of the fleet ends up in Orlando and the remaining $\frac{4}{7}$ in Tampa. This information is helpful to the company. Knowing the demand patterns in each city, the company can now estimate how many cars they will have to ship. In the problem set, we ask you to explore the system to determine whether it is sensitive to the coefficients in the equations.

Example 2   *Competitive Hunter Models—Spotted Owls and Hawks*

In Section 3.2 we developed several models for a spotted owl population. One of the models assumed that a second species was competing for the resources (such as food) in the habitat. The effect of the presence of the second species is to diminish the growth rate of the spotted owl. We now consider a specific example.

Let O_n represent the value of the spotted owl population at the end of day n and let H_n denote the competing hawk population. Based on the competing species

model in Example 2 of Section 3.2, we have

$$O_{n+1} = (1 + k_1)O_n - k_3 O_n H_n$$
$$H_{n+1} = (1 + k_2)H_n - k_4 O_n H_n$$

where k_1, k_2, k_3, and k_4 are positive constants.

Note that in the absence of the other species, each species exhibits a positive growth rate, and the presence of the second species diminishes the growth rate of the first. Now let's choose specific values for the constants of proportionality and consider the system:

$$O_{n+1} = 1.2O_n - 0.001 O_n H_n \qquad \text{(3.3)}$$
$$H_{n+1} = 1.3H_n - 0.002 O_n H_n$$

Equilibrium values If we call the equilibrium values (O, H), then we must have $O = O_{n+1} = O_n$ and $H = H_{n+1} = H_n$ simultaneously. Substituting into the system yields

$$O = 1.2O - 0.001 OH$$
$$H = 1.3H - 0.002 OH$$

or

$$0 = 0.2O - 0.001 OH = O(0.2 - 0.001H)$$
$$0 = 0.3H - 0.002 OH = H(0.3 - 0.002O)$$

The first equation indicates that there is no change in the owl population if $O = 0$ or $H = 0.2/0.001 = 200$. The second equation indicates there is no change in the hawk population if $H = 0$ or $O = 0.3/0.002 = 150$, as depicted in Figure 3.21. Note that equilibrium values exist at $(O, H) = (0, 0)$ and $(O, H) = (150, 200)$ because *neither* population will change at those points. (Substitute the equilibrium values into Equation (3.3) to check that the system indeed remains at $(0, 0)$ and $(150, 200)$ if either of those points are the starting values.)

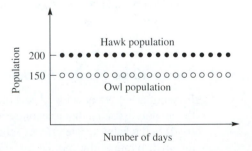

FIGURE 3.21 If the owl population begins at 150 and the hawk population begins at 200, the two populations remain at their starting values

Now let's analyze what happens in the vicinity of the equilibrium values we have found. Let's build numerical solutions for the three starting populations given here. Note that the first two values are close to the equilibrium value $(150, 200)$, whereas the third is near the origin.

	Owls	Hawks
Case 1	151	199
Case 2	149	201
Case 3	10	10

Iterating Equation (3.3) beginning with the starting values given results in the numerical solutions graphed in Figure 3.22. Note that in each case eventually one of the two species drives the other to extinction.

Sensitivity to initial conditions and long-term behavior   Suppose 350 owls and hawks are to be placed in a habitat modeled by Equation (3.3). If 150 of the birds are

n	Owls	Hawks
1	151	199
2	151.151	198.602
3	151.3623	198.1448
4	151.6431	197.6049
5	152.0063	196.9556
6	152.4691	196.1653
7	153.0538	195.1966
8	153.7889	194.0044
9	154.711	192.5343
10	155.866	190.7202
11	157.3124	188.4827
12	159.1242	185.7261
13	161.3956	182.3369
14	164.2463	178.1812
15	167.83	173.1044
16	172.3438	166.9315
17	178.043	159.4717
18	185.2588	150.5276
19	194.424	139.9128
20	206.1064	127.4818
21	221.0528	113.1767
22	240.2454	97.09366
23	264.9681	79.56915
24	296.8785	61.27332
25	338.0634	43.27385
26	391.0468	26.99739
27	458.6989	13.98212
28	544.0252	5.349589
29	649.9199	1.133844
30	779.1669	0.000182

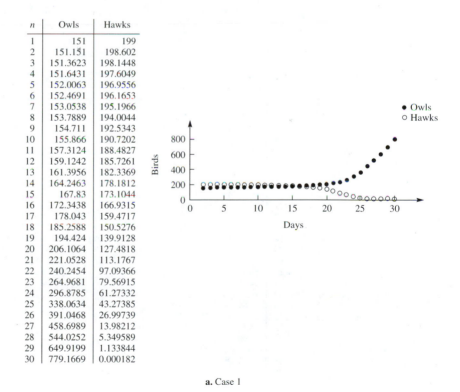

a. Case 1

FIGURE 3.22   Either the owls or hawks dominate the competition (*continues*)

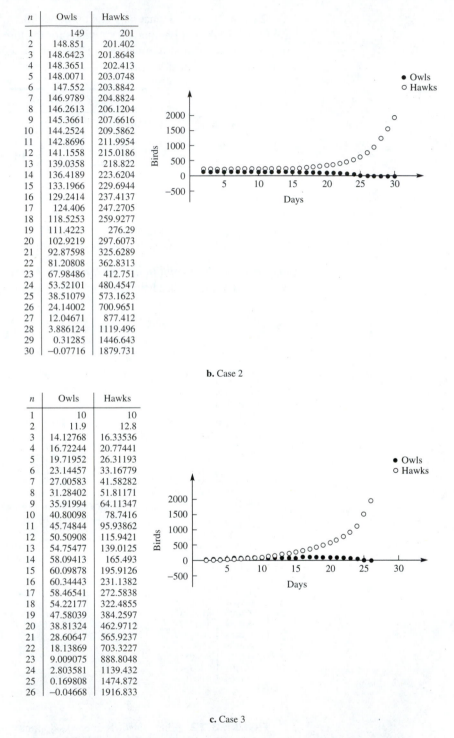

n	Owls	Hawks
1	149	201
2	148.851	201.402
3	148.6423	201.8648
4	148.3651	202.413
5	148.0071	203.0748
6	147.552	203.8842
7	146.9789	204.8824
8	146.2613	206.1204
9	145.3661	207.6616
10	144.2524	209.5862
11	142.8696	211.9954
12	141.1558	215.0186
13	139.0358	218.822
14	136.4189	223.6204
15	133.1966	229.6944
16	129.2414	237.4137
17	124.406	247.2705
18	118.5253	259.9277
19	111.4223	276.29
20	102.9219	297.6073
21	92.87598	325.6289
22	81.20808	362.8313
23	67.98486	412.751
24	53.52101	480.4547
25	38.51079	573.1623
26	24.14002	700.9651
27	12.04671	877.412
28	3.886124	1119.496
29	0.31285	1446.643
30	−0.07716	1879.731

b. Case 2

n	Owls	Hawks
1	10	10
2	11.9	12.8
3	14.12768	16.33536
4	16.72244	20.77441
5	19.71952	26.31193
6	23.14457	33.16779
7	27.00583	41.58282
8	31.28402	51.81171
9	35.91994	64.11347
10	40.80098	78.7416
11	45.74844	95.93862
12	50.50908	115.9421
13	54.75477	139.0125
14	58.09413	165.493
15	60.09878	195.9126
16	60.34443	231.1382
17	58.46541	272.5838
18	54.22177	322.4855
19	47.58039	384.2597
20	38.81324	462.9712
21	28.60647	565.9237
22	18.13869	703.3227
23	9.009075	888.8048
24	2.803581	1139.432
25	0.169808	1474.872
26	−0.04668	1916.833

c. Case 3

FIGURE 3.22   Either the owls or hawks dominate the competition

owls, our model predicts the owls will remain at 150 forever. If one owl is removed from the habitat (i.e., 149), the model predicts that the owl population will die out. If 151 owls are placed in the habitat, however, the model predicts that the owls will grow without bound while the hawks will disappear. This model is extremely sensitive to the initial conditions. The equilibrium values are unstable in the sense that if we start close to either equilibrium value, we do not remain close. Note how the model predicts that coexistence of the two species in a single habitat is highly unlikely because one of the two species will eventually dominate the habitat. In the problem set, you are asked to explore this system further by examining other starting points and by changing the coefficients of the model.

Example 3   Predator–Prey: Owls and Mice

In Section 3.2 we developed a model for the spotted owl assuming mice were the sole source of food. We now consider a specific example.

Let O_n represent the size of the spotted owl population at the end of day n, and let M_n denote the size of the mouse population. Then from Section 3.2,

$$M_{n+1} = (1 + k_1)M_n - k_3 O_n M_n$$
$$O_{n+1} = (1 - k_2)O_n + k_4 O_n M_n$$

where k_1, k_2, k_3, and k_4 are positive constants.

Now let's consider specific values of the constants of proportionality to obtain the system

$$M_{n+1} = 1.2M_n - 0.001 O_n M_n \qquad \text{(3.4)}$$
$$O_{n+1} = -0.7 O_n + 0.002 O_n M_n$$

Note that in the absence of owls, the mice grow without bound. Similarly, in the absence of mice, the owls die out. Notice also that the form of the system is similar to the model used to model competing species (such as the owls and hawks just analyzed) except that the signs are different for some of the coefficients. How much of a difference do you think this will make in the model's predictions? Let's find and analyze the equilibrium values.

Equilibrium values   If we call the equilibrium values (M, O), then we must have $M = M_{n+1} = M_n$ and $O = O_{n+1} = O_n$ simultaneously. Substituting into the system, we have,

$$M = 1.2M - 0.001MO$$
$$O = -0.7O + 0.002MO$$

or

$$0 = 0.2M - 0.001MO = M(0.2 - 0.001O)$$
$$0 = -1.7O + 0.002MO = O(-1.7 + 0.002M)$$

The first equation indicates there is no change in the mouse population if $M = 0$ or $O = \frac{0.2}{0.001} = 200$; the second equation indicates there is no change in the owl population if $O = 0$ or $M = \frac{1.7}{0.002} = 850$. Note that equilibrium values exist at $(M, O) = (0, 0)$ and $(M, O) = (850, 200)$ since *neither* population changes at those points. (Substitute the equilibrium values into Equation (3.4) to check that the system indeed remains at $(0, 0)$ and $(850, 200)$ if either of those points represents the starting values.)

Now let's see what happens in the vicinity of the equilibrium values. Let's build numerical solutions of the starting populations listed here. Note that we have picked three cases near the equilibrium point $(850, 200)$.

Four Starting Values for Mice and Owls

	Mice	Owls
Case 1	850	200
Case 2	849	201
Case 3	851	199
Case 4	860	190

Iterating Equation (3.4) beginning with the starting values given in the preceding table, results in the numerical solutions graphed in Figure 3.23. Note that in Cases 2, 3, and 4 there is oscillation in the sizes of the two populations. The amplitudes of the oscillations seem to increase with time. Does one of the two species eventually become extinct?

In the problem set you are asked to continue this exploration and determine the long-term behavior and sensitivity of the model to initial conditions (as we did for the competing species model).

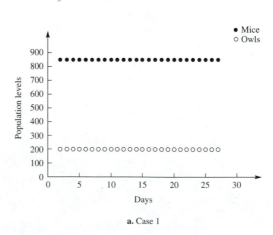

a. Case 1

FIGURE 3.23 Mice and owls *(continues)*

n	Mice	Owls
1	849	201
2	848.151	200.598
3	847.6438	199.8562
4	847.7657	198.9144
5	848.6861	198.0255
6	850.3618	197.5051
7	852.4833	197.648
8	854.4883	198.6297
9	855.6592	200.4127
10	855.3061	202.6811
11	853.013	204.832
12	848.8913	206.0663
13	843.7416	205.6093
14	839.0088	203.0358
15	836.4618	198.5726
16	837.6558	193.1959
17	843.3552	188.4262
18	853.116	185.9221
19	865.1261	187.0808
20	876.3028	192.7404
21	882.6644	202.8797
22	880.1227	216.1336
23	865.9232	229.1546
24	840.6775	236.4523
25	810.0329	232.0437
26	784.0765	213.4954
27	773.495	185.3467
28	784.8293	156.9868
29	818.5873	136.5249
30	870.5472	127.9477

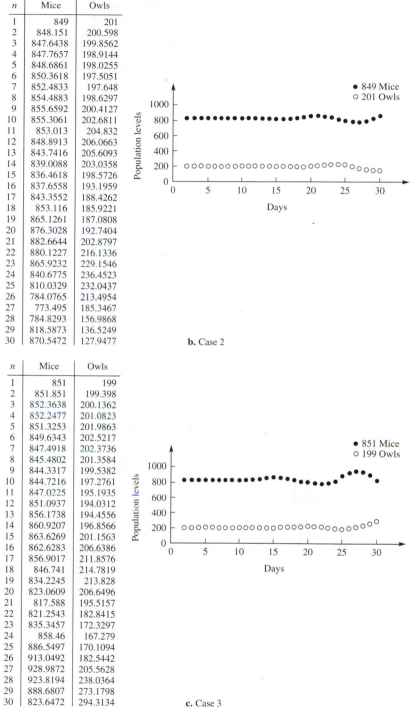

b. Case 2

n	Mice	Owls
1	851	199
2	851.851	199.398
3	852.3638	200.1362
4	852.2477	201.0823
5	851.3253	201.9863
6	849.6343	202.5217
7	847.4918	202.3736
8	845.4802	201.3584
9	844.3317	199.5382
10	844.7216	197.2761
11	847.0225	195.1935
12	851.0937	194.0312
13	856.1738	194.4556
14	860.9207	196.8566
15	863.6269	201.1563
16	862.6283	206.6386
17	856.9017	211.8576
18	846.741	214.7819
19	834.2245	213.828
20	823.0609	206.6496
21	817.588	195.5157
22	821.2543	182.8415
23	835.3457	172.3297
24	858.46	167.279
25	886.5497	170.1094
26	913.0492	182.5442
27	928.9872	205.5628
28	923.8194	238.0364
29	888.6807	273.1798
30	823.6472	294.3134

c. Case 3

FIGURE 3.23   Mice and owls (*continues*)

n	Mice	Owls
1	860	190
2	868.6	193.8
3	873.9853	201.0094
4	873.1032	210.6519
5	863.8029	220.3854
6	846.194	226.4693
7	823.7959	224.7454
8	803.4107	212.9669
9	792.993	193.1229
10	798.4464	171.1042
11	821.5182	153.4621
12	859.7499	144.7204
13	907.2765	147.5424
14	954.8701	164.4438
15	988.8216	198.9343
16	989.8754	254.1671
17	936.2568	325.2705
18	818.9714	381.3841
19	670.423	357.7165
20	564.6863	229.2412
21	548.1742	98.42985
22	603.8523	39.01251
23	701.065	19.80683
24	827.3921	13.90697
25	981.364	13.27815
26	1164.606	16.7667
27	1378.001	27.31651
28	1615.959	56.16278
29	1848.394	142.1995
30	1955.232	426.1417

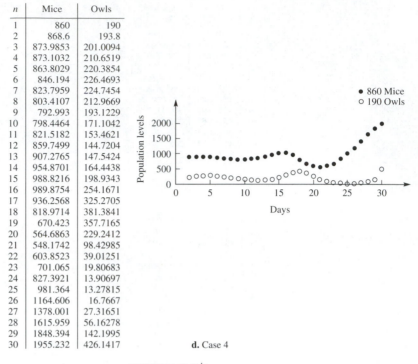

d. Case 4

FIGURE 3.23   Mice and owls

Example 4   *Voting Tendencies of the Political Parties*

Consider a three-party system with Republicans, Democrats, and Independents. Assume that in the next election, 75% of those who voted Republican again vote Republican, 5% vote Democrat, and 20% vote Independent. Of those who voted Democrat before, 20% vote Republican, 60% again vote Democrat, and 20% vote Independent. Of those who voted Independent, 40% vote Republican, 20% Democrat, and 40% again vote Independent. Assume these tendencies continue from election to election and that no additional voters enter or leave the system. These tendencies are depicted in Figure 3.24.

To formulate a system of difference equations, let n represent the nth election and define

$$R_n = \text{the number of Republican voters in the } n\text{th election}$$
$$D_n = \text{the number of Democrat voters in the } n\text{th election}$$
$$I_n = \text{the number of Independent voters in the } n\text{th election}$$

Formulating the system of difference equations, we have the following dynamical system:

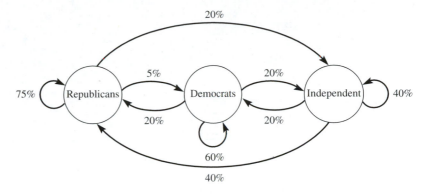

FIGURE 3.24   Voting tendencies among Republicans, Democrats, and Independents

$$R_{n+1} = 0.75R_n + 0.20D_n + 0.40I_n$$
$$D_{n+1} = 0.05R_n + 0.60D_n + 0.20I_n$$
$$I_{n+1} = 0.20R_n + 0.20D_n + 0.40I_n$$

Equilibrium values   If we call the equilibrium values (R, D, I), then we must have $R = R_{n+1} = R_n$, $D = D_{n+1} = D_n$, and $I = I_{n+1} = I_n$ simultaneously. Substituting into the dynamical system yields

$$-0.25R + 0.20D + 0.40I = 0$$
$$0.05R - 0.40D + 0.20I = 0$$
$$0.20R + 0.20D - 0.60I = 0$$

There is an infinite number of solutions to this system of equations. Letting $I = 1$, the system is satisfied if $R = 2.2221$ and $D = 0.7777694$ (approximately). Suppose the system has 399,998 voters. Then $R = 222,221$, $D = 77,777$, and $I = 100,000$ voters should approximate the equilibrium values. Let's use a spreadsheet to check the equilibrium and several other values. The total voters in the system is 399,998 with the initial voting as follows:

	Republicans	Democrats	Independents
Case 1	222,221	77,777	100,000
Case 2	227,221	82,777	90,000
Case 3	100,000	100,000	199,998
Case 4	0	0	399,998

The numerical solutions for the starting values are graphed in Figure 3.25.

Sensitivity to initial conditions and long-term behavior   Suppose initially there are 399,998 voters in the system and all remain in the system. At least for the

n	Republicans	Democrats	Independents
0	222221	77777	100000
1	222221.2	77777.25	99999.6
2	222221.2	77777.33	99999.52
3	222221.1	77777.36	99999.5
4	222221.1	77777.37	99999.5
5	222221.1	77777.38	99999.5
6	222221.1	77777.38	99999.5
7	222221.1	77777.39	99999.5
8	222221.1	77777.39	99999.5
9	222221.1	77777.39	99999.5
10	222221.1	77777.39	99999.5

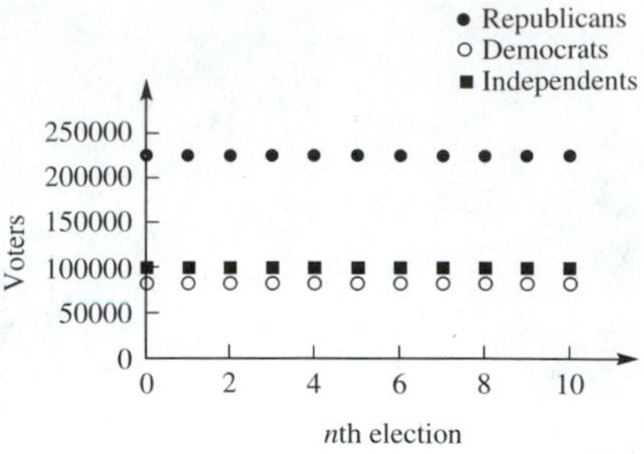

a. Case 1

n	Republicans	Democrats	Independents
0	227221	82777	90000
1	222971.2	79027.25	97999.6
2	222233.7	78164.83	99599.52
3	222148	77930.48	99919.5
4	222164.9	77849.59	99983.5
5	222187	77814.7	99996.3
6	222201.7	77797.43	99998.86
7	222210.3	77788.32	99999.37
8	222215.1	77783.38	99999.47
9	222217.8	77780.68	99999.49
10	222219.3	77779.2	99999.5

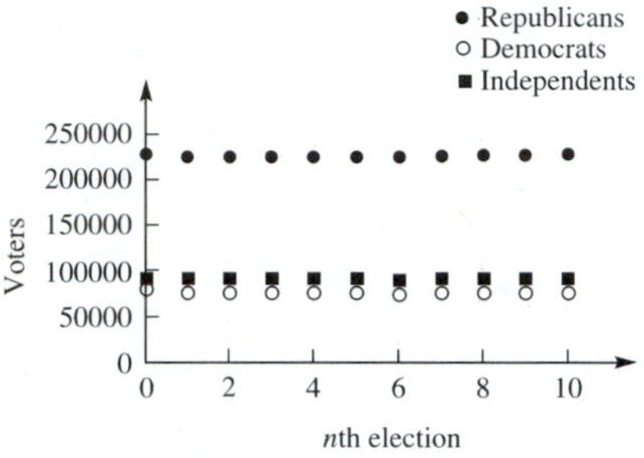

b. Case 2

n	Republicans	Democrats	Independents
0	100000	100000	199998
1	174999.2	104999.6	119999.2
2	200249	95749.56	103999.4
3	210936.4	88262.07	100799.5
4	216174.5	83663.96	100159.5
5	218927.5	81039	100031.5
6	220416	79576.08	100005.9
7	221229.6	78767.63	100000.8
8	221676	78322.21	99999.76
9	221921.4	78077.08	99999.55
10	222056.3	77942.23	99999.51

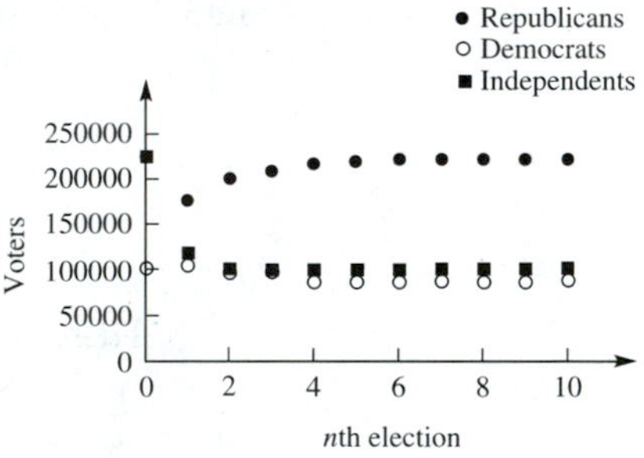

c. Case 3

n	Republicans	Democrats	Independents
0	0	0	399998
1	159999.2	79999.6	159999.2
2	199999	87999.56	111999.4
3	212398.9	85199.57	102399.5
4	217298.9	82219.59	100479.5
5	219609.9	80292.6	100095.5
6	220804.1	79175.15	100018.7
7	221445.6	78549.04	100003.3
8	221795.4	78202.37	100000.3
9	221987.1	78011.25	99999.65
10	222092.4	77906.03	99999.53

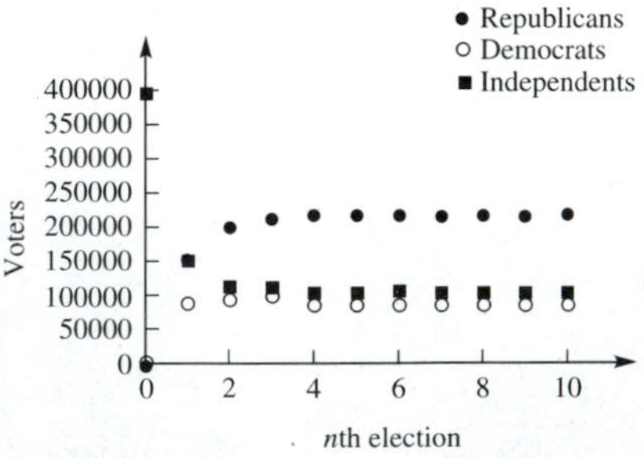

d. Case 4

FIGURE 3.25 Voting tendencies

starting values we investigated, the system approaches the same result, even if initially there are no Republicans or Democrats in the system. The equilibrium investigated appears to be stable. Starting values in the vicinity of the equilibrium appear to approach the equilibrium. What about the origin: Is it stable? In the problem set you are asked to explore this system further by examining other starting points and changing the coefficients of the model. Additionally, you are asked to investigate systems where voters enter and leave the system.

Summary

In this section we have explored numerically several dynamical systems and observed some interesting, perhaps surprising, results. We will return to analyze these systems further in Chapter 11 where we assume the behavior under study is taking place continuously. In many cases, the equilibrium points for continuous systems

nonlinear   can be classified. The nature of the equilibrium points for **nonlinear discrete** systems (such as the competitive hunter and predator–prey models) is highly dependent on the coefficients of the model. You are asked to explore examples of such systems in the following problem set.

3.4 PROBLEMS

1. Consider Example 1, A Car Rental Company. Experiment with different values for the coefficients. Iterate the resulting dynamical system for the given initial values. Then experiment with different starting values. Do your experimental results indicate that the model is sensitive

   a. to the coefficients?

   b. to the starting values?

2. Consider Example 2, Spotted Owls and Hawks. Experiment with different values for the coefficients using the starting values given. Then try different starting values. What is the long-term behavior? Do your experimental results indicate that the model is sensitive

   a. to the coefficients?

   b. to the starting values?

3. Consider Example 3, Owls and Mice. For the model and starting values given, continue to iterate the system until you are confident that one of the species becomes extinct or that an equilibrium is achieved. Now experiment with different values for the coefficients using the starting values given. Then try different starting values. What is the long-term behavior? Do your experimental results indicate that the model is sensitive

   a. to the coefficients?

   b. to the starting values?

4. Consider Example 4, Voting Tendencies. Experiment with starting values near the origin. Does the origin appear to be a stable equilibrium? Explain. Experiment with different values for the coefficients using the starting values given. Then try

different starting values. What is the long-term behavior? Do your experimental results indicate that the model is sensitive

a. to the coefficients?

b. to the starting values?

Now assume that each party recruits new party members. Initially assume that the total number of voters increases as each party recruits unregistered citizens. Experiment with different values for new party members. What is the long-term behavior. Does it seem to be sensitive to the recruiting rates? How would you adjust your model to reflect that the total number of citizens in the voting district is constant? Adjust the model to reflect what you think is happening in your voting district. What do you think will happen in your district over the long haul?

5. *Continuation of Problem 11, Section 3.2*: Build a numerical solution to the resulting model for different starting values. Does your model indicate that coexistence of the two species is possible? Likely? Vary the k_i. Interpret the results in terms of the given situation.

6. *Continuation of Problem 10, Section 3.2*: Pick values for your coefficients and try several starting values. What is the long-term behavior predicted by your model? Vary the coefficients. Do your experimental results indicate that the model is sensitive

a. to the coefficients?

b. to the starting values?

3.4 PROJECTS

1. Complete the requirements of the UMAP module, "Graphical Analysis of Some Difference Equations in Biology," by Martin Eisen, UMAP 553. The growth of many biological populations can be modeled by difference equations. This module shows how the behavior of the solutions to certain equations can be predicted by graphical techniques.

2. Prepare a summary of a paper by May et al. listed in the references for this section.

3.4 Further Reading

CLARK, Colin W. *Mathematical Bioeconomics: The Optimal Management of Renewable Resources*. New York: Wiley, 1976.

MAY, R. M. *Stability and Complexity in Model Ecosystems*, Monographs in Population Biology VI. Princeton, NJ: Princeton University Press, 1973.

MAY, R. M., ed. *Theoretical Ecology: Principles and Applications*. Philadelphia: Saunders, 1976.

MAY, R. M., J. R. Beddington, C. W. Clark, S. J. Holt, & R. M. Lewis, "Management of Multispecies Fisheries." *Science* 205 (July 1979): 267–277.

Chapter Four

MODELING USING PROPORTIONALITY

INTRODUCTION

In the previous chapter, we used the concept of proportionality to model change. In this chapter we formalize the concept of proportionality and use it to uncover relationships among variables we have selected in the model-building process. We then present the technique of geometric similarity, a powerful concept for uncovering relationships among variables. In Chapter 5 we will see how proportionality can be used in fitting the model with specific criteria to data that have been collected.

4.1 PROPORTIONALITY

Two positive quantities x and y are said to be **proportional** (to each other) if one quantity is a constant positive multiple of the other; that is, if

$$y = kx$$

for some positive constant k. We write $y \propto x$ in that situation and say that y is proportional to x. Thus

$$y \propto x \text{ if and only if } y = kx \text{ for some constant } k > 0 \qquad \textbf{(4.1)}$$

Of course, if $y \propto x$ then $x \propto y$ also, because the constant k in Equation (4.1) is greater than zero. Then $x = \left(\frac{1}{k}\right)y$. Other examples of proportionality relationships

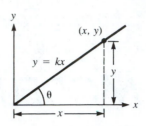

FIGURE 4.1   Geometrical interpretation of $y \propto x$

are the following:

$$y \propto x^2 \quad \text{if and only if } y = k_1 x^2 \quad \text{for } k_1 \text{ a constant} \tag{4.2}$$

$$y \propto \ln x \quad \text{if and only if } y = k_2 \ln x \quad \text{for } k_2 \text{ a constant} \tag{4.3}$$

$$y \propto e^x \quad \text{if and only if } y = k_3 e^x \quad \text{for } k_3 \text{ a constant} \tag{4.4}$$

In Equation (4.2), $y = kx^2, k > 0$ so we also have $x \propto y^{1/2}$ because $x = \left(\frac{1}{\sqrt{k}}\right) y^{1/2}$. This leads us to consider how to link proportionalities together, a transitive rule for proportionality:

$$\text{if } y \propto x \text{ and } x \propto z, \text{ then } y \propto z$$

Thus any variables proportional to the same variables are proportional to one another.

Now let's explore a geometric interpretation of proportionality. In Model (4.1), $y = kx$ yields $k = y/x$. Thus k may be interpreted as the tangent of the angle θ depicted in Figure 4.1, and the relation $y \propto x$ defines a set of points along a line in the plane with angle of inclination θ.

Comparing the general form of a proportionality relationship $y = kx$ with the equation for a straight line $y = mx + b$, we can see that the graph of a proportionality relationship is a line (possibly extended) passing through the origin. If we plot the proportionality variables for Models (4.2), (4.3), and (4.4), we obtain the straight-line graphs presented in Figure 4.2.

It is important to note that not just any straight line represents a proportionality relationship: The y-intercept *must* be zero so that the line passes through the origin.

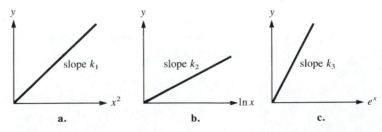

FIGURE 4.2   Geometrical interpretation of Models (a) (4.2), (b) (4.3), and (c) (4.4)

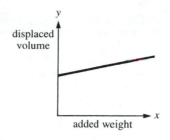

FIGURE 4.3 A straight-line relationship exists between displaced volume and total weight, but it is not a proportionality because the line fails to pass through the origin

Failure to recognize this point can lead to erroneous results when using our model. For example, suppose we are interested in predicting the volume of water displaced by a boat as it is loaded with cargo. Because a floating object displaces a volume of water equal to its weight, we might be tempted to assume that the total volume y of displaced water is proportional to the weight x of the added cargo. However, there is a difficulty with that assumption because the unloaded boat already displaces a volume of water equal to its weight. Although the graph of total volume of displaced water versus weight of added cargo is given by a straight line, it is not given by a line passing through the origin (see Figure 4.3), so the proportionality argument is incorrect.

A proportionality relationship may, however, be a reasonable simplifying assumption, depending on the size of the y-intercept and the slope of the line. The domain of the independent variable also can be significant because the relative error

$$\frac{y_a - y_p}{y_a}$$

is greater for small values of x. These features are depicted in Figure 4.4. If the slope is nearly zero, proportionality may be a poor assumption because the initial displacement dwarfs the effect of the added weight. For example, there would be virtually no effect in placing 400 lb on an aircraft carrier already weighing many

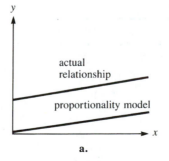

a.

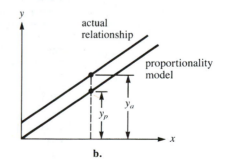

b.

FIGURE 4.4 Proportionality as a simplifying assumption

TABLE 4.1 Famous proportionalities

- Hooke's Law $F = kS$
  where F is the restoring force in a spring stretched or compressed a distance S.
- Newton's Law $F = ma$ or $a = \frac{1}{m}F$
  where a is the acceleration of a mass m subjected to a net external force F.
- Ohm's Law $V = iR$
  where i is the current induced by a voltage V across a resistance R.
- Boyle's Law $V = \frac{k}{p}$
  where under a constant temperature k the volume V is inversely proportional to the pressure p.
- Einstein's Theory of Relativity $E = c^2M$
  where under the constant speed of light squared c^2 the energy E is proportional to the mass M of the object.
- Kepler's Third Law $T = cR^{3/2}$
  where T is the period (days) and R is the mean distance to the sun.

tons. On the other hand, if the initial displacement is relatively small and the slope is large, the effect of the initial displacement is dwarfed quickly, and proportionality will be a good simplifying assumption. Some famous proportionalities are displayed in Table 4.1.

Example 1   To assist in further understanding the idea of proportionality, let's examine one of the famous proportionalities from Table 4.1, Kepler's Third Law. In 1601 the German astronomer Johannes Kepler became director of the Prague Observatory. Kepler had been helping Tycho Brahe collect 13 years of observations on the relative motion of the planet Mars. By 1609 Kepler had formulated his first two laws:

1. Each planet moves along an ellipse with the sun at one focus.
2. For each planet, the line from the sun to the planet sweeps out equal areas in equal times.

Kepler spent many years verifying these laws and formulating the third law given in Table 4.1, which relates the orbital periods and mean distances of the planets from the sun. The data shown in Table 4.2 are from the 1993 World Almanac:

TABLE 4.2 Orbital periods and mean distances of planets from the sun

Planet	Period (days)	Mean Distance (millions of miles)
Mercury	88.0	36
Venus	224.7	67.25
Earth	365.3	93
Mars	687.0	141.75
Jupiter	4,331.8	483.80
Saturn	10,760.0	887.97
Uranus	30,684.0	1,764.50
Neptune	60,188.3	2,791.05
Pluto	90,466.8	3,653.90

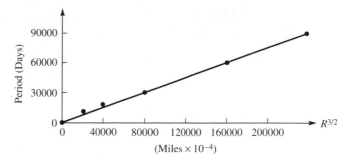

FIGURE 4.5 Kepler's Third Law as a proportionality

In Figure 4.5, we plot the period versus the mean distance to the $\frac{3}{2}$ power. The plot approximates a line that projects through the origin. We can easily estimate the slope (constant of proportionality) by picking any two points that lie on the line passing through the origin:

$$slope = \frac{90,466.8 - 88}{220,869.1 - 216} \approx 0.410$$

We estimate the model to be $T = 0.410R^{3/2}$.

Example 2 Consider a spring-mass system, such as the one shown in Figure 4.6. We conduct an experiment to measure the stretch of the spring as a function of the mass (measured as weight) placed on the spring. Consider the data collected for this experiment, displayed in Table 4.3. The graph showing an approximate straight line (passing through the origin) is presented in Figure 4.7.

The data appear to follow the proportionality rule that elongation (e) is proportional to the mass (m), $e \propto m$. The straight line appears to pass through the origin. This geometric understanding allows us to look at the data to determine if proportionality is a reasonable simplifying assumption from which we can estimate the slope, k. In this case, the assumption appears valid, so we estimate the constant

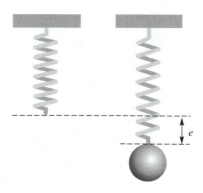

FIGURE 4.6 Spring-mass system

TABLE 4.3
Spring-mass
system

Mass	Elong
50	1.000
100	1.875
150	2.750
200	3.250
250	4.375
300	4.875
350	5.675
400	6.500
450	7.250
500	8.000
550	8.750

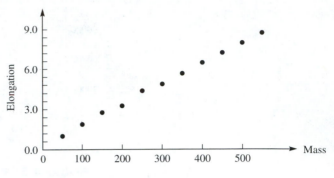

FIGURE 4.7   Data from spring-mass system

of proportionality by picking any two points that lie on our straight line as

$$slope = \frac{\Delta \text{elongation}}{\Delta \text{mass}} = \frac{3.25 - 1.875}{200 - 100} \approx 0.0135$$

We estimate our model as

$$e = 0.0135m \tag{4.5}$$

We then examine how close our model fits the data (see Figure 4.8) and conclude that Model (4.5) is reasonable.

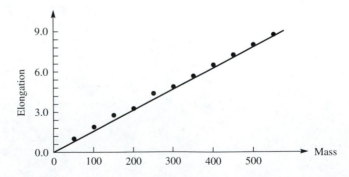

FIGURE 4.8   Data from spring-mass system with proportionality line

4.1 PROBLEMS

1. Show graphically the meaning of the proportionality $y \propto u/v$.

2. If a spring is stretched 0.37 in. by a 14-lb force, what stretch will be produced by a 9-lb force? by a 22-lb force? Assume Hooke's Law, which asserts the distance stretched is proportional to the force applied.

3. If an architectural drawing is scaled so that 0.75 in. represents 4 ft, what length represents 27 ft?

4. Determine whether the following data support a proportionality argument for $y \propto z^{1/2}$. If so, estimate the slope.

y	3.5	5	6	7	8
z	3	6	9	12	15

5. A new planet is discovered beyond Pluto at a mean distance to the sun of 400.4 million miles. Using Kepler's Third Law, determine an estimate for the time T to travel around the sun in an orbit.

6. In Example 2, if a 1000-g mass is added to the spring, how far would the spring stretch? If a 1,000,000-g weight is added, how far would the spring stretch? Is the model valid for any added mass?

7. In Example 2, convert the weights from grams into pounds. Find the model that now relates elongation to mass.

4.1 PROJECTS

1. Consider an automobile suspension system. Build a model that relates the stretch (or compression) of the spring to the mass it supports. If available, obtain a car shock absorber and collect data by measuring the change in spring size to the mass supported by the shock absorber. Graphically test your proportionality argument. If it is reasonable, find the constant of proportionality.

2. Research and prepare a 10-min report on Hooke's Law.

4.2   VEHICULAR STOPPING DISTANCE

Consider again the scenario posed in Example 1 of Section 2.2. Recall the general rule that allows one car length for every 10 mph of speed. It was also stated that this rule is the same as allowing for 2 sec between cars. The rules are in fact different from each other (at least for most cars). For the rules to be the same, at 10 mph both should allow one car length:

$$1 \text{ car length} = \text{distance} = \frac{\text{speed in ft}}{\text{sec}}(2 \text{ sec})$$

$$= \left(\frac{10 \text{ mi}}{\text{hr}}\right)\left(\frac{5280 \text{ ft}}{\text{mi}}\right)\left(\frac{1 \text{ hr}}{3600 \text{ sec}}\right)(2 \text{ sec})$$

$$= 29.33 \text{ ft}$$

This is an unreasonable result for an average car length of 15 ft, so the rules are *not* the same.

Let's interpret the one-car-length rule geometrically. If we assume a car length of 15 ft and plot this rule, we obtain the graph shown in Figure 4.9, which shows that the distance allowed by the rule is proportional to the velocity. In fact, if we plot the speed in feet per second, the constant of proportionality has the units seconds and represents the total reaction time for the equation $D = kv$ to make sense. Moreover, in the case of a 15-ft car, we obtain a constant of proportionality as follows:

$$k = \frac{15 \text{ ft}}{10 \text{ mph}} = \frac{15 \text{ ft}}{52,800 \text{ ft}/3600 \text{ sec}} = \frac{90}{88} \text{ sec}$$

In our previous discussion of this problem, we presented the model

$$\text{total stopping distance} = \text{reaction distance} + \text{braking distance}$$

Let's consider the submodels for reaction distance and braking distance.

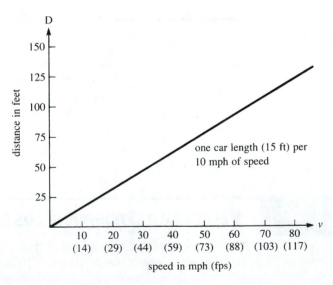

FIGURE 4.9 Geometrical interpretation of the one-car-length rule

Recall that

$$\text{reaction distance} = f \text{ (response time, speed)}$$

Now assume that the vehicle continues at constant speed from the time the driver determines the need to stop until the brakes are applied. Under this assumption, reaction distance d_r is simply the product of response time t_r and velocity v:

$$d_r = t_r v \tag{4.6}$$

To test the Submodel (4.6), plot measured reaction distance versus velocity. If the resultant graph approximates a straight line through the origin, we could estimate the slope t_r and feel fairly confident in the submodel. Alternatively, we could test a group of drivers representative of the assumptions made in Example 1 of Section 2.2 and estimate t_r directly.

Next, consider the braking distance:

$$\text{braking distance} = h \text{ (weight, speed)}$$

Suppose there is a panic stop and the maximum brake force F is applied throughout the stop. The brakes are basically an energy dissipating device; that is, the brakes do work on the vehicle, producing a change in the velocity that results in a loss of kinetic energy. Now, the work done is the force F times the braking distance d_b. This work must equal the change in kinetic energy, which, in this situation, is simply $0.5\,mv^2$. Thus, we have

$$\text{work done} = Fd_b = 0.5mv^2 \tag{4.7}$$

Next, we consider how the force F relates to the mass of the car. A reasonable design criterion would be to build cars in such a way that the maximum deceleration is constant when the maximum brake force is applied regardless of the mass of the car. This assumption means that the panic deceleration of a larger car, such as a Cadillac, is the same as that of a small car, such as a Honda, owing to the design of the braking system. Moreover, constant deceleration occurs throughout the panic stop. From Newton's Second Law, $F = ma$, it follows that the force F is proportional to the mass. Combining this result with Equation (4.7) gives the proportionality relation

$$d_b \propto v^2$$

At this point we might want to design a test for the two submodels, or we could test the submodels against the data provided by the U.S. Bureau of Public Roads given in Table 4.4.

Figure 4.10 depicts the plot of driver reaction distance against velocity using the data in Table 4.4. The graph is a straight line of approximate slope 1.1 passing through the origin; our results are *too good!* Because we always expect some

TABLE 4.4 Observed reaction and braking distances

Speed (mph)	Driver reaction distance (ft)	Braking distance* (ft)		Total stopping distance (ft)	
20	22	18–22	(20)	40–44	(42)
25	28	25–31	(28)	53–59	(56)
30	33	36–45	(40.5)	69–78	(73.5)
35	39	47–58	(52.5)	86–97	(91.5)
40	44	64–80	(72)	108–124	(116)
45	50	82–103	(92.5)	132–153	(142.5)
50	55	105–131	(118)	160–186	(173)
55	61	132–165	(148.5)	193–226	(209.5)
60	66	162–202	(182)	228–268	(248)
65	72	196–245	(220.5)	268–317	(292.5)
70	77	237–295	(266)	314–372	(343)
75	83	283–353	(318)	366–436	(401)
80	88	334–418	(376)	422–506	(464)

*Interval given includes 85% of the observations based on tests conducted by the U.S. Bureau of Public Roads. Figures in parentheses represent average values.

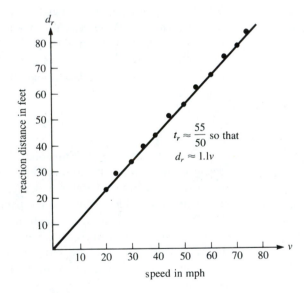

$$t_r \approx \frac{55}{50} \text{ so that}$$

$$d_r \approx 1.1v$$

FIGURE 4.10 Proportionality of reaction distance and speed

deviation in experimental results, we should be suspicious. In fact, the results of Table 4.4 are based on the Submodel (4.7), where an average response time of 3/4 sec was obtained independently. So we might later decide to design another test for the submodel.

To test the submodel for braking distance, we plot the observed braking distance recorded in Table 4.4 against v^2, as shown in Figure 4.11. Proportionality

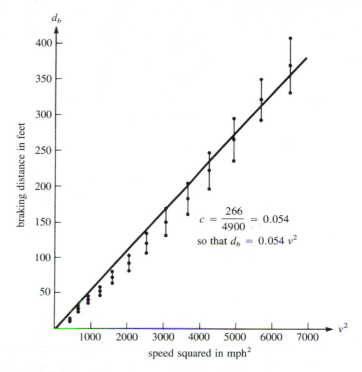

FIGURE 4.11   Proportionality of braking distance and the speed squared

seems to be a reasonable assumption at the lower speed, although it does seem to be less convincing at the higher speeds. By graphically fitting a straight line to the data, we estimate the slope and obtain the submodel

$$d_b = 0.054v^2 \tag{4.8}$$

We will learn how to fit the model to the data analytically in Chapter 5.

Connecting the two Submodels (4.6) and (4.8), we obtain the following model for the total stopping distance d:

$$d = 1.1v + 0.054v^2 \tag{4.9}$$

The predictions of Model (4.9) and the actual observed stopping distance recorded in Table 4.4 are plotted in Figure 4.12. Considering the grossness of the assumptions and the inaccuracies of the data, the model seems to agree fairly reasonably with the observations up to 70 mph. The rule of one 15-ft car length for every 10 mph of speed is also plotted in Figure 4.12. We can see that the rule significantly underestimates the total stopping distance at speeds exceeding 40 mph.

Let's suggest an alternative rule that is easy to understand and use. Assume the driver of the trailing vehicle must be fully stopped by the time he or she reaches

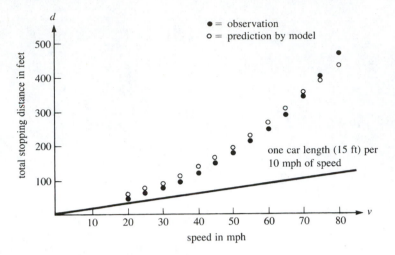

FIGURE 4.12 Total stopping distance

the point occupied by the lead vehicle at the exact time of the observation. Thus the driver must trail the lead vehicle by the total stopping distance, either based on our Model (4.9) or on the observed data in Table 4.4. The maximum stopping distance can readily be converted to a trailing time. The results of these computations for the observed distances, in which 85% of the drivers were able to stop, are given in Table 4.5. These computations suggest the following general rule.

TABLE 4.5 Time required to allow the proper stopping distance

Speed (mph)	(fps)	Stopping distance* (ft)		Trailing time required for maximum stopping distance (sec)
20	(29.3)	42	(44)†	1.5
25	(36.7)	56	(59)	1.6
30	(44.0)	73.5	(78)	1.8
35	(51.3)	91.5	(97)	1.9
40	(58.7)	116	(124)	2.1
45	(66.0)	142.5	(153)	2.3
50	(73.3)	173	(186)	2.5
55	(80.7)	209.5	(226)	2.8
60	(88.0)	248	(268)	3.0
65	(95.3)	292.5	(317)	3.3
70	(102.7)	343	(372)	3.6
75	(110.0)	401	(436)	4.0
80	(117.3)	464	(506)	4.3

*Includes 85% of the observations based on tests conducted by the U.S. Bureau of Public Roads.
†Figures in parentheses under stopping distance represent maximum values and are used to calculate trailing times.

Speed (mph)	Guideline (sec)
0–10	1
10–40	2
40–60	3
60–75	4

This alternative rule is plotted in Figure 4.13. An alternative to using such a rule might be to convince manufacturers to modify existing speedometers to compute stopping distance and time for the car's speed v based on Equation (4.9). We will revisit the braking distance problem in Section 10.3 with a model based on the derivative.

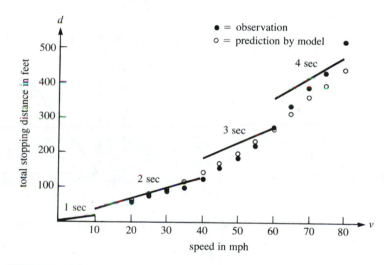

FIGURE 4.13   Total stopping distance and alternate general rule. The plotted observations are the maximum values from Table 4.5

4.2 PROBLEMS

1. Design a test to determine the average response time. Design a test to determine average reaction distance. Discuss the difference between the two statistics. If you were going to use the results of these tests to predict the total stopping distance, would you want to use the average reaction distance?

2. For the submodel concerning braking distance, how would you design a brake system so that the maximum deceleration is constant for all vehicles regardless of their mass? Consider the surface area of the brake pads and the capacity of the hydraulic system to apply a force.

4.3   **AUTOMOBILE GASOLINE MILEAGE**

In Example 2 of Section 2.2 we posed the following problem:

> For a particular driver, driving her car on a given day on a level highway at constant highway speeds near the optimum speed for fuel economy, provide a qualitative explanation of how fuel economy varies with small increases in speed.

The newspaper article from which this problem was derived gave the general rule that for every 5 mph over 50 mph, there is a loss of 1 mile to the gallon. Let's graph this rule. If you plot the loss in miles per gallon against speed minus 50, the graph is a straight line passing through the origin as depicted in Figure 4.14. Let's see if this linear graph is qualitatively correct.

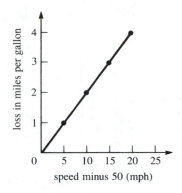

FIGURE 4.14   For every 5 mph over 50 mph, there is a loss of 1 mile to the gallon

Because the automobile is to be driven at constant speeds, the acceleration is zero. Newton's Second Law then yields that the resultant force must be zero or that the forces of propulsion and resistance must be in equilibrium. That is,

$$F_p = F_r$$

where F_p represents the propulsion force and F_r the resisting force.

First consider the force of propulsion. Each gallon of gasoline contains an amount of energy, say K. If C_r represents the amount of fuel burned per unit time, then $C_r K$ represents the power available to the car. Assuming a constant rate of power conversion, it follows that the converted power is proportional to $C_r K$. Because power is the product of force and velocity for a constant force, this argument yields the proportionality relation

$$F_p \propto \frac{C_r K}{v}$$

By further assuming a constant fuel rating K, this last proportionality simplifies to

$$F_p \propto \frac{C_r}{v} \tag{4.10}$$

Next consider the resisting force. Because we restricted the problem to highway speeds, it is reasonable to assume that frictional forces are small when compared with the drag forces caused by air resistance. At highway speeds, one sensible submodel for these drag forces is

$$F_r \propto Sv^2$$

where S is the cross-sectional area perpendicular to the direction of the moving car. (This assumption is commonly used in engineering for moderate sizes of S.) Because S is constant in our restricted problem, it follows that

$$F_r \propto v^2$$

Application of the condition $F_p = F_r$ and the proportionality (4.10) then yields

$$\frac{C_r}{v} \propto v^2$$

or

$$C_r \propto v^3 \tag{4.11}$$

The proportionality (4.11) gives the qualitative information that the fuel consumption rate should increase as the cube of the velocity. However, fuel consumption rate is not an especially good indicator of fuel efficiency: Although the proportionality (4.11) says that the car is using more fuel per unit time at higher speeds, the car is also traveling farther. Therefore, we define gasoline mileage as follows:

$$\text{mileage} = \frac{\text{distance}}{\text{consumption}}$$

Substitution of vt for distance and $C_r t$ for consumption gives the proportionality

$$\text{mileage} = \frac{v}{C_r} \propto v^{-2} \tag{4.12}$$

Thus gasoline mileage is inversely proportional to the square of the velocity.

Model (4.12) provides some useful qualitative information to assist us in explaining automobile fuel consumption. For one thing, we should be very suspicious of the rule suggesting a linear graph as depicted in Figure 4.14. We should be careful about the conclusions we do draw. Although the power relationship in (4.12) appears impressive, it is valid only over a restricted range of speeds. Over that restricted range the relationship could be nearly linear, depending on the size of the constant of proportionality. Moreover, don't forget that we have ignored many factors in our analysis and have assumed that several important factors are constant. Thus our model is quite fragile, and its use is limited to a qualitative explanation over a restricted range of speeds. But then, that's precisely how we identified the problem.

4.3 PROBLEMS

1. In the automobile gasoline mileage example, suppose you plot miles per gallon against speed as a graphical representation of the general rule (instead of the graph depicted in Figure 4.14). Explain why it would be difficult to deduce a proportionality relationship from that graph.

2. In the automobile gasoline mileage example, assume the drag forces are proportional to Sv, where S is the cross-sectional area perpendicular to the direction of the moving car and v is its speed. What conclusions can you draw? Discuss the factors that might influence the choice of Sv^2 over Sv for the drag forces submodel. How could you test the submodel?

3. Discuss several factors that were completely ignored in our analysis of the gasoline mileage problem.

4.4   GEOMETRIC SIMILARITY

Geometric similarity is a concept related to proportionality and can be useful to simplify the mathematical modeling process. The definition is as follows: Two

geometrically similar

objects are said to be **geometrically similar** if there is a one-to-one correspondence between points of the objects such that the ratio of distances between corresponding points is constant for all possible pairs of points.

For example, consider the two boxes depicted in Figure 4.15. Let l denote the distance between the points A and B in Figure 4.15a, and let l' be the distance

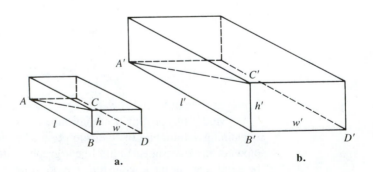

FIGURE 4.15   Two geometrically similar objects

between the corresponding points A' and B' in Figure 4.15b. Other corresponding points in the two figures, and the associated distances between the points, are marked in the same way. For the boxes to be geometrically similar, it must be true that

$$\frac{l}{l'} = \frac{w}{w'} = \frac{h}{h'} = k \text{ for some constant } k > 0$$

Let's interpret the last result geometrically. In Figure 4.15, consider the triangles ABC and $A'B'C'$. If the two boxes are geometrically similar, these two triangles must be similar. The same argument can be applied to any corresponding pair of triangles, such as CBD and $C'B'D'$. Thus *corresponding angles are equal for objects that are geometrically similar.* In other words, the shape is the same for two geometrically similar objects, and one object is simply an enlarged copy of the other. You may think of geometrically similar objects as scaled replicas of one another, as in an architectural drawing where all the dimensions are simply scaled by some constant factor k.

One advantage that results when two objects are geometrically similar is a simplification in certain computations, such as volume and surface area. For the boxes depicted in Figure 4.15, consider the following argument for the ratio of the volumes V and V':

$$\frac{V}{V'} = \frac{lwh}{l'w'h'} = k^3 \tag{4.13}$$

Similarly, the ratio of their total surface areas S and S' is given by

$$\frac{S}{S'} = \frac{2lh + 2wh + 2wl}{2l'h' + 2w'h' + 2w'l'} = k^2 \tag{4.14}$$

Not only are these ratios immediately known once the scaling factor k has been specified, but the surface area and volume may be expressed as a proportionality in terms of some selected **characteristic dimension**. For example, with $l/l' = k$, we have

characteristic dimension

$$\frac{S}{S'} = k^2 = \frac{l^2}{l'^2}$$

Therefore,

$$\frac{S}{l^2} = \frac{S'}{l'^2} = \text{constant}$$

holds for any two geometrically similar objects. That is, surface area is always proportional to the square of the characteristic dimension length:

$$S \propto l^2$$

Likewise, volume is proportional to the length cubed:

$$V \propto l^3$$

Thus if we are interested in some function depending on an object's length, surface area, and volume, for example,

$$y = f(l, S, V)$$

we could express all the function arguments in terms of some selected characteristic dimension, such as length, giving

$$y = g(l, l^2, l^3)$$

Geometric similarity is a powerful simplifying assumption. To illustrate the concepts of both proportionality and geometric similarity, we present two examples: raindrops falling from a motionless cloud and the weight of a human heart.

Example 1 Raindrops from a Motionless Cloud

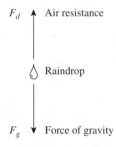

Suppose we are interested in the terminal velocity of a raindrop from a motionless cloud. Examining the free-body diagram, the only forces acting on the raindrop are gravity and drag. Assume that the atmospheric drag on the raindrop is proportional to its surface area S times the square of its speed v. The mass m of the raindrop is proportional to the weight of the raindrop (assuming constant gravity in Newton's Second Law):

$$F = F_g - F_d = ma$$

Under terminal velocity (assuming v equals v_t), we have $a = 0$ so Newton's Second Law is reduced to

$$F_g - F_d = 0$$

or

$$F_g = F_d$$

We are assuming that $F_d \propto Sv^2$ and that F_g is proportional to weight w. Because $m \propto w$, we have $F_g \propto m$.

Next we assume all the raindrops are geometrically similar. This assumption allows us to relate area and volume so that

$$S \propto l^2 \quad \text{and} \quad V \propto l^3$$

for any characteristic dimension l. Thus $l \propto S^{1/2} \propto V^{1/3}$, which implies

$$S \propto V^{2/3}$$

Because weight and mass are proportional to volume, we can substitute, yielding

$$S \propto m^{2/3}$$

From the equation $F_g = F_d$, we now have $m \propto m^{2/3}v_t^2$. Solving for the terminal velocity, we have

$$m^{1/3} \propto v_t^2 \quad \text{or} \quad m^{1/6} \propto v_t$$

Therefore, the terminal velocity of the raindrop is proportional to its mass raised to the one-sixth power.

Example 2 The Weight of a Human Heart

Suppose we are interested in constructing a model of a heart weight versus some measurable dimension of the heart. We will assume that the heart is essentially spherical and choose the circumference c of any great circle as the characteristic dimension. Assuming all hearts are geometrically similar yields the volume V of the heart being proportional to c^3. Now, assuming a constant weight per unit volume (weight density), we have

$$w \propto V \propto c^3$$

Likewise, if the characteristic dimension had been the diameter d of the spherical heart, or the radius r, then the model would have been

$$w \propto V \propto c^3 \propto d^3 \propto r^3$$

Testing Geometric Similarity

The principle of geometric similarity suggests a convenient method for testing to determine whether it holds among a collection of objects. Note that the definition requires that the ratio of distances between corresponding pairs of points be the same for all pairs of points. The satisfaction of this assumption can be tested to see if the ratio is constant.

Example 3 Testing Geometric Similarity

Consider the two circles displayed in Figure 4.16. Let c_1 denote the distance from A around the circle back to A, and let r_1 denote the distance from A to B along the chord $\overline{AB}$. Let c_2 and r_2 be defined similarly, using the corresponding points in Figure 4.16b. The definition of geometric similarity requires that

$$\frac{c_1}{c_2} = \frac{r_1}{r_2}$$

Further, this ratio must hold for all possible pairs of points. If the objects are geometrically similar, the ratio represents a *scaling* of the objects.

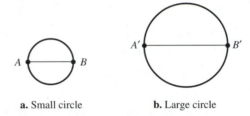

a. Small circle   **b.** Large circle

FIGURE 4.16   Circles are geometrically similar

The following data were collected for the distances c_i and r_i corresponding to 20 various scaled circles:

Data for geometric circles

cir1	r_1	cir2	r_2
3.14	1	6.28	2
6.28	2	12.56	4
9.42	3	18.84	6
12.56	4	25.12	8
15.70	5	31.40	10
18.84	6	37.68	12
21.98	7	43.96	14
25.12	8	50.24	16
28.26	9	56.52	18
31.40	10	62.80	20

A plot of c_i versus r_i is displayed in Figure 4.17. The plot appears to be a straight line through the origin and thus supports the reasonableness of the geometric similarity argument for the circles.

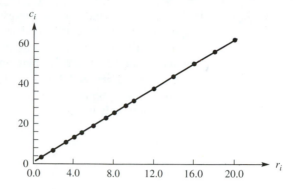

FIGURE 4.17   Testing the assumption of geometric similarity

4.4 PROBLEMS

1. Suppose two maps of the same country on different scales are drawn on tracing paper and superimposed, with possibly one of the maps being turned over before being superimposed on the other. Show there is just one place that is represented by the same spot on both maps.

2. Consider a 20-lb pink flamingo that stands 3 ft in height and has legs that are 2 ft in length. Model the height and leg length of a 100-lb flamingo. What assumptions are necessary? Are they reasonable assumptions?

3. An object is sliding down a ramp inclined at an angle of θ radians and attains a terminal velocity before reaching the bottom. Assume that the drag force caused by the air is proportional to Sv^2, where S is the cross-sectional area perpendicular to the direction of motion and v is the speed. Further assume that the sliding friction between the object and the ramp is proportional to the normal weight of the object. Determine the relationship between the terminal velocity and the mass of the object. If two different boxes, weighing 600 and 800 lb, are pushed down the ramp, find the relationship between their terminal velocities.

4. Assume that under certain conditions the heat loss of an object is proportional to the exposed surface area. Relate the heat loss of a cubic object with side length 6 in. to one with a side length 12 in. Now consider two irregularly shaped objects, such as two submarines. Relate the heat loss of a 70-ft submarine with a 7-ft scale model. Suppose you are interested in the amount of energy needed to maintain a constant internal temperature in the submarine. Relate the energy needed in the actual submarine to that required by the scaled model. Specify the assumptions you have made.

5. Consider the situation of two warm-blooded adult animals essentially at rest and under the same conditions (as in a zoo). Assume the animals maintain the same

body temperature and that the energy available to maintain this temperature is proportional to the amount of food provided to them. Challenge this assumption. If you are willing to assume that the animals are geometrically similar, relate the amounts of food necessary to maintain their body temperatures to their lengths and volumes. (*Hint:* See Problem 4.) List any assumptions you have made. What additional assumptions are necessary to relate the amount of food necessary to maintain their body weight?

4.4 PROJECTS

1. *Superstars*—In the TV show "Superstars" the top athletes from various sports compete against one another in a variety of events. The athletes vary considerably in height and weight. To compensate for this in the weight-lifting competition, the body weight of the athlete is subtracted from his lift. What kind of relationship does this suggest? Use the following table, which displays the winning lifts at the 1976 Montreal Olympic Games (weights are in pounds), to show this relationship:

Bodyweight class		Total winning lifts (lb)		
	Max. weight (lb)	Snatch	Jerk	Total weight
Flyweight	114.5	231.5	303.1	534.6
Bantamweight	123.5	259.0	319.7	578.7
Featherweight	132.5	275.6	352.7	628.3
Lightweight	149.0	297.6	380.3	677.9
Middleweight	165.5	319.7	418.9	738.5
Light-heavyweight	182.0	358.3	446.4	804.7
Middle-heavyweight	198.5	374.8	468.5	843.3
Heavyweight	242.5	385.8	496.0	881.8

Physiological arguments have been proposed that suggest that the strength of a muscle is proportional to its cross-sectional area. Using this submodel for strength, construct a model relating lifting ability and body weight. List all assumptions. Do you have to assume that all weight lifters are geometrically similar? Test your model with the data provided.

Now consider a refinement to the model. Suppose there is a certain amount of body weight that is independent of size in adults. Suggest a model that incorporates this refinement and test it against the data provided.

Criticize the use of the preceding data. What data would you really like to handicap the weight lifters? Who is the best weight lifter according to your models? Suggest a general rule for the "Superstars" show to handicap the weight lifters.

2. *Heart Rate of Birds*—Warm-blooded animals use large quantities of energy to maintain body temperature because of heat loss through the body surface. In fact, biologists believe that the primary energy drain on a resting warm-blooded animal is the maintenance of body temperature.

   a. Construct a model relating blood flow through the heart to body weight. Assume the amount of energy available is proportional to the blood flow through the lungs, which is the source of oxygen. Assuming the least amount of blood needed is circulated, the amount of available energy will equal the amount of energy used to maintain the body temperature.

   b. The following data relate weights of some birds to their heart rate measured in beats per minute. Construct a model that relates heart rate to body weight. Discuss the assumptions of your model. Use the data to check your model.

Bird	Body weight (g)	Pulse rate
Canary	20	1000
Pigeon	300	185
Crow	341	378
Buzzard	658	300
Duck	1100	190
Hen	2000	312
Goose	2300	240
Turkey	8750	193
Ostrich	71,000	60–70

Data from A. J. Clark, *Comparative Physiology of the Heart* (New York: Macmillan, 1927), p. 99.

3. *Heart Rate of Mammals*—The following data relate the weights of some mammals to their heart rate in beats per minute. Based on the discussion relating

Mammal	Body weight (g)	Pulse rate
Vesperugo pipistrellus	4	660
Mouse	25	670
Rat	200	420
Guinea Pig	300	300
Rabbit	2000	205
Little dog	5000	120
Big dog	30,000	85
Sheep	50,000	70
Man	70,000	72
Horse	450,000	38
Ox	500,000	40
Elephant	3,000,000	48

Data from A. J. Clark, *Comparative Physiology of the Heart* (New York: Macmillan, 1927), p. 99.

blood flow through the heart to body weight, as just presented in Problem 2, construct a model that relates heart rate to body weight. Discuss the assumptions of your model. Use the data to check your model.

4. *Lumber Cutters*—Lumber cutters wish to use readily available measurements to estimate the number of board feet of lumber in a tree. Assume they measure the diameter of the tree in inches at waist height. Develop a model that predicts board feet as a function of diameter in inches.

Use the following data for your test:

x	17	19	20	23	25	28	32	38	39	41
y	19	25	32	57	71	113	123	252	259	294

The variable x is the diameter in inches of a ponderosa pine and y is the number of board feet divided by 10.

a. Consider two separate assumptions, allowing each to lead to a model. Completely analyze each model.

   i. Assume all trees are right-circular cylinders and are about the same height.

   ii. Assume all trees are right-circular cylinders and the height of the tree is proportional to the diameter.

b. Which model appears better? Why? Justify your conclusions.

5. *Racing Shells*—If you have been to a rowing regatta, you might have observed that the more oarsmen there are in a boat, the faster the boat travels. Investigate whether there is a mathematical relationship between the speed of a boat and the number of crew members. Consider the follwing assumptions (partial list) in formulating a model:

a. The total force exerted by the crew is constant for a particular crew throughout the race.

b. The drag force experienced by the boat as it moves through the water is proportional to the square of the velocity times the wetted surface area of the hull.

c. Work is defined as force times distance. Power is defined as work per unit time.

Number in crew	Race 1 (sec)	Race 2 (sec)
1	429.6	430.2
2	412.2	406.2
4	379.8	367.8
8	346.8	343.8

Hint: When additional oarsmen are added to a shell it is not obvious whether the amount of *force* is proportional to the number in the crew or the amount of *power* is proportional to the number in the crew. Which assumption appears the most reasonable? Which yields a more accurate model?

6. *Scaling a braking system*—Suppose after years of experience your auto company has designed an optimum braking system for its prestigious full-sized car. That is, the distance required to brake the car is the best in its weight class, and the occupants feel the system is very smooth. Your firm has decided to build cars in the lighter weight classes. Discuss how you would scale the braking system of your present car to have the same performance in the smaller versions. Be sure to consider the hydraulic system and the size of the brake pads. Would a simple geometric similarity suffice? Let's suppose the wheels are scaled in such a manner that the pressure (at rest) on the tires is constant in all car models. Would the brake pads seem proportionally larger or smaller in the scaled-down cars?

4.5   A BASS FISHING DERBY

Consider a sport fishing club that for conservation purposes wishes to encourage its membership to release their fish immediately after catching them. The club also wishes to make awards based on the total weight of fish caught: honorary membership in The 1000 Pound Club, Greatest Total Weight Caught During a Derby Award, and so forth. How does someone fishing determine the weight of a fish he or she has caught? You might suggest that each individual carry a small portable scale. However, portable scales tend to be inconvenient and inaccurate, especially for smaller fish.

Problem identification   We can identify a problem as follows: *Predict the weight of a fish in terms of some easily measurable dimensions.*

Assumptions   Many factors that affect the weight of a fish can easily be identified. Different species have different shapes and different average weights per unit volume (weight density) based on the proportions and densities of meat, bone, and so on. Gender also plays an important role, especially during spawning season. The various seasons themselves probably have a considerable effect on weight.

Because a general rule for sport fishing is sought, let's initially restrict attention to a single species of fish, say bass, and assume that within the species the average weight density is constant. Later, it may be desirable to refine our model if the results prove unsatisfactory or if it is determined that considerable variability in density does exist. Furthermore, let's also neglect gender and season. Thus initially we will predict weight as a function of size (volume) and constant average weight density.

Assuming that all bass are geometrically similar, the volume of any bass is proportional to the cube of some characteristic dimension. Note that we are not assuming any particular shape but only that the bass are scaled models of one another. The basic shape can be quite irregular as long as the ratio between corresponding pairs of points in two distinct bass remains constant for all possible pairs of points. This idea is illustrated in Figure 4.18.

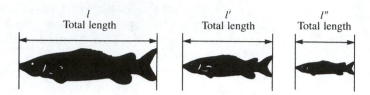

FIGURE 4.18 Fish that are geometrically similar are simply scaled models of one another

Now choose the length *l* of the fish as the characteristic dimension. This choice is depicted in Figure 4.18. Thus the volume of a bass satisfies the proportionality

$$V \propto l^3$$

Because weight *W* is volume times average weight density and a constant average density is being assumed, it follows immediately that

$$W \propto l^3$$

Model verification Let's test our model. Consider the following data collected during a fishing derby:

Length, l (in.)	14.5	12.5	17.25	14.5	12.625	17.75	14.125	12.625
Weight, W (oz)	27	17	41	26	17	49	23	16

If our model is correct, the graph of *W* versus l^3 should be a straight line passing through the origin. The graph showing an approximating straight line is presented in Figure 4.19. (Note the judgment here is qualitative. In Chapter 5 we develop analytical methods to determine a best-fitting model for collected data.)

Let's accept the model, at least for further testing, based on the small amount of data presented so far. Because the data point $(14.5^3, 26)$ happens to lie along the line we have drawn in Figure 4.19, we can estimate the slope of the line as $26/3049 = 0.00853$, yielding the model

$$W = 0.00853 \, l^3 \tag{4.15}$$

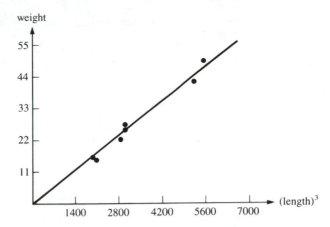

FIGURE 4.19 If the model is valid, the graph of W versus l^3 should be a straight line passing through the origin

Of course, if we had drawn our line a little differently, we would have obtained a slightly different slope. In Chapter 5 you will be asked to show analytically that the coefficients that minimize the sum of squared deviations between the model $W = kl^3$ and the given data points is $k = 0.008437$. A graph of Model (4.15) is presented in Figure 4.20, showing also a plot of the original data points.

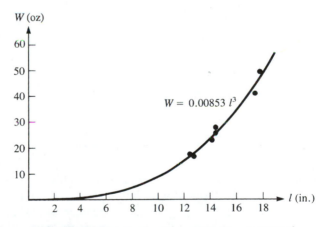

FIGURE 4.20 Graph of the model $W = 0.00853\, l^3$

Model (4.15) provides a convenient general rule. For example, from Figure 4.20 we might estimate that a 12-in. bass weighs approximately 1 lb. This means that an 18-in. bass should weigh approximately $(1.5)^3 = 3.4$ lb and a 24-in. bass approximately $2^3 = 8$ lb. For the fishing derby, a card converting the length of a caught fish to its weight in ounces or pounds could be given to each fisher, or a cloth tape or a retractable metal tape could be marked with a conversion scale if the use of the rule becomes popular enough. A conversion scale for Model (4.15) is as follows:

Length (in.)	12	13	14	15	16	17	18	19	20	21	22	23	24	25	26
Weight (oz)	15	19	23	29	35	42	50	59	68	79	91	104	118	133	150
Weight (lb)	0.9	1.2	1.5	1.8	2.2	2.6	3.1	3.7	4.3	4.9	5.7	6.5	7.4	8.3	9.4

Even though our rule seems reasonable based on the limited data we have obtained, a fisher may not like the rule because it doesn't reward the catching of a fat fish: The model treats fat and skinny fish alike. Let's address this dissatisfaction. Instead of assuming that the fish are geometrically similar, assume that only their cross-sectional areas are similar. This does not imply any particular shape for the cross section, only that the definition of geometric similarity is satisfied. For example, consider the ellipses depicted in Figure 4.21. From the definition of geometric similarity, if the ellipses are geometrically similar, then the ratio of the distance $\overline{P_1P_2}$ and the corresponding distance $\overline{P_1'P_2'}$ must be the same as the ratio between $\overline{P_3P_4}$ and its corresponding distance $\overline{P_3'P_4'}$. That is,

$$\overline{P_1P_2} : \overline{P_1'P_2'} = \overline{P_3P_4} : \overline{P_3'P_4'}$$

or

$$\frac{a_1}{a_2} = \frac{b_1}{b_2} \tag{4.16}$$

Now a characteristic dimension such as d in Figure 4.21 and a factor such as $r = a/b$, which is the ratio of the major axis to the minor axis, can be used to describe **shape factor** any two-dimensional ellipse. This factor is called the **shape factor**. Shape factors are important in similitude, as you will find in Section 13.5. From Equation (4.16), $a_1/b_1 = a_2/b_2$, so that geometrically similar ellipses have identical shape factors. Thus once the shape factor for a given ellipse is known, all geometrically similar ellipses can be described by giving the characteristic dimension d. Likewise, once the shape of the cross section of a bass is described, all bass with geometrically similar cross sections are described by giving the chosen characteristic dimension because they have the same shape factors. We assume that the transverse cross sections of the bass are geometrically similar and choose the characteristic dimension to be the girth, g, defined subsequently.

Now assume that the major portion of the weight of the fish is from the main body. Thus the head and tail contribute relatively little to the total weight. Constant

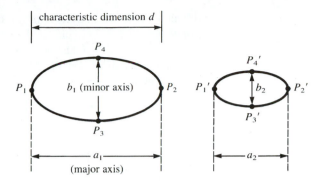

FIGURE 4.21   An ellipse is completely described by the shape factor $r = a/b$ and a convenient characteristic dimension d. Two geometrically similar ellipses have the same shape factor and are scaled by the ratio of their characteristic dimensions

terms can be added later if our model proves worthy of refinement. Next assume that the main body is of varying cross-sectional area. Then the volume can be found by multiplying the average cross-sectional area A_{avg} by the effective length l_{eff}:

$$V \approx l_{eff}(A_{avg})$$

But how shall the effective length l_{eff} and the average cross-sectional area A_{avg} be measured? Have the fishing contestants measure the length l of the fish as before, and assume the proportionality $l_{eff} \propto l$. To estimate the average cross-sectional area, have each fisher take a cloth measuring tape and measure the circumference of the fish at its widest point. Call this measurement the *girth*, g. Assume the average cross-sectional area is proportional to the square of the girth. Combining these two proportionality assumptions gives

$$V \propto lg^2$$

Finally, assuming constant density, $W \propto V$ as before so that

$$W = klg^2 \tag{4.17}$$

for some positive constant k.

Several assumptions have been made, so let's get an initial test of our model. Consider again the following data:

Length, l (in.)	14.5	12.5	17.25	14.5	12.625	17.75	14.125	12.625
Girth, g (in.)	9.75	8.375	11.0	9.75	8.5	12.5	9.0	8.5
Weight, W (oz)	27	17	41	26	17	49	23	16

Because our model suggests a proportionality between W and lg^2, we consider a plot of W versus lg^2. This plot is depicted in Figure 4.22. The plotted data lie approximately along a straight line passing through the origin, so the proportionality assumption seems reasonable. Now the point corresponding to the 41-oz fish happens to lie along the line shown in Figure 4.22, so the slope can be estimated as

$$\frac{41}{(17.25)(11)^2} \approx 0.0196$$

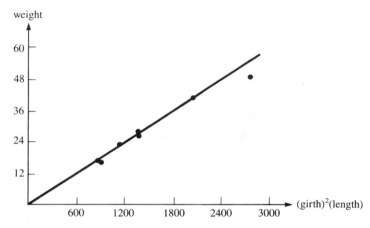

FIGURE 4.22   Testing the proportionality between W and lg^2

This computation leads to the model

$$W = 0.0196 \, lg^2 \tag{4.18}$$

In Chapter 5 we will show analytically that choosing the slope in such a way that the sum of the squared deviations from the given data points is minimized leads to the model

$$W = 0.0187 \, lg^2$$

A fisher would probably be happier with the new rule (4.18), because doubling the girth leads to a fourfold increase in the weight of the fish. However, the model appears more inconvenient to apply. Because $1/0.0196 \approx 50.9$, we could round this coefficient to 50 and have a fisher apply one of the simple rules:

$$W = \frac{lg^2}{50} \quad \text{for } W \text{ in ounces and } l, g \text{ measured in inches}$$

$$W = \frac{lg^2}{800} \quad \text{for } W \text{ in pounds and } l, g \text{ measured in inches}$$

However, the application of either of the preceding rules would probably require the fisher to record the length and the girth of each fish, and then compute its weight on a four-function calculator. Or perhaps he or she could be given a two-dimensional card showing the correct weights for different values of length and girth. The fishing competitors probably would prefer a simple plastic disk on which girth and length measurements can be entered in such a way that the weight of the bass appears magically in a window. You are asked to design such a disk in the following problem section.

4.5 PROBLEMS

1. Design a plastic disk to perform the calculations given by Model (4.18).

2. Consider Models (4.15) and (4.18); which do you think is better? Why? Discuss the models qualitatively. In Chapter 5 you will be asked to compare the two models analytically.

3. Under what circumstances, if any, will Models (4.15) and (4.18) coincide? Explain fully.

4. Consider the models $W \propto l^2 g$ and $W \propto g^3$. Interpret each of these models geometrically. Explain how these two models differ from Models (4.15) and (4.18), respectively. Under what circumstances, if any, would the four models coincide? Which model do you think would do the best job of predicting W? Why? In Chapter 5 you will be asked to compare the four models analytically.

5. **a.** Let $A(x)$ denote a typical cross-sectional area of a bass, $0 \le x \le l$, where l denotes the length of the fish. Use the mean value theorem from calculus to show that the volume V of the fish is given by

$$V = l \cdot \overline{A}$$

where $\overline{A}$ is the average value of $A(x)$.

b. Assuming that $\overline{A}$ is proportional to the square of the girth g and that weight density for the bass is constant, establish that

$$W \propto l g^2$$

4.6   BODY WEIGHT AND HEIGHT, STRENGTH AND AGILITY[1]

Body Weight and Height

A question of interest to almost all Americans is, How much should I weigh? A rule often given to people desiring to run a marathon is 2 lb of body weight per inch

---

[1]Optional section

of height, but shorter marathon runners seem to have a much easier time meeting this rule than taller ones. Tables have been designed to suggest weights for different purposes: Doctors are concerned about a reasonable weight for health purposes, and Americans seek weight standards based on physical appearance. Moreover, some organizations, such as the Army, are concerned about physical conditioning and define an upper weight allowance of acceptability. Quite often these weight tables are categorized in some manner. For example, consider Table 4.6, which gives upper weight limits of acceptability for males between the ages of 17 and 21. (The table has no further delineators such as bone structure.)

TABLE 4.6 Weight versus height for males aged 17–21

Height (in.)	Weight (lb)	Height (in.)	Weight (lb)
60	132	71	185
61	136	72	190
62	141	73	195
63	145	74	201
64	150	75	206
65	155	76	212
66	160	77	218
67	165	78	223
68	170	79	229
69	175	80	234
70	180		

If the differences between successive weight entries in Table 4.6 are computed to determine how much weight is allowed for each additional inch of height, it will be seen that throughout a large portion of the table a constant 5 lb per inch is allowed (with some fours and sixes appearing at the lower and upper ends of the scales, respectively). Certainly this last rule is more liberal than the previous 2 lb per inch rule recommended for the marathoners, but just how reasonable a constant weight per height rule is remains to be seen. In this section we examine qualitatively how weight and height should vary.

Body weight depends on a number of factors, some of which we have mentioned. In addition to height, bone density could be a factor. Is there a significant variation in bone density, or is it essentially constant? What about the relative volume occupied by the bones? Is the volume essentially constant, or are there heavy, medium, and light bone structures? And what about a body density factor? How can differences in the densities of bone, muscle, and fat be accounted for? Do these densities vary? Is body density typically a function of age and gender in the sense that the relative composition of muscle, bone, and fat vary as a person becomes older? Are there different compositions of muscle and fat between males and females at the same age?

Let's define the problem so that bone density is considered constant (by accepting an upper limit) and predict weight as a function of height, gender, age, and body density. The purposes or basis of the weight table must also be specified, so we base the table on physical appearance.

Problem identification   We identify the problem as follows: *For various heights, genders, and age groups, determine upper weight limits that represent maximum levels of acceptability based on physical appearance.*

Assumptions   Assumptions about body density are needed if we are to predict successfully weight as a function of height. As one simplifying assumption, suppose some parts of the body are composed of an inner core of a different density. Assume too that the inner core is composed primarily of bones and muscle material and that the outer core is primarily a fatty material, giving rise to the different densities (see Figure 4.23). We next construct submodels to explain how the weight of each core might vary with height.

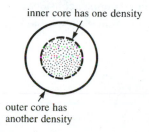

inner core has one density

outer core has
another density

FIGURE 4.23   Assume that parts of the body are composed of an inner and outer core of distinct densities

How does body weight vary with height? To begin, assume that for adults certain parts of the body, such as the head, have the same volume and density for different people. Thus the weight of an adult is given by

$$W = k_1 + W_{\text{in}} + W_{\text{out}} \tag{4.19}$$

where $k_1 > 0$ is the constant weight of those parts having the same volume and density for different individuals, and W_{in} and W_{out} are the weights of the inner and outer cores, respectively.

Next, let's turn our attention to the inner core. How does the volume of the extremities and trunk vary with height? We know that people are not geometrically similar because they don't appear to be scaled models of one another, having different shapes and different relative proportions of the trunk and extremities. By definition of the problem at hand, however, we are concerned with an upper weight limit based on physical appearance. Even though this may be somewhat subjective, it would seem reasonable that whatever image might be visualized as an upper limit standard of acceptability for a 74-in. person would be a scaled image of a 65-in. person. Thus for purposes of our problem, geometric similarity of individuals is a reasonable assumption. Note that no particular shape is being assumed but only that the ratio of distances between corresponding points in individuals is the same. Under this assumption the volume of each component we are considering is proportional to the cube of a characteristic dimension, which we select to be height h. Hence the sum of the components must be proportional to the cube of the height, or

$$V_{in} \propto h^3 \tag{4.20}$$

Now, what should we assume about the average weight density of the inner core? Assuming that the inner core is composed of muscle and bone, each of which has a different density, what percentage of the total volume of the inner core is occupied by the bones? If bone diameter is assumed to be proportional to the height, then the total volume occupied by the bones is proportional to the cube of the height. This implies that the percentage of the total volume of the inner core occupied by the bones in geometrically similar individuals is constant. It follows from the ensuing argument that the average weight density ρ_{in} is constant as well. For example, consider the average weight density ρ_{avg} of a volume V consisting of two components V_1 and V_2, each with a density ρ_1 and ρ_2. Then

$$V = V_1 + V_2$$

and

$$\rho_{avg} V = W = \rho_1 V_1 + \rho_2 V_2$$

yield

$$\rho_{avg} = \rho_1 \frac{V_1}{V} + \rho_2 \frac{V_2}{V}$$

Therefore as long as the ratios V_1/V and V_2/V do not change, the average weight density ρ_{avg} is constant. Application of this result to the inner core implies that the average weight density ρ_{in} is constant, yielding

$$W_{in} = V_{in}\rho_{in} \propto h^3$$

or

$$W_{in} = k_2 h^3 \quad \text{for } k_2 > 0 \tag{4.21}$$

Note that the preceding submodel includes any case of materials with densities different from muscles and bone (such as tendons, ligaments, and organs) as long as the percentage of the total volume of the inner core occupied by those materials is constant.

Now consider the outer core of fatty material. Because the table is to be based on personal appearance, it can be argued that the thickness of the outer core should be constant regardless of the height (see Problem 3). If τ represents this thickness, then the weight of the outer core is

$$W_{\text{out}} = \tau\rho_{\text{out}}S_{\text{out}}$$

where S_{out} is the surface area of the outer core and ρ_{out} is the density of the outer core. Again assuming that the subjects are geometrically similar, it follows that the surface area is proportional to the square of the height. If the density of the outer core of fatty material is assumed to be constant for all individuals, we have

$$W_{\text{out}} \propto h^2$$

It may be argued, however, that taller people can carry a greater thickness for the fatty layer. If it is assumed that the thickness of the outer core is proportional to the height, then

$$W_{\text{out}} \propto h^3$$

Allowing both these assumptions to reside in a single submodel gives

$$W_{\text{out}} = k_3h^2 + k_4h^3, \text{ where } k_3, k_4 \geq 0 \qquad \textbf{(4.22)}$$

Here the constants k_3 and k_4 are allowed to assume a zero value.

Summing the submodels represented by Equations (4.19), (4.21), and (4.22) to determine a model for weight yields

$$W = k_1 + k_3h^2 + k_5h^3 \quad \text{for } k_1, k_5 > 0 \text{ and } k_3 \geq 0 \qquad \textbf{(4.23)}$$

where $k_5 = k_2 + k_4$. Note that Model (4.23) suggests variations in weight of a higher order than the first power of h. If the model is valid, then taller people will indeed have a difficult time satisfying the linear rules given earlier. At the moment, however, our judgment can only be qualitative because we have not verified our submodels. Some ideas on how to test the model are discussed in the problem set. Furthermore, we have not placed any relative significance on the higher-order terms occurring in (4.23). In the study of statistics, regression techniques are given to provide insight to the significance of each term in the model.

Model interpretation   Let's interpret the general rules given earlier, which allowed a constant weight increase for each additional inch of height, in terms of our submodel. Consider the amount of weight attributable to an increase in the length of the trunk. Because the total allowable weight increase per inch is assumed constant by the given rules, the portion allowed for the trunk increase may also be assumed constant. To allow a constant weight increase, the trunk must increase in length while maintaining *the same cross-sectional area*. This implies, for example, that the waist size remains constant. Lets suppose that in the age 17–21 category, a 30-in. waist is judged the upper limit acceptable for the sake of personal appearance in a male with a height of 66 in. The 2 lb per inch rule would allow a 30-in. waist for a male with a height of 72 in., as well. On the other hand, the model based on geometric similarity suggests that all distances between corresponding points should increase by the same ratio. Thus the male with a height of 72 in. should have a waist of 30(72/66), or approximately 32.7 in., to be proportioned similarly. Comparing the two models, we get the following data.

Height (in.)	Linear models (in., waist measure)	Geometric similarity model (in., waist measure)
66	30	30.0
72	30	32.7
78	30	35.5
84	30	38.2

Now we can see why tall marathoners who follow the 2 lb per inch rule appear *very thin*.

Strength and Agility

Consider a competitive sports contest in which men or women of various sizes compete in events emphasizing strength (such as weight lifting) or agility (such as an obstacle course). How would you handicap such events? Let's define a problem as follows.

Problem identification *For various heights, weights, genders, and age groups, determine a relationship with agility in competitive sports.*

Assumptions Let's initially neglect gender and age. We assume that agility is proportional to the ratio strength/weight. We further assume that strength is proportional to the size of the muscles being used in the event, and we measure this size in terms of the muscles' cross-sectional area. (See Project 1 in Section 4.4.) Recall too that weight is proportional to volume (assuming constant weight density). If we assume all participants are geometrically similar, we have

$$\text{agility} \propto \frac{\text{strength}}{\text{weight}} \propto \frac{l^2}{l^3} \propto \frac{1}{l}$$

This is clearly a *nonlinear relationship* between agility and the characteristic dimension l. We also see that under these assumptions agility has a *nonlinear relationship* with weight. How would you collect data for your model? How would you test and verify your model?

4.6 PROBLEMS

1. Describe in detail the data you would like to obtain to test the various submodels supporting Model (4.23). How would you go about collecting the data?

2. Tests exist to measure the percent of body fat. Assume that such tests are accurate and that a large amount of carefully collected data are available. You may specify any other statistics, such as waist size and height, that you would like collected. Explain how the data could be arranged to check the assumptions underlying the submodels in this section. For example, suppose the data for males between ages 17 and 21 with constant body fat and height are examined. Explain how the assumption of constant density of the inner core could be checked.

3. A popular measure of physical condition and personal appearance is the pinch test. To administer this test, you measure the thickness of the outer core at selected locations on the body by pinching. Where and how should the pinch be made? What thickness of pinch should be allowed? Should the pinch thickness be allowed to vary with height?

4. It has been said that gymnastics is a sport that requires great agility. Use the model and assumptions developed for agility in this section to argue why there are few tall gymnasts.

4.6 PROJECTS

1. Consider an endurance test that measures only aerobic fitness. This test could be a swimming test, running test, or bike test. Assume we want all competitors to do an equal amount of work. Build a mathematical model that relates work done by the competitor to some measurable characteristic such as height or weight. Next consider a refinement using kinetic energy in your model. Collect data for one of these aerobic tests and determine the reasonableness of these models.

Chapter Five

MODEL FITTING

INTRODUCTION

In the mathematical modeling process we encounter situations that cause us to analyze data for different purposes. We have already seen how our assumptions can lead to a model of a particular type. For example, in Chapter 4, when we analyzed the distance required to bring a car to a safe stop once the brakes are applied, our assumptions led to a submodel of the form

$$d_b = Cv^2$$

where d_b is the distance required to stop the car, v is the velocity of the car at the time the brakes are applied, and C is some arbitrary constant of proportionality. At this point we can collect and analyze sufficient data to determine if the assumptions are reasonable. If they are, we want to determine the constant C that selects the particular member from the family $y = Cv^2$ corresponding to the braking distance submodel.

We may encounter situations in which there are different assumptions leading to different submodels. For example, when studying the motion of a projectile through a medium such as air, we can make different assumptions about the nature of a drag force, such as the drag force being proportional to v or v^2. We might even choose to neglect the drag force completely. As another example, when we are determining how fuel consumption varies with automobile velocity, our different assumptions about the drag force can lead to models that predict mileage varies as $C_1 v^{-1}$ or as $C_2 v^{-2}$. The resulting problem can be thought of in the following way: First, use collected data to choose C_1 and C_2 in a way that selects the curve from each family that best fits the data, and then choose whichever resultant model is most appropriate for the particular situation under investigation.

A different case arises when the problem is so complex as to prevent the formulation of a model explaining the situation. For instance, if the submodels involve partial differential equations that are not solvable in closed form, there is little hope for constructing a master model that can be solved and analyzed without the aid of a computer. Or there may be so many significant variables involved that we would not even attempt to construct an explicative model. In such cases experiments may have to be conducted to investigate the behavior of the independent variable(s) within the range of the the data points.

The preceding discussion identifies three possible tasks when analyzing a collection of data points:

1. Fitting a selected model type or types to the data.
2. Choosing the most appropriate model from competing types that have been fitted. For example, we may need to determine whether the best-fitting exponential model is a better model than the best-fitting polynomial model.
3. Making predictions from the collected data.

In the first two tasks a model or competing models exist that seem to *explain* the behavior being observed. We address these two cases in this chapter under the general heading of *model fitting*. In the third case, however, a model does not exist to explain the behavior being observed. Rather, there exists a collection of data points that can be used to *predict* the behavior within some range of interest. In essence we wish to construct an *empirical model* based on the collected data. In Chapter 6 we study such empirical model construction under the general heading of *interpolation*. It is important to understand both the philosophical and the mathematical distinctions between model fitting and interpolation.

Relationship Between Model Fitting and Interpolation

Let's analyze the three tasks to determine what must be done in each case. In Task 1 the precise meaning of *best* model must be identified and the resulting mathematical problem resolved. In Task 2 a criterion is needed for comparing models of different types. In Task 3 a criterion must be established for determining how to make predictions in between the observed data points.

Note the difference in the modeler's attitude in each of these situations. In the two model-fitting tasks, a relationship of a particular type is strongly suspected, and the modeler is willing to accept some deviation between the model and the collected data points to have a model that satisfactorily *explains* the situation under investigation. In fact, the modeler expects errors to be present in both the model and the data. On the other hand, when interpolating, the modeler is strongly guided by the data that have been carefully collected and analyzed, and a curve is sought that captures the trend of the data to *predict* in between the data points. Thus the modeler generally attaches little explicative significance to the interpolating curves. In all situations the modeler may ultimately want to make predictions from the model. However, the modeler tends to emphasize the proposed *models* over the data when model fitting, whereas he or she places greater confidence in the *collected data* when interpolating and attaches less significance to the form of the model. In a sense, explicative models are *theory* driven, whereas predictive models are *data* driven.

Let's illustrate the preceding ideas with an example. Suppose we are attempting to relate two variables, y and x, and have gathered the data plotted in Figure 5.1. If the modeler is going to make predictions based solely on the data in the figure, he or she might use a technique such as *spline interpolation* (which we study in Chapter 6) to pass a smooth curve through the points (see Figure 5.2). Note that in Figure 5.2 the interpolating curve passes through the data points and captures the trend of the behavior over the range of observations.

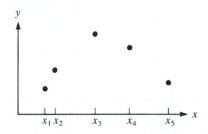

FIGURE 5.1 Observations relating the variables y and x

Suppose, however, that in studying the particular behavior depicted in Figure 5.1 the modeler makes assumptions leading to the expectation of a quadratic model, or parabola, of the form $y = C_1 x^2 + C_2 x + C_3$. In this case, the data of Figure 5.1 would be used to determine the arbitrary constants C_1, C_2, and C_3 to select the best parabola (see Figure 5.3). The fact that the parabola may deviate from some or all of the data points would be of no concern. Note the difference in the values of the predictions made by the curves in Figures 5.2 and 5.3 in the vicinity of the values x_1 and x_5.

A modeler may find it necessary both to fit a model and to interpolate in the same problem. The best-fitting model of a given type may prove to be unwieldy, or even impossible, for subsequent analysis involving operations such as integration or differentiation. In such situations the model may be replaced with an interpolating curve (such as a polynomial) that is more readily differentiated or integrated. For example, a step function used to model a square wave might be replaced by a trigonometric approximation to facilitate subsequent analysis. In these instances

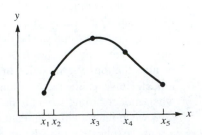

FIGURE 5.2 Interpolating the data using a smooth polynomial

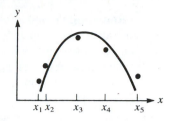

FIGURE 5.3   Fitting a parabola $y = C_1 x^2 + C_2 x + C_3$ to the data points

approximation

the modeler desires the interpolating curve to approximate closely the essential characteristics of the function it replaces. This type of interpolation is usually called **approximation** and is typically addressed in introductory numerical analysis courses.

Sources of Error in the Modeling Process

Before discussing criteria on which to base curve-fitting and interpolation decisions, we need to examine the modeling process to ascertain where errors can arise. If error considerations are neglected, undue confidence may be placed on intermediate results, causing faulty decisions in subsequent steps. Our goals are to ensure that all parts of the modeling process are computationally compatible and to consider the effects of cumulative errors likely to exist from previous steps.

For purposes of easy reference, we classify errors under the following category scheme:

1. Formulation error
2. Truncation error
3. Round-off error
4. Measurement error

formulation error

Formulation errors result from the assumption that certain variables are negligible or from simplifications in describing interrelationships among the variables in the various submodels. For example, when we determined a submodel for braking distance in Chapter 4, we completely neglected road friction and assumed a simple relationship for the nature of the drag force caused by air resistance. Formulation errors are present in even the best models.

truncation error

Truncation errors are attributable to the numerical method used to solve a mathematical problem. For example, we may find it necessary to approximate $\sin x$ with a polynomial representation obtained from the power series

$$\sin x = x - \frac{x^3}{3!} + \frac{x^5}{5!} - \cdots$$

An error will be introduced when the series is truncated to produce the polynomial.

round-off error   **Round-off errors** are caused by using a finite digit machine for computation. Because all numbers cannot be represented exactly using only finite representations, we must always expect round-off errors. For example, consider a calculator or computer that uses eight-digit arithmetic. The number $\frac{1}{3}$ is represented by .33333333 so that 3 times $\frac{1}{3}$ is .99999999 rather than the actual value 1. The error 10^{-8} is caused by round off. The ideal real number $\frac{1}{3}$ is an *infinite* string of decimal digits .3333..., but calculators and computers can do arithmetic only with numbers having finite precision. When many arithmetic operations are performed in succession, each with its own round-off, the accumulated effect of round-off can significantly alter the numbers that are supposed to be the answer. Round-off is just one of the things we have to live with—*and be aware of*—when we use computing machines.

measurement error   **Measurement errors** are caused by imprecision in data collection. This imprecision may include such diverse factors as human errors in recording or reporting the data or physical limitations of the laboratory equipment. For example, considerable measurement error would be expected in the data reflecting the response distance and the braking distance in the braking distance problem.

5.1   FITTING MODELS TO DATA GRAPHICALLY

Assume the modeler has made certain assumptions leading to a model of a particular type. The model generally contains one or more parameters, and sufficient data must be gathered to determine them. Let's consider the problem of data collection.

The determination of how many data points to collect involves a trade-off between the cost of obtaining them and the accuracy required of the model. As a minimum, the modeler needs as many data points as there are arbitrary constants in the model curve. Additional points are required to determine any arbitrary constants involved with some technique of "best fit" being used. The *range* over which the model is to be used determines the end points of the interval for the independent variable(s).

The *spacing* of the data points within that interval is also important because any part of the interval over which the model must fit particularly well can be weighted by using unequal spacing. We may choose to take more data points where maximum use of the model is expected, or we may collect more data points where we anticipate abrupt changes in the dependent variable(s).

Even if the experiment has been carefully designed and the trials meticulously conducted, the modeler needs to appraise the accuracy of the data before attempting to fit the model. How were the data collected? What is the accuracy of the measuring devices used in the collection process? Do any points appear suspicious? Following such an appraisal and elimination (or replacement) of spurious data, it is useful to think of each data point as an interval of relative confidence rather than as a single point. This idea is shown in Figure 5.4. The length of each interval should be commensurate with the appraisal of the errors present in the data collection process.

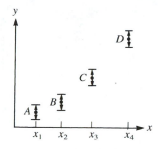

FIGURE 5.4   Each data point is thought of as an interval of confidence

Visual Model Fitting with the Original Data

Suppose we wish to fit the model $y = ax + b$ to the data shown in Figure 5.4. How might the constants a and b be chosen to determine the line that best fits the data? Generally, when more than two data points exist, not all of them can be expected to lie exactly along a single straight line, even if such a line accurately models the relationship between the two variables x and y. That is, ordinarily there will be some vertical discrepancy between some of the data points and any particular line being considered. We refer to these vertical discrepancies as **absolute deviations** (see Figure 5.5). For the best-fitting line, we might try to minimize the sum of these absolute deviations, leading to the model depicted in Figure 5.5. Although success may be achieved in minimizing the sum of the absolute deviations, the absolute deviation from individual points may be quite large. For example, consider point D in Figure 5.5. If the modeler has confidence in the accuracy of this data point, there would be concern for the predictions made from the fitted line near the point. As an alternative, suppose a line is selected that minimizes the largest deviation from any point. Applying this criterion to the data points might give the line shown in Figure 5.6.

Although these visual methods for fitting a line to data points may appear imprecise, the methods are often quite *compatible* with the accuracy of the modeling process itself. The grossness of the assumptions and the imprecision involved in the data collection may not warrant a more sophisticated analysis. In such situations

absolute deviations

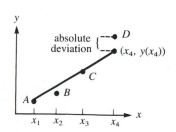

FIGURE 5.5   Minimizing the sum of the absolute deviations from the fitted line

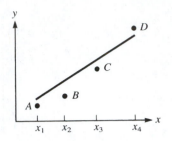

FIGURE 5.6 Minimizing the largest absolute deviation from the fitted line

the blind application of one of the analytic methods to be presented in Section 5.2 may lead to models far less appropriate than one obtained graphically. Furthermore, a visual inspection of the model fitted graphically to the data immediately gives an impression of *how good* the fit is and *where* it appears to fit well. Unfortunately, these important considerations are often overlooked in problems with large amounts of data analytically fitted via computer codes. Because the model-fitting portion of the modeling process seems to be more precise and analytical than some of the other steps, there is a tendency to place undue faith in the numerical computations.

Transforming the Data

Most of us are limited visually to fitting only lines. So how can we graphically fit other curves as models? Suppose, for example, a relationship of the form $y = Ce^x$ is suspected for some submodel and the data shown in Table 5.1 have been collected.

TABLE 5.1 Collected data

x	1	2 .	3	4
y	8.1	22.1	60.1	165

The model states that y is proportional to e^x. Thus, if we plot y versus e^x, we should obtain approximately a straight line. The situation is depicted in Figure 5.7. Because the plotted data points do lie approximately along a line that projects

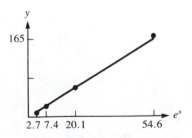

FIGURE 5.7 Plot of y versus e^x for the data given in Table 5.1

through the origin, we conclude that the assumed proportionality is reasonable. From the figure, the slope of the line is approximated as

$$C = \frac{165 - 60.1}{54.6 - 20.1} \approx 3.0$$

Now let's consider an alternate technique that is useful in a variety of problems. Take the logarithm of each side of the equation $y = Ce^x$ to obtain

$$\ln y = \ln C + x$$

Note that this expression is an equation of a line in the variables $\ln y$ and x. The number $\ln C$ is the intercept when $x = 0$. The transformed data are shown in Table 5.2 and plotted in Figure 5.8. Semilog paper or a computer is useful when plotting large amounts of data.

TABLE 5.2 The transformed data from Table 5.1

x	1	2	3	4
$\ln y$	2.1	3.1	4.1	5.1

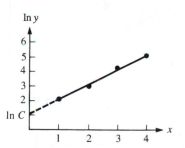

FIGURE 5.8   Plot of $\ln y$ versus x using Table 5.2

From Figure 5.8 we can determine that the intercept $\ln C$ is approximately 1.1, giving $C = e^{1.1} \approx 3.0$ as before.

A similar transformation can be performed on a variety of other curves to produce linear relationships among the resulting transformed variables. For example, if $y = x^a$, then

$$\ln y = a \ln x$$

is a linear relationship in the transformed variable $\ln y$ and $\ln x$. Here log-log paper or a computer is useful when plotting large amounts of data.

Let's pause and make an important observation. Suppose we do invoke a transformation and plot ln y versus x, as in Figure 5.8, and find the line that successfully minimizes the sum of the absolute deviations of the transformed data points. The line then determines ln C, which in turn produces the proportionality constant C. Although it is not obvious, the resulting model $y = Ce^x$ is not the member of the family of exponential curves of the form ke^x that minimizes the sum of the absolute deviations from the original data points (when we plot y versus x). This important idea will be demonstrated both graphically and analytically in the ensuing discussion. When transformations of the form $y = \ln x$ are made, the distance concept is distorted. Whereas a fit that is compatible with the inherent limitations of a graphical analysis may be obtained, the modeler must be aware of this distortion and *verify the model using the graph from which it is intended to make predictions or conclusions—namely, the* y *versus* x *graph in the original data rather than the graph of the transformed variables.*

We now present an example illustrating how a transformation may distort distance in the xy-plane. Consider the data plotted in Figure 5.9 and assume the data are expected to fit a model of the form $y = Ce^{1/x}$. Using a logarithmic transformation as before, we find

$$\ln y = \frac{1}{x} + \ln C$$

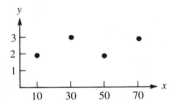

FIGURE 5.9 A plot of some collected data points

A plot of the points ln y versus $1/x$ based on the original data is shown in Figure 5.10. Note from the figure how the transformation distorts the distances between the original data points and squeezes them all together. Consequently, if a straight line is made to fit the transformed data plotted in Figure 5.10, the absolute deviations appear relatively small (i.e., small computed on the Figure 5.10 scale rather than on the Figure 5.9 scale). If we were to plot the fitted model $y = Ce^{1/x}$ to

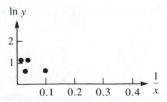

FIGURE 5.10 A plot of the transformed data points

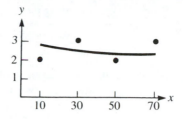

FIGURE 5.11   A plot of the curve $y = Ce^{1/x}$ based on the value $\ln C \approx 0.9$ from Figure 5.10

the data in Figure 5.9, however, we would see that it fits the data relatively poorly, as shown in Figure 5.11.

From the preceding example we see that if a modeler is not careful when using transformations, he or she can be tricked into selecting a relatively poor model. This realization becomes especially important when comparing alternative models. Serious errors can be introduced when selecting the best model unless all comparisons are made with the original data (plotted in Figure 5.9 in our example). Otherwise, the choice of best model may be determined by a peculiarity of the transformation rather than on the merits of the model and how well it fits the original data. Whereas the danger of making transformations is evident in this graphical illustration, a modeler may be fooled if he or she is not especially observant, because many computer codes fit models by first making a transformation. If the modeler intends to use indicators such as the sum of the absolute deviations to make decisions about the adequacy of a particular submodel or choose among competing submodels, the modeler must first ascertain how those indicators were computed.

5.1 PROBLEMS

1. The model in Figure 5.2 would normally be used to predict behavior between x_1 and x_5. What would be the danger of using the model to predict y for values of less than x_1 or greater than x_5? Suppose we are modeling the trajectory of a thrown baseball.

2. The following table gives the elongation e in inches per inch (in./in.) for a given stress S on a steel wire measured in pounds per square inch (lb/in.^{2}). Test the model $e = c_1 S$ by plotting the data. Estimate c_1 graphically.

$S\ (\times 10^{-3})$	5	10	20	30	40	50	60	70	80	90	100
$e\ (\times 10^5)$	0	19	57	94	134	173	216	256	297	343	390

3. In the following data, x is the diameter in inches of a ponderosa pine measured at breast height and y is a measure of volume—number of board feet divided by 10. Test the model $y = ax^b$ by plotting the transformed data. If the model seems reasonable, estimate the parameters a and b of the model graphically.

x	17	19	20	22	23	25	28	31	32	33	36	37	38	39	41
y	19	25	32	51	57	71	113	141	123	187	192	205	252	259	294

4. In the following data, V represents a mean walking velocity and P represents the population size. We wish to know if we can predict the population size P by observing how fast people walk. Plot the data. What kind of a relationship is suggested? Test the following models by plotting the appropriate transformed data:

a. $P = aV^b$

b. $P = a \ln V$

V	2.27	2.76	3.27	3.31	3.70	3.85	4.31	4.39	4.42
P	2500	365	23700	5491	14000	78200	70700	138000	304500

V	4.81	4.90	5.05	5.21	5.62	5.88
P	341948	49375	2602000	867023	1340000	1092759

5. The following data represent the growth of a population of fruit flies over a 6-week period. Test the following models by plotting an appropriate set of data. Estimate the parameters of the models:

a. $P = c_1 t$

b. $P = ae^{bt}$

t (days)	7	14	21	28	35	42
P (number of observed flies)	8	41	133	250	280	297

6. The following data represent (hypothetical) energy consumption normalized to the year 1900. Plot the data. Test the model $Q = ae^{bx}$ by plotting the transformed data. Estimate the parameters of the model graphically.

x	Year	Consumption Q
0	1900	1.00
10	1910	2.01
20	1920	4.06
30	1930	8.17
40	1940	16.44
50	1950	33.12
60	1960	66.69
70	1970	134.29
80	1980	270.43
90	1990	544.57
100	2000	1096.63

7. In 1601 the German astronomer Johannes Kepler became director of the Prague Observatory. Kepler had been helping Tycho Brahe in collecting 13 years of observations on the relative motion of the planet Mars. By 1609 Kepler had formulated his first two laws:

   i. Each planet moves on an ellipse with the sun at one focus.

   ii. For each planet, the line from the sun to the planet sweeps out equal areas in equal times.

   Kepler spent many years verifying these laws and formulating a third law, which relates the orbital periods and mean distances from the sun.

   a. Plot the period time T versus the mean distance r using the following updated observational data:

Planet	Period (days)	Mean distance from sun (km $\times 10^{-6}$)
Mercury	88	57.9
Venus	225	108.2
Earth	365	149.6
Mars	687	227.9
Jupiter	4329	778.3
Saturn	10,753	1427
Uranus	30,660	2870
Neptune	60,150	4497
Pluto	90,670	5907

   b. Assuming a relationship of the form

$$T = Cr^a$$

   determine the parameters C and a by plotting $\ln T$ versus $\ln r$. Does the model seem reasonable? Try to formulate Kepler's Third Law.

5.2

ANALYTIC METHODS OF MODEL FITTING

In this section we investigate several criteria for fitting curves to a collection of data points. Each criterion gives a way of selecting the best curve from a given family in the sense that the curve most accurately represents the data according to the criterion. We also discuss how the various criteria are related.

Chebyshev Approximation Criterion

In the preceding section we graphically fit lines to a given collection of data points. One of the best-fit criteria used was to minimize the largest distance from the line to any corresponding data point. Let's analyze this geometric construction. Given a collection of m data points $(x_i, y_i), i = 1, 2, \ldots, m$, fit the collection to the line $y = ax + b$, determined by the parameters a and b, that minimizes the distance between any data point (x_i, y_i) and its corresponding point on the line $(x_i, ax_i + b)$. That is, minimize the largest absolute deviation $|y_i - y(x_i)|$ over the entire collection of data points. Now let's generalize this criterion.

Given some function type $y = f(x)$ and a collection of m data points (x_i, y_i), minimize the largest absolute deviation $|y_i - f(x_i)|$ over the entire collection. That is, determine the parameters of the function type $y = f(x)$ that minimizes the number

$$\text{Maximum } |y_i - f(x_i)| \quad i = 1, 2, \ldots, m \tag{5.1}$$

Chebyshev approx-
imation criterion

This important criterion is often called the **Chebyshev approximation criterion**. The difficulty with the Chebyshev criterion is that it is often complicated to apply in practice, at least using only elementary calculus. The optimization problems that result from applying the criterion may require advanced mathematical procedures or numerical algorithms requiring the use of a computer.

For example, suppose we want to measure the line segments AB, BC, and AC represented in Figure 5.12. Assume our measurements yield the estimates $AB = 13$, $BC = 7$, and $AC = 19$. As we should expect in any physical measuring process, discrepancy results. In this situation, the values of AB and BC add up to 20 rather than the estimated $AC = 19$. Let's resolve the discrepancy of 1 unit using the Chebyshev criterion. That is, we will assign values to the three line segments in such a way that the largest absolute deviation between any corresponding pair of assigned and observed values is minimized. Assume the same degree of confidence in each measurement so that each measurement has equal weight. In that case, the discrepancy should be distributed equally across each segment, resulting in the predictions $AB = 12\frac{2}{3}$, $BC = 6\frac{2}{3}$, and $AC = 19\frac{1}{3}$. Thus, each absolute deviation is $\frac{1}{3}$. Convince yourself that reducing any of these deviations causes one of the other deviations to increase. (Remember that $AB + BC$ must equal AC.) Let's formulate the problem symbolically.

A B C

FIGURE 5.12 The line segment AC is divided into two segments AB and BC

Let x_1 represent the true value of the length of the segment AB and x_2 the true value of BC. For ease of our presentation, let r_1, r_2, and r_3 represent the discrepancies between the true and measured values as follows:

$$x_1 - 13 = r_1 \quad \text{(line segment } AB)$$
$$x_2 - 7 = r_2 \quad \text{(line segment } BC)$$
$$x_1 + x_2 - 19 = r_3 \quad \text{(line segment } AC)$$

residuals   The numbers r_1, r_2, and r_3 are called **residuals**.

If the Chebyshev approximation criterion is applied, values would be assigned to x_1 and x_2 in such a way as to minimize the largest of the three numbers $|r_1|$, $|r_2|$, $|r_3|$. If we call that largest number r, then we want to

Minimize r

Subject to the three conditions:

$$|r_1| \le r \quad \text{or} \quad -r \le r_1 \le r$$
$$|r_2| \le r \quad \text{or} \quad -r \le r_2 \le r$$
$$|r_3| \le r \quad \text{or} \quad -r \le r_3 \le r$$

Each of these conditions can be replaced by two inequalities. For example, $|r_1| \le r$ can be replaced by $r - r_1 \ge 0$ and $r + r_1 \ge 0$. If this is done for each condition, the problem can be stated in the form of a classical mathematical problem:

Minimize r

Subject to

$$r - x_1 \qquad + 13 \ge 0 \quad (r - r_1 \ge 0)$$
$$r + x_1 \qquad - 13 \ge 0 \quad (r + r_1 \ge 0)$$
$$r \qquad - x_2 + 7 \ge 0 \quad (r - r_2 \ge 0)$$
$$r \qquad + x_2 - 7 \ge 0 \quad (r + r_2 \ge 0)$$
$$r - x_1 - x_2 + 19 \ge 0 \quad (r - r_3 \ge 0)$$
$$r + x_1 + x_2 - 19 \ge 0 \quad (r + r_3 \ge 0)$$

linear program   This problem is called a **linear program**. We will discuss linear programs further in Chapters 8 and 9. Even large linear programs can be solved by computer implementation of an algorithm known as the Simplex Method. In the preceding line segment example, the Simplex Method yields a minimum value of $r = \frac{1}{3}$, and $x_1 = 12\frac{2}{3}$ and $x_2 = 6\frac{2}{3}$.

We now generalize this procedure. Given some function type $y = f(x)$, whose parameters are to be determined, and a collection of m data points (x_i, y_i), define the residuals $r_i = y_i - f(x_i)$. If r represents the largest absolute value of these residuals, then the problem is to

$$\text{Minimize } r$$

Subject to

$$\left.\begin{array}{l} r - r_i \geq 0 \\ r + r_i \geq 0 \end{array}\right\} \text{ for } i = 1, 2, \ldots, m$$

Although we discuss linear programs in Chapter 8, we should note here that the model resulting from this procedure is not always a linear program; for example, consider fitting the function $f(x) = \sin kx$. Also note that many computer codes of the simplex algorithm require using variables that are allowed to assume only nonnegative values. This requirement can be accomplished with a simple substitution (see Problem 5).

As we will see, alternative criteria lead to optimization problems that often can be resolved more conveniently. Primarily for this reason, the Chebyshev criterion is not used often for fitting a curve to a finite collection of data points. However, its application should be considered whenever minimizing the largest absolute deviation is important. (We consider several applications of the criterion in Chapters 8 and 9.) Furthermore, the principle underlying the Chebyshev criterion is extremely important when we are replacing a function defined over an interval by another function and the largest difference between the two functions over the interval must be minimized. This topic is studied in approximation theory and is typically covered in introductory numerical analysis.

Minimizing the Sum of the Absolute Deviations

When we were graphically fitting lines to the data in Section 5.1, one of our criteria minimized the total sum of the absolute deviations between the data points and their corresponding points on the fitted line. This criterion can be generalized: Given some function type $y = f(x)$ and a collection of m data points (x_i, y_i), minimize the sum of the absolute deviations $|y_i - f(x_i)|$. That is, determine the parameters of the function type $y = f(x)$ to minimize

$$\sum_{i=1}^{m} |y_i - f(x_i)| \tag{5.2}$$

If we let $R_i = |y_i - f(x_i)|, i = 1, 2, \ldots, m$ represent each absolute deviation, then criterion (5.2) can be interpreted as minimizing the length of the line formed by adding together the numbers R_i. This is illustrated for the case $m = 2$ in Figure 5.13.

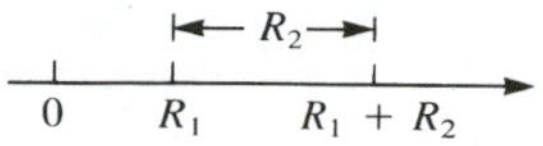

FIGURE 5.13 A geometrical interpretation of minimizing the sum of the absolute deviations

Although we geometrically applied this criterion in Section 5.1 when the function type $y = f(x)$ is a line, the general criterion presents severe problems. To solve this optimization problem using the calculus, we need to differentiate the sum (5.2) with respect to the parameters of $f(x)$ to find the critical points. However, the various derivatives of the sum fail to be continuous because of the presence of the absolute values, so we will not pursue this criterion any further. In Chapters 8 and 9 we consider other applications of the criterion and present a technique for approximating solutions numerically.

Least-Squares Criterion

least-squares criterion Currently the most frequently used curve-fitting criterion is the **least-squares criterion**. Using the same notation as shown earlier, the problem is to determine the parameters of the function type $y = f(x)$ to minimize the sum

$$\sum_{i=1}^{m} |y_i - f(x_i)|^2 \tag{5.3}$$

Part of the popularity of this criterion stems from the ease with which the resulting optimization problem can be solved using only the calculus of several variables. However, relatively recent advances in mathematical programming techniques (such as the Simplex Method for solving many applications of the Chebyshev criterion) and advances in numerical methods for approximating solutions to criterion (5.2) promise to dissipate this advantage. The justification for using the least-squares method increases when considering probabilistic arguments that assume the errors are distributed randomly. We will not discuss such probabilistic arguments until later in the text.

We now give a geometric interpretation of the least-squares criterion. Consider the case of three data points and let $R_i = |y_i - f(x_i)|$ denote the absolute deviation between the observed and predicted values for $i = 1, 2, 3$. You can think of R_i as the scalar components of a deviation vector as depicted in Figure 5.14. Thus the vector $\mathbf{R} = R_1\mathbf{i} + R_2\mathbf{j} + R_3\mathbf{k}$ represents the resultant deviation between the observed and predicted values. The magnitude of the deviation vector is given by

$$|\mathbf{R}| = \sqrt{R_1^2 + R_2^2 + R_3^2}$$

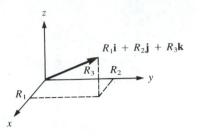

FIGURE 5.14 A geometrical interpretation of the least-squares criterion

To minimize $|\mathbf{R}|$ we can minimize $|\mathbf{R}|^2$ (see Problem 1 at the end of this section). Thus the least-squares problem is to determine the parameters of the function type $y = f(x)$ such that

$$|\mathbf{R}|^2 = \sum_{i=1}^{3} R_i^2 = \sum_{i=1}^{3} \left[y_i - f(x_i) \right]^2$$

is minimized. That is, we may interpret the least-squares criterion as minimizing the magnitude of the vector whose coordinates represent the absolute deviation between the observed and predicted values.

Relating the Criteria

The geometric interpretations of the three curve-fitting criteria help in providing a qualitative description comparing the criteria. Minimizing the sum of the absolute deviations tends to treat each data point with equal weight and to average the deviations. The Chebyshev criterion gives more weight to a single point potentially having a large deviation. The least-squares criterion is somewhere in between as far as weighting individual points with significant deviations is concerned. But let's be more precise. Because the Chebyshev and least-squares criteria are the most convenient to apply analytically, we now derive a method for relating the deviations resulting from using these two criteria.

Suppose the Chebyshev criterion is applied and the resulting optimization problem solved to yield the function $f_1(x)$. The absolute deviations resulting from the fit are defined as follows:

$$|y_i - f_1(x_i)| = c_i, \quad i = 1, 2, \ldots, m$$

Now define $c_{\max}$ as the largest of the absolute deviations c_i. There is a special significance attached to $c_{\max}$. Because the parameters of the function $f_1(x)$ are determined to minimize the value of $c_{\max}$, it is the minimal largest absolute deviation obtainable.

On the other hand, suppose the least-squares criterion is applied and the resulting optimization problem solved to yield the function $f_2(x)$. The absolute deviations resulting from the fit are then given by

$$|y_i - f_2(x_i)| = d_i, \quad i = 1, 2, \ldots, m$$

Define d_{max} as the largest of the absolute deviations d_i. At this point it can only be said that d_{max} is at least as large as c_{max} because of the special significance of the latter as discussed previously. However, let's attempt to relate d_{max} and c_{max} more precisely.

The special significance the least-squares criterion attaches to d_i is that the sum of their squares is the smallest such sum obtainable. Thus it must be true that

$$d_1^2 + d_2^2 + \cdots + d_m^2 \le c_1^2 + c_2^2 + \cdots + c_m^2$$

Because $c_i \le c_{max}$ for every i, these inequalities imply

$$d_1^2 + d_2^2 + \cdots + d_m^2 \le mc_{max}^2$$

or

$$\sqrt{\frac{d_1^2 + d_2^2 + \cdots + d_m^2}{m}} \le c_{max}$$

For ease of discussion define

$$D = \sqrt{\frac{d_1^2 + d_2^2 + \cdots + d_m^2}{m}}$$

Thus

$$D \le c_{max} \le d_{max}$$

This last relationship is revealing. Suppose it is more convenient to apply the least-squares criterion in a particular situation, but there is concern about the largest absolute deviation c_{max} that may result. If we compute D, a lower bound on c_{max} is obtained and d_{max} gives an upper bound. Thus, if there is considerable difference between D and d_{max}, the modeler should consider applying the Chebyshev criterion.

5.2 PROBLEMS

1. Using elementary calculus, show that the minimum and maximum points for $y = f(x)$ occur among the minimum and maximum points for $y = f^2(x)$. Assuming $f(x) \ge 0$, why can we minimize $f(x)$ by minimizing $f^2(x)$?

2. For each of the following data sets, formulate the mathematical model that minimizes the largest deviation between the data and the line $y = ax + b$. If a computer is available, solve for the estimates of a and b.

a.

x	1.0	2.3	3.7	4.2	6.1	7.0
y	3.6	3.0	3.2	5.1	5.3	6.8

b.

x	29.1	48.2	72.7	92.0	118	140	165	199
y	0.0493	0.0821	0.123	0.154	0.197	0.234	0.274	0.328

c.

x	2.5	3.0	3.5	4.0	4.5	5.0	5.5
y	4.32	4.83	5.27	5.74	6.26	6.79	7.23

3. For the data given, formulate the mathematical model that minimizes the largest deviation between the data and the model $y = c_1x^2 + c_2x + c_3$. If a computer is available, solve for the estimates of c_1, c_2, and c_3.

x	0.1	0.2	0.3	0.4	0.5
y	0.06	0.12	0.36	0.65	0.95

4. For the following data, formulate the mathematical model that minimizes the largest deviation between the data and the model $P = ae^{bt}$. If a computer is available, solve for the estimates of a and b.

t	7	14	21	28	35	42
P	8	41	133	250	280	297

5. Suppose the variable x_1 can assume any real value. Show that the following substitution using nonnegative variables x_2 and x_3 permits x_1 to assume any real value:

$$x_1 = x_2 - x_3, \quad \text{where } x_1 \text{ is unconstrained}$$

and

$$x_2 \geq 0 \quad \text{and} \quad x_3 \geq 0$$

Thus if a computer allows only nonnegative variables, the substitution allows for solving the linear program in the variables x_2 and x_3, and then recovering the value of the variable x_1.

5.3   APPLYING THE LEAST-SQUARES CRITERION

Suppose our assumptions lead us to expect a model of a certain type and that data have been collected and analyzed. In this section the least-squares criterion is applied to estimate the parameters for several types of curves.

Fitting a Straight Line

Suppose a model of the form $y = Ax + B$ is expected and it's been decided to use the m data points (x_i, y_i), $i = 1, 2, \ldots, m$, to estimate A and B. Denote the least-squares estimate of $y = Ax + B$ by $y = ax + b$. Applying the least-squares criterion (5.3) to this situation requires the minimization of

$$S = \sum_{i=1}^{m} \left[y_i - f(x_i) \right]^2 = \sum_{i=1}^{m} (y_i - ax_i - b)^2$$

A necessary condition for optimality is that the two partial derivatives $\partial S / \partial a$ and $\partial S / \partial b$ equal zero, yielding the equations

$$\frac{\partial S}{\partial a} = -2 \sum_{i=1}^{m} (y_i - ax_i - b)x_i = 0$$

$$\frac{\partial S}{\partial b} = -2 \sum_{i=1}^{m} (y_i - ax_i - b) = 0$$

These equations can be rewritten to give

normal equations

$$\left. \begin{array}{l} a \sum_{i=1}^{m} x_i^2 + b \sum_{i=1}^{m} x_i = \sum_{i=1}^{m} x_i y_i \\[2mm] a \sum_{i=1}^{m} x_i + mb = \sum_{i=1}^{m} y_i \end{array} \right\} \tag{5.4}$$

The preceding equations can be solved for a and b once all the values for x_i and y_i are substituted into them. The solutions for the parameters a and b are easily obtained by elimination and are found to be (see Problem 1 at the end of this section)

slope

$$a = \frac{m \sum x_i y_i - \sum x_i \sum y_i}{m \sum x_i^2 - \left(\sum x_i \right)^2}, \quad \text{the slope} \tag{5.5}$$

and

intercept

$$b = \frac{\sum x_i^2 \sum y_i - \sum x_i y_i \sum x_i}{m \sum x_i^2 - \left(\sum x_i \right)^2}, \quad \text{the intercept} \tag{5.6}$$

Computer algorithms are easily written to compute these values for a and b for any collection of data points. Equations (5.4) are called the **normal equations**.

Fitting a Power Curve

Now let's use the least-squares criterion to fit a curve of the form $y = Ax^n$, where n is fixed, to a given collection of data points. Call the least-squares estimate of the model $f(x) = ax^n$. Application of the criterion then requires minimization of

$$S = \sum_{i=1}^{m} \left[y_i - f(x_i) \right]^2 = \sum_{i=1}^{m} \left[y_i - ax_i^n \right]^2$$

A necessary condition for optimality is that the derivative dS/da equals zero, giving the equation

$$\frac{dS}{da} = -2 \sum_{i=1}^{m} x_i^n \left[y_i - ax_i^n \right] = 0$$

Solving the equation for a yields

$$a = \frac{\sum x_i^n y_i}{\sum x_i^{2n}} \tag{5.7}$$

Remember, the number n is *fixed* in Equation (5.7).

The least-squares criterion can be applied to other models as well. The limitation in applying the method lies in calculating the various derivatives required in the optimization process, setting these derivatives to zero, and solving the resulting equations for the parameters in the model type.

For example, let's fit $y = Ax^2$ to the data shown in Table 5.3 and predict the value of y when $x = 2.25$.

TABLE 5.3 Data collected to fit $y = Ax^2$

x	0.5	1.0	1.5	2.0	2.5
y	0.7	3.4	7.2	12.4	20.1

In this case, the least-squares estimate a is given by

$$a = \frac{\sum x_i^2 y_i}{\sum x_i^4}$$

We compute $\sum x_i^4 = 61.1875$, $\sum x_i^2 y_i = 195.0$ to yield $a = 3.1869$ (to four decimal places). This computation gives the least-squares approximate model

$$y = 3.1869x^2$$

When $x = 2.25$, the predicted value for y is 16.1337.

Transformed Least-Squares Fit

Although the least-squares criterion appears easy to apply in theory, in practice it may be difficult. For example, consider fitting the model $y = Ae^{Bx}$ using the least-squares criterion. Call the least-squares estimate of the model $f(x) = ae^{bx}$. Application of the criterion then requires the minimization of

$$S = \sum_{i=1}^{m} \left[y_i - f(x_i) \right]^2 = \sum_{i=1}^{m} \left[y_i - ae^{bx_i} \right]^2.$$

A necessary condition for optimality is that $\partial S / \partial a = \partial S / \partial b = 0$. Formulate the conditions and convince yourself that solving the resulting system of nonlinear equations would not be easy. Many simple models result in derivatives that are complex or in systems of equations that are difficult to solve. For this reason we use transformations that allow us to *approximate* the least-squares model.

In graphically fitting lines to data in Section 5.1, we sometimes found it convenient to transform the data first, and then fit a line to the transformed data. For example, in graphically fitting $y = Ce^x$, we found it convenient to plot $\ln y$ versus x, and then fit a line to the transformed data. The same idea can be used with the least-squares criterion to simplify the computational aspects of the process. In particular, if a convenient substitution can be found so that the problem takes the form $Y = AX + B$ in the transformed variables X and Y, then Equations (5.4) can be used to fit a line to the transformed variables. We illustrate the technique with the example we just worked out.

Suppose we wish to fit the power curve $y = Ax^N$ to a collection of data points. Let's denote the estimate of A by α and the estimate of N by n. Taking the logarithm of both sides of the equation $y = \alpha x^n$ yields

$$\ln y = \ln \alpha + n \ln x \tag{5.8}$$

Note that in plotting the variables $\ln y$ versus $\ln x$, Equation (5.8) yields a straight line. On that graph, $\ln \alpha$ is the intercept when $\ln x = 0$ and the slope of the line is n. Using Equations (5.5) and (5.6) to solve for the slope n and intercept $\ln \alpha$ with the transformed variables and $m = 5$ data points, we have

$$n = \frac{5 \sum (\ln x_i)(\ln y_i) - \left(\sum \ln x_i \right) \left(\sum \ln y_i \right)}{5 \sum (\ln x_i)^2 - \left(\sum \ln x_i \right)^2}$$

$$\ln \alpha = \frac{\sum (\ln x_i)^2 (\ln y_i) - \left(\sum \ln x_i \right)(\ln y_i) \sum \ln x_i}{5 \sum (\ln x_i)^2 - \left(\sum \ln x_i \right)^2}$$

For the data displayed in Table 5.3 we get $\sum \ln x_i = 1.3217558$, $\sum \ln y_i = 8.359597801$, $\sum (\ln x_i)^2 = 1.9648967$, and $\sum (\ln x_i)(\ln y_i) = 5.542315175$, yielding $n = 2.062809314$ and $\ln \alpha = 1.126613508$ or $\alpha = 3.085190815$. Thus our least-squares best fit of Equation (5.8) is (rounded to four decimal places)

$$y = 3.0852x^{2.0628}$$

This model predicts $y = 16.4348$ when $x = 2.25$. Note, however, that this model fails to be a quadratic such as the one we fit previously.

Suppose we still wish to fit a *quadratic* $y = Ax^2$ to the collection of data. Denote the estimate of A by a_1 to distinguish this constant from the constants a and α computed previously. Taking the logarithm of both sides of the equation $y = a_1 x^2$ yields

$$\ln y = \ln a_1 + 2 \ln x$$

In this situation the graph of $\ln y$ versus $\ln x$ is a straight line of slope 2 and intercept $\ln a_1$. Using the second equation in (5.4) to compute the intercept, we have

$$2 \sum \ln x_i + 5 \ln a_1 = \sum \ln y_i$$

For the data displayed in Table 5.3 we get $\sum \ln x_i = 1.3217558$ and $\sum \ln y_i = 8.359597801$. Therefore, this last equation gives $\ln a_1 = 1.14321724$ or $a_1 = 3.136844129$, yielding the least-squares best fit (rounded to four decimal places)

$$y = 3.1368x^2$$

The model predicts $y = 15.8801$ when $x = 2.25$, which differs significantly from the value 16.1337 predicted by the first quadratic $y = 3.1869x^2$ obtained as the least-squares best fit of $y = Ax^2$ without transforming the data. We compare these two quadratic models (as well as a third model) in the next section.

The preceding example illustrates two facts. First, if an equation can be transformed to yield an equation of a straight line in the transformed variables, Equations (5.4) can be used directly to solve for the slope and intercept of the transformed graph. Second, the least-squares best fit to the transformed equations *does not* coincide with the least-squares best fit of the original equations. The reason for this discrepancy is that the resulting optimization problems are different. In the case of the original problem we are finding the curve that minimizes the sum of the squares of the deviations using the original data, whereas in the case of the transformed problem we are minimizing the sum of the squares of the deviations using the *transformed* variables.

5.3 PROBLEMS

1. Solve the two equations given by (5.4) to obtain the values of the parameters given by Equations (5.5) and (5.6), respectively.

2. Use Equations (5.5) and (5.6) to estimate the coefficients of the line $y = ax + b$ such that the sum of the squared deviations between the line and the following data points is minimized.

    a.

x	1.0	2.3	3.7	4.2	6.1	7.0
y	3.6	3.0	3.2	5.1	5.3	6.8

    b.

x	29.1	48.2	72.7	92.0	118	140	165	199
y	0.0493	0.0821	0.123	0.154	0.197	0.234	0.274	0.328

    c.

x	2.5	3.0	3.5	4.0	4.5	5.0	5.5
y	4.32	4.83	5.27	5.74	6.26	6.79	7.23

    For each problem, compute D and d_{max} to bound c_{max}. Compare the results with your solutions to Problem 2 in Section 5.2.

3. Derive the equations that minimize the sum of the squared deviations between a set of data points and the quadratic model $y = c_1 x^2 + c_2 x + c_3$. Use the equations to find estimates of c_1, c_2, and c_3 for the following set of data.

x	0.1	0.2	0.3	0.4	0.5
y	0.06	0.12	0.36	0.65	0.95

    Compute D and d_{max} to bound c_{max}. Compare the results with your solution to Problem 3 in Section 5.2.

4. Make an appropriate transformation to fit the model $P = ae^{bt}$ using Equations (5.4). Estimate a and b.

t	7	14	21	28	35	42
P	8	41	133	250	280	297

5. Examine closely the system of equations that result when you fit the quadratic in Problem 3. Suppose $c_2 = 0$. What would be the corresponding system of equations? Repeat for the cases $c_1 = 0$ and $c_3 = 0$. Suggest a system of equations for a cubic. Check your result. Explain how you would generalize the system of Equations (5.4) to fit any polynomial. Explain what you would do if one or more of the coefficients in the polynomial is zero.

6. A general rule for computing a person's ideal weight is as follows: For a female, multiply the height in inches by 3.5 and subtract 108; for a male, multiply the height in inches by 4.0 and subtract 128. If the person is small bone–structured, adjust this computation by subtracting 10%; for a large bone–structured person, add 10%. No adjustment is made for an average-size person. Gather some data on the weight versus height of people of differing age, size, and gender. Using Equations (5.4), fit a straight line to your data for males and another straight line to your data for females. What are the slopes and intercepts of those lines? How do the results compare with the general rule?

5.3 PROJECTS

1. Complete the requirements of the module "Curve Fitting via the Criterion of Least Squares," by John W. Alexander, Jr., UMAP 321. This unit provides an easy introduction to correlations, scatter diagrams (polynomial, logarithmic, and exponential scatters), and lines and curves of regression. Students construct scatter diagrams, choose appropriate functions to fit specific data, and use a computer program to fit curves. Recommended for students who wish an introduction to statistical measures of correlation.

5.3 Further Reading

BURDEN, Richard L. & J. Douglas Faires. *Numerical Analysis*, 6th ed. Pacific Grove, CA: Brooks/Cole, 1997.

HAMMING, R. W. *Numerical Methods for Scientists and Engineers*. New York: McGraw-Hill, 1973.

CHENEY, E. Ward & David Kincaid. *Numerical Mathematics and Computing*, 3rd ed. Pacific Grove, CA: Brooks/Cole, 1994.

STANTON, Ralph G. *Numerical Methods for Science and Engineering*. Englewood Cliffs, NJ: Prentice-Hall, 1961.

STIEFEL, Edward L. *An Introduction to Numerical Mathematics*. New York: Academic Press, 1963.

5.4 CHOOSING A BEST MODEL

Let's consider the adequacy of the various models of the form $y = Ax^2$ that we fit using the least-squares and transformed least-squares criteria in the previous section. Using the least-squares criterion, we obtained the model $y = 3.1869x^2$. One way of evaluating how well the model fits the data is to compute the deviations between the model and the actual data. If we compute the sum of the squares of the deviations, we can bound c_{max} as well. For the model $y = 3.1869x^2$ and the data in Table 5.3, we compute the deviations shown in Table 5.4.

TABLE 5.4 Deviations between the data in Table 5.3 and the fitted model $y = 3.1869x^2$

x_i	0.5	1.0	1.5	2.0	2.5
$y_i - y(x_i)$	-0.0967	0.2131	0.02948	-0.3476	0.181875

From Table 5.4 we compute the sum of the squares of the deviations as 0.20954, so that $D = (0.20954/5)^{1/2} = 0.204714$. Because the largest absolute deviation is 0.3476 when $x = 2.0$, c_{max} can be bounded as follows:

$$D = 0.204714 \le c_{max} \le 0.3476 = d_{max}$$

Let's find c_{max}. Because there are five data points, the mathematical problem is to minimize the largest of the five numbers $|r_i| = |y_i - y(x_i)|$. Calling that largest number r, we want to minimize r subject to $r \geq r_i$ and $r \geq -r_i$ for each $i = 1, 2, 3, 4, 5$. Denote our model by $y(x) = a_2 x^2$. Then substitution of the observed data points in Table 5.3 into the inequalities $r \geq r_i$ and $r \geq -r_i$ for each $i = 1, 2, 3, 4, 5$ yields the following linear program:

$$\text{Minimize } r$$

Subject to

$$r - r_1 = r - (0.7 - 0.25a_2) \geq 0$$
$$r + r_1 = r + (0.7 - 0.25a_2) \geq 0$$
$$r - r_2 = r - (3.4 - a_2) \geq 0$$
$$r + r_2 = r + (3.4 - a_2) \geq 0$$
$$r - r_3 = r - (7.2 - 2.25a_2) \geq 0$$
$$r + r_3 = r + (7.2 - 2.25a_2) \geq 0$$
$$r - r_4 = r - (12.4 - 4a_2) \geq 0$$
$$r + r_4 = r + (12.4 - 4a_2) \geq 0$$
$$r - r_5 = r - (20.1 - 6.25a_2) \geq 0$$
$$r + r_5 = r + (20.1 - 6.25a_2) \geq 0$$

The solution of the preceding linear program yields $r = 0.28293$ and $a_2 = 3.17073$. Thus we have reduced our largest deviation from $d_{max} = 0.3476$ to $c_{max} = 0.28293$. Note that we can reduce the largest deviation no further than 0.28293 for the model type $y = Ax^2$.

Now that we have determined three estimates of the parameter A for the model type $y = Ax^2$, which estimate is best? For each model we can readily compute the deviations from each data point as recorded in Table 5.5.

TABLE 5.5 Summary of the deviations for each model $y = Ax^2$

x_i	y_i	$y_i - 3.1869x_i^2$	$y_i - 3.1368x_i^2$	$y_i - 3.17073x_i^2$
0.5	0.7	-0.0967	-0.0842	-0.0927
1.0	3.4	0.2131	0.2632	0.2293
1.5	7.2	0.029475	0.1422	0.0659
2.0	12.4	-0.3476	-0.1472	-0.2829
2.5	20.1	0.181875	0.4950	0.28293

For each of the three models we can compute the sum of the squares of the deviations and the maximum absolute deviation. The results are shown in Table 5.6.

As you would expect, each model has something to commend it. However, notice the increase in the sum of the squares of the deviations in the transformed least-squares model. It is tempting to apply a simple rule, such as choose the model

TABLE 5.6 Summary of the results for the three models

Criterion	Model	$\sum [y_i - y(x_i)]^2$	Max $|y_i - y(x_i)|$
Least-squares	$y = 3.1869x^2$	0.2095	0.3476
Transformed least-squares	$y = 3.1368x^2$	0.3633	0.4950
Chebyshev	$y = 3.17073x^2$	0.2256	0.28293

with the smallest absolute deviation. (Other statistical indicators of goodness of fit exist as well. For example, see *Probability and Statistics in Engineering and Management Science*, by William W. Hines and Douglas C. Montgomery, New York: Wiley, 1972.) These indicators are useful for eliminating obviously poor models, but there is no easy answer to the question Which model is best? The model with the smallest absolute deviation or the smallest sum of squares may fit poorly over the range where you most intend to use it. Furthermore, as you will see in Chapter 6, models that pass through each data point thereby yielding a zero sum of squares and zero maximum deviation can easily be constructed. So we need to answer the question of which model is best on a case-by-case basis, taking into account such things as the purpose of the model, the precision demanded by the scenario, the accuracy of the data, and the range of values for the independent variable over which the model will be used.

When choosing among models or judging the adequacy of a model, we may find it tempting to rely on the value of the best-fit criterion being used. For example, it is tempting to choose the model that has the smallest sum of squared deviations for the given data set or conclude that a sum of squared deviations less than a predetermined value indicates a good fit. However, in isolation these indicators may be misleading. For example, consider the data displayed in Figure 5.15. In each of the four cases, the model $y = x$ results in exactly the same deviations. Without

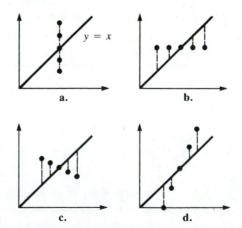

FIGURE 5.15 In each of these graphs the model $y = x$ has the same sum of squared deviations

the benefit of the graphs, therefore, we might conclude that in each case the model fits the data about the same. As the graphs show, however, there is a significant variation in each model's ability to capture the trend of the data. The following example illustrates how the various indicators can be used to help in reaching a decision on the adequacy of a particular model. Normally, a graphical plot is of great benefit.

Example 1   *Vehicular Stopping Distance*

Let's reconsider the problem of predicting a motor vehicle's stopping distance as a function of its speed. (This problem was addressed in Sections 2.2 and 4.2.) In Section 4.2 the submodel in which reaction distance d_r was proportional to the velocity v was tested graphically and the constant of proportionality was estimated to be 1.1. Similarly, the submodel predicting a proportionality between braking distance d_b and the square of the velocity was tested. We found reasonable agreement with the submodel and estimated the proportionality constant to be 0.054. Hence the model for stopping distance was given by

$$d = 1.1v + 0.054v^2 \tag{5.9}$$

We now fit these submodels analytically and compare the various fits.

To fit the model $d_r = A_v$ using the least-squares criterion, we use the formula from (5.7):

$$A = \frac{\sum x_i y_i}{\sum x_i^2}$$

where y_i denotes the driver reaction distance and x_i denotes the speed at each data point. For the 13 data points given in Table 4.4, we compute $\sum x_i y_i = 40905$ and $\sum x_i^2 = 37050$ giving $A = 1.104049$.

For the model type $d_b = Bv^2$, we use the formula

$$B = \frac{\sum x_i^2 y_i}{\sum x_i^4}$$

where y_i denotes the average braking distance and x_i denotes the speed at each data point. For the 13 data points given in Table 4.4, we compute $\sum x_i^2 y_i = 8258350$ and $\sum x_i^4 = 152343750$, giving $B = 0.054209$. Because the data are relatively imprecise and the modeling done qualitatively, we round the coefficients to obtain the model

$$d = 1.104v + 0.0542v^2 \tag{5.10}$$

Model (5.10) does not differ significantly from that obtained graphically in Chapter 4.

Next, let's analyze how well the model fits. We can readily compute the deviations between the observed data points in Table 4.4 and the values predicted by Models (5.9) and (5.10). The deviations are summarized in Table 5.7. The fits of both models are similar. The largest absolute deviation for Model (5.9) is 30.4 and for Model (5.10) it is 28.8. Note that both models overestimate the stopping distance up to 70 mph, and then they begin to underestimate the stopping distance. We should point out that a better fitting model would be obtained by directly fitting the data for total stopping distance to

$$d = k_1v + k_2v^2$$

instead of fitting the submodels individually as we did. The advantage of fitting the submodels individually and then testing each submodel is that we can measure how well they explain the behavior.

TABLE 5.7 Deviations from the observed data points and Models (5.9) and (5.10)

Speed	Graphical model (5.9)	Least-squares model (5.10)
20	1.6	1.76
25	5.25	5.475
30	8.1	8.4
35	13.15	13.535
40	14.4	14.88
45	16.35	16.935
50	17	17.7
55	14.35	15.175
60	12.4	13.36
65	7.15	8.255
70	−1.4	−0.14
75	−14.75	−13.325
80	−30.4	−28.8

A plot of the proposed model(s) and the observed data points is useful to determine how well the model fits the data. Model (5.10) and the observations are plotted in Figure 5.16. It is evident from the figure that a definite trend exists in the data and that Model (5.10) does a reasonable job of capturing that trend, especially at the lower speeds.

A powerful technique for quickly determining where the model is breaking down is to plot the deviations (residuals) as a function of the independent variable(s). For Model (5.10) a plot of the deviations is given in Figure 5.17, showing that the model is indeed reasonable up to 70 mph. Beyond 70 mph, however, there is a breakdown in the model's ability to predict the observed behavior.

Let's examine Figure 5.17 more closely. Note that although the deviations up to 70 mph are relatively small, they are all positive. If the model fully explains the behavior, the deviations should not only be small, but they should be positive and

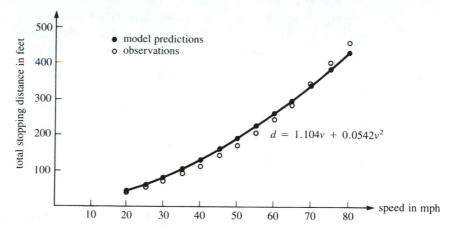

FIGURE 5.16   A plot of the proposed model and the observed data points provides a visual check on the adequacy of the model

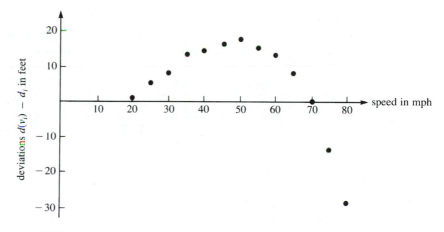

FIGURE 5.17   A plot of the deviations (residuals) reveals those regions where the model does not fit well

negative as well. Why? In Figure 5.17 we note a definite pattern in the nature of the deviations, which might cause us to reexamine the model and/or the data. The nature of the pattern in the deviations can give us clues on how to refine the model further. In this case the imprecision in the data collection process probably does not warrant further model refinement.

5.4 PROBLEMS

For each of the following problems, find a model using the least-squares criterion either on the data or the transformed data (as appropriate). Compare your results with the graphical fits obtained in the 5.1 problem set by computing the deviations, maximum absolute deviation, and sum of the squared deviations for each model. Find a bound on c_{max} if the model was fit using the least-squares criterion.

1. Problem 3 in Section 5.1
2. Problem 4a in Section 5.1
3. Problem 4b in Section 5.1
4. Problem 5a in Section 5.1
5. Problem 2 in Section 5.1
6. Problem 6 in Section 5.1
7. **a.** In the following data, W represents the weight of a fish (bass) and l represents its length. Fit the model $W = kl^3$ to the data using the least-squares criterion.

Length, l (in.)	14.5	12.5	17.25	14.5	12.625	17.75	14.125	12.625
Weight, W (oz)	27	17	41	26	17	49	23	16

   b. In the following data, g represents the girth of a fish. Fit the model $W = klg^2$ to the data using the least-squares criterion.

Length, l (in.)	14.5	12.5	17.25	14.5	12.625	17.75	14.125	12.625
Girth, g (in.)	9.75	8.375	11.0	9.75	8.5	12.5	9.0	8.5
Weight, W (oz)	27	17	41	26	17	49	23	16

   c. Which of the two models fits the data better? Justify fully. Which model do you prefer? Why?
8. Use the data presented in Problem 7b to fit the models $W = cg^3$ and $W = kgl^2$. Interpret these models. Compute appropriate indicators and determine which model is best. Explain.

5.4 PROJECTS

1. Write a computer program that finds the least-squares estimates of the coefficients in the following models:
   a. $y = ax^2 + bx + c$
   b. $y = ax^n$
2. Write a computer program that computes the deviation from the data points and any model that the user enters. Assuming that the model was fitted using the least-squares criterion, compute D and d_{max}. Output each data point, the deviation from each data point, D, d_{max}, and the sum of the squared deviations.

3. Write a computer program that uses Equations (5.4) and the appropriate transformed data to estimate the parameters of the following models:

a. $y = bx^a$

b. $y = be^{ax}$

c. $y = a \ln x + b$

d. $y = ax^2$

e. $y = ax^3$

Chapter Six

EXPERIMENTAL MODELING

INTRODUCTION

In Chapter 5 we discussed the philosophical differences between curve fitting and interpolation. When fitting a curve, the modeler is using assumptions that select a particular type of model that explains the behavior being observed. If collected data then corroborate the reasonableness of those assumptions, the modeler's task is to choose the parameters of the selected curve that best fits the data according to some criterion (such as least squares). In this situation the modeler expects, and willingly accepts, some deviations between the fitted model and the collected data to obtain a model explaining the behavior. The problem with this approach is that in many cases the modeler is unable to construct a tractable model form that satisfactorily explains the behavior. Thus the modeler doesn't know what kind of curve actually describes the behavior. If it is necessary to predict the behavior nevertheless, the modeler may conduct experiments (or otherwise gather data) to investigate the behavior of the dependent variable(s) for selected values of the independent variable(s) within some range. In essence, the modeler desires to construct an **empirical model** *based on the collected data* rather than select a model based on certain assumptions. In such cases the modeler is strongly influenced by the data that have been carefully collected and analyzed, so he or she seeks a curve that captures the trend of the data to *predict* in between the data points.

empirical model

For example, consider the data shown in Figure 6.1a. If the modeler's assumptions lead to the expectation of a quadratic model, a parabola would be fit to the data points as illustrated in Figure 6.1b. If the modeler has no reason to expect a particular type model, however, a smooth curve may be passed through the data points instead, as illustrated in Figure 6.1c.

In this chapter we address the construction of empirical models. In Section 6.1 we study the selection process for simple one-term models that capture the trend of the data. A few scenarios are addressed for which few modelers would attempt to construct an explicative model—namely, predicting the harvest of sea

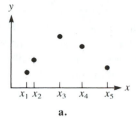 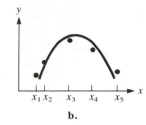 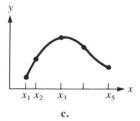

a. b. c.

FIGURE 6.1 If the modeler expects a quadratic relationship, a parabola may be fit to the data, as in **b**. Otherwise, a smooth curve may be passed through the points, as in **c**.

life in the Chesapeake Bay. In Section 6.2, we discuss the construction of higher-order polynomials that pass through the collected data points. In Section 6.3, we investigate smoothing of data using low-order polynomials. Finally, in Section 6.4, we present the technique of cubic spline interpolation, where a distinct cubic polynomial is used across successive pairs of data points.

6.1 HARVESTING IN THE CHESAPEAKE BAY AND OTHER ONE-TERM MODELS

Let's consider a situation in which a modeler has collected some data but is unable to construct an explicative model. In 1992, the *Daily Press* (a newspaper in Virginia) reported some observations (data) collected over the past 50 years on harvesting sea life in the Chesapeake Bay. We will examine several scenarios using observations from (a) harvesting bluefish and (b) harvesting blue crabs by the commercial fishing industry of the Chesapeake Bay. Table 6.1 shows the data we will use in our one-term models.

TABLE 6.1 Harvesting the Bay 1940–1990

Year	Bluefish (lbs)	Blue crabs (lbs)
1940	15,000	100,000
1945	150,000	850,000
1950	250,000	1,330,000
1955	275,000	2,500,000
1960	270,000	3,000,000
1965	280,000	3,700,000
1970	290,000	4,400,000
1975	650,000	4,660,000
1980	1,200,000	4,800,000
1985	1,500,000	4,420,000
1990	2,750,000	5,000,000

A scatterplot of harvesting bluefish versus time is shown in Figure 6.2, and a scatterplot of harvesting blue crabs is shown in Figure 6.3. Figure 6.2 clearly shows a tendency to harvest more bluefish over time, indicating or suggesting the availability of bluefish. A more precise description is not so obvious. In Figure 6.3, the tendency is for the harvesting of the blue crabs to be increasing. Again, a precise model is not so obvious.

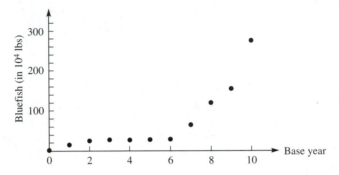

FIGURE 6.2 Scatterplot of harvesting bluefish versus base year (5-year periods from 1940 to 1990)

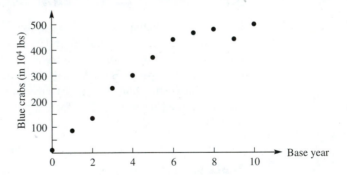

FIGURE 6.3 Scatterplot of harvesting blue crabs versus base year (5-year periods from 1940 to 1990)

In the rest of this section, we suggest how we might begin to predict the availability of bluefish over time. Our strategy will be to transform the data of Table 6.1 in such a way that the resulting graph approximates a line, thus achieving a working model. But how do we determine the transformation? We will use the Ladder of Powers[1] of a variable z to help select the appropriate linearizing transformation.

[1] See Paul F. Velleman and David C. Hoaglin, *Applications, Basics, and Computing of Exploratory Data Analysis* (Boston: Duxbury Press, 1981), p. 49.

Ladder of powers

$$\vdots$$
$$z^2$$
$$z$$
$$\sqrt{z}$$
$$\log z$$
$$\frac{1}{\sqrt{z}}$$
$$\frac{1}{z}$$
$$\frac{1}{z^2}$$
$$\vdots$$

Figure 6.4 shows a set of five data points (x, y) together with the line $y = x$, for $x > 1$. Suppose we change the y value of each point to $\sqrt{y}$. This procedure yields a new relation $y = \sqrt{x}$ whose y values are closer together over the domain in question. Note that all the y values are reduced, but the larger values are reduced more than the smaller ones.

Changing the y value of each point to $\log y$ has a similar but more pronounced effect, and each additional step down the ladder produces a stronger version of the same effect.

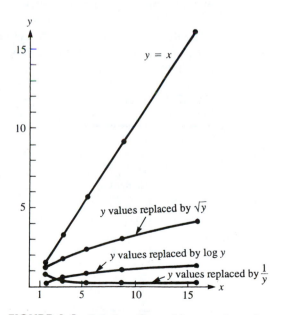

FIGURE 6.4   Relative effects of five transformations

We started in Figure 6.4, in the simplest way—with a linear function. But that was only a convenience. Given any positive-valued function $y = f(x)$, $x > 1$, that is concave up, like this

then some transformation in the ladder below y—changing the y values to $\sqrt{y}$, or log y, or a more drastic change—squeezes the right-hand tail downward and has a chance of generating a new function more nearly linear than the original. Which transformation should be used is a matter of trial and error and of experience. Another possibility is to stretch the right-hand tail to the right (try changing the x values to x^2, x^3 values, and so on).

Given a positive-valued function $y = f(x)$, $x > 1$, that is increasing and concave down, like this

we can hope to (more nearly) linearize it by stretching the right-hand tail upward (try changing y values to y^2, y^3 values, and so on). Another possibility is to squeeze the right-hand tail to the left (try changing the x values to $\sqrt{x}$, or log x, or by a more drastic choice from the ladder).

A final comment: Note that, although replacing z by $1/z$ or $1/z^2$, and so on, may sometimes have a desirable effect, such replacements also have an undesirable one—an increasing function is converted to a decreasing one. As a result, when using the transformation in the ladder of powers below log z, data analysts usually use a negative sign to keep the transformation data in the same order as the original data. Table 6.2 shows the Ladder of Transformations as it is usually used.

TABLE 6.2 Ladder of Transformations

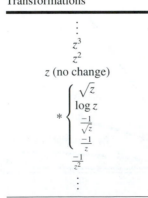

*Most often used transformations

With this information on the Ladder of Transformations, let's return to the Chesapeake Bay harvesting data.

Example 1   Harvesting Bluefish

Recall from the scatterplot in Figure 6.2 that the trend of the data appears to be increasing and concave up. Using the ladder of powers to squeeze the right-hand tail downward, we can change y values by replacing y with log y or other transformations down the ladder. Another choice would be to replace x values with x^2 or x^3 values or other powers up the ladder. We will use the data in Table 6.3, where we make 1940 the base year of $x = 0$ for numerical convenience, with each base year representing a 5-year period.

TABLE 6.3 Harvesting the Bay: Bluefish 1940–1990

Year	Base year	Bluefish (lbs)
	x	y
1940	0	15,000
1945	1	150,000
1950	2	250,000
1955	3	275,000
1960	4	270,000
1965	5	280,000
1970	6	290,000
1975	7	650,000
1980	8	1,200,000
1985	9	1,550,000
1990	10	2,750,000

We begin by squeezing the right-hand tail downward by changing x to various values going up the ladder (x^2, x^3, and so on). None of these transformations resulted in a linear graph. We next change y to values of $\sqrt{y}$ and log y, going down the ladder. Both $\sqrt{y}$ and log y plots versus x appeared more linear than the transformations to the x variable. (The plot of y versus x^2 and x versus $\sqrt{y}$ appeared nearly identical in their linearity.) We choose the log y versus x model. We fit with the least-squares criterion the model of the form

$$\log y = mx + b$$

and obtain the following estimated curve:

$$\log y = 0.7231 + 0.1654x$$

where x is the base year and log y is to the base 10.

Using the property that $y = \log n$ if and only if $10^y = n$, we can rewrite this equation (with the aid of a calculator) as

$$y = 5.2857(1.4635)^x \tag{6.1}$$

where y is measured in pounds of bluefish and x is the base year. The plot of the model appears to fit the data reasonably well. Figure 6.5 shows the graph of this curve superimposed on the scatterplot. We will accept some error to have a simple, one-term model.

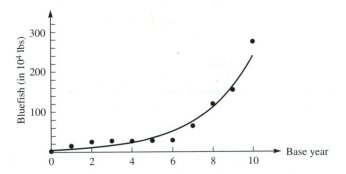

FIGURE 6.5   Superimposed data and model $y = 5.2857(1.4635)^x$

Example 2   Harvesting Blue Crabs

Recall from our original scatterplot, Figure 6.3, that the trend of the data is increasing and concave down. With this information, we can use the Ladder of Transformations. We will use the data in Table 6.4, modified by making 1940 (year $x = 0$) the base year, with each base year representing a 5-year period.

TABLE 6.4 Harvesting the Bay: Blue crabs 1940–1990

Year	Base year	Blue crabs (lbs)
	x	y
1940	0	100,000
1945	1	850,000
1950	2	1,330,000
1955	3	2,500,000
1960	4	3,000,000
1965	5	3,700,000
1970	6	4,400,000
1975	7	4,660,000
1980	8	4,800,000
1985	9	4,420,000
1990	10	5,000,000

As previously stated, we can attempt to linearize these data by changing y values to y^2 or y^3 values or to others moving up the ladder. After several experiments, we chose to replace the x values with $\sqrt{x}$. This squeezes the right-hand tail to the left. We provide a plot of y versus $\sqrt{x}$ (Figure 6.6). In Figure 6.7 we superimpose a line $y = k\sqrt{x}$ projected through the origin (no y-intercept). We use least squares from Chapter 5 to find k, yielding

$$y = 158.344\sqrt{x} \tag{6.2}$$

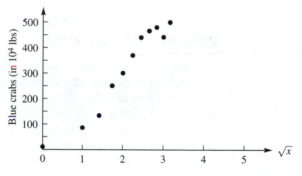

FIGURE 6.6   Blue crabs (in 10^4 lbs) versus $\sqrt{x}$

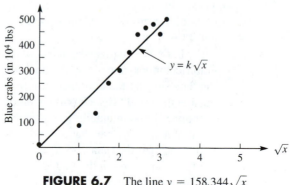

FIGURE 6.7   The line $y = 158.344\sqrt{x}$

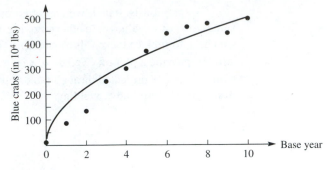

FIGURE 6.8 Superimposed data and model $y = 158.344\sqrt{x}$

Figure 6.8 shows the graph of the preceding model superimposed on the scatterplot. Note that the curve seems to be reasonable because it fits the data about as well as we would expect for a simple one-term model.

Verifying the Models

How good are the predictions based on the models? Part of the answer lies in comparing the observed values with those predicted. We can calculate the residuals and the relative errors for each pair of data. In many cases the question asked of the modeler is to predict or extrapolate to the future. How would these models hold up to predicting the amounts of harvests from the bay in the year 2000?

The following result for the bluefish may be larger than one might predict, whereas the result for the blue crabs may be a little lower than one might predict:

Bluefish $y = 5.2857(1.4635)^{12} = 510.2879 \ (10^4 \text{ lbs}) \approx 5.1 \text{ million lbs}$

Blue crabs $y = 158.344\sqrt{12} = 548.5197 \ (10^4 \text{ lbs}) \approx 5.48 \text{ million lbs}$

These simple, one-term models should be used for interpolation and not extrapolation. We discuss modeling population growth more thoroughly in Chapters 10 and 11.

Let's summarize the ideas of this section. When we are constructing an empirical model, we always begin with a careful analysis of the collected data. Do the data suggest the existence of a trend? Are there data points that obviously lie outside the trend? If such outliers do exist, it may be desirable to discard them or, if obtained experimentally, to repeat the experiment as a check for a data collection error. When it is clear that a trend exists, we next attempt to find a function that will transform the data into a straight line (approximately). In addition to trying the functions listed in the Ladder of Transformations presented in this section, we can also attempt the transformations discussed in Chapter 5. Thus if the model

$y = ax^b$ is selected, we would plot $\ln y$ versus $\ln x$ to see if a straight line results. Likewise, when investigating the appropriateness of the model $y = ae^{bx}$, we would plot $\ln y$ versus x to see if a straight line results. Keep in mind our discussion in Chapter 5 about how the use of transformations may be deceiving, especially if the data points are squeezed together. Our judgment is strictly qualitative; the idea is to determine if a particular model type appears promising. When we are satisfied that a certain model type does seem to capture the trend of the data, we can estimate the parameters of the model graphically or by using the analytical techniques discussed in Chapter 5. Eventually we must analyze the goodness of fit using the indicators discussed in Chapter 5. Remember to graph the proposed model against the original data points, not the transformed data. If we are dissatisfied with the fit, we can investigate other one-term models. Because of their inherent simplicity, however, one-term models cannot fit all data sets. In such situations other techniques can be used; we will discuss these methods in the next several sections.

6.1 PROBLEMS

In 1976 Marc and Helen Bornstein studied the pace of life.[2] To see if life becomes more hectic as the size of the city becomes larger, they systematically observed the mean time required for pedestrians to walk 50 ft on the main streets of their cities and towns. In Table 6.5 we present some of the data they collected. The variable P represents the population of the town or city, and the variable V represents the mean velocity of pedestrians walking the 50 ft. Problems 1–5 are based on the data in Table 6.5.

1. Fit the model $V = CP^a$ to the pace of life data in Table 6.5. Use the transformation: $\log V = a \log P + \log C$. Plot $\log V$ versus $\log P$. Does the relationship seem reasonable?

   a. Make a table of $\log P$ versus $\log V$.

   b. Construct a scatterplot of your log–log data.

   c. Eyeball a line l onto your scatterplot.

   d. Estimate the slope and the intercept.

   e. Find the linear equation that relates $\log V$ and $\log P$.

   f. Find the equation of the form $V = CP^a$ that expresses V in terms of P.

2. Graph the equation you found in Problem **1f** superimposed on the original scatterplot.

[2]Bornstein, Marc H. and Helen G. Bornstein. "The Pace of Life." *Nature* 259 (19 February 1976): 557–559.

TABLE 6.5 Population and mean velocity over a 50-foot course, for 15 locations*

Location	Population P	Mean velocity V (ft/sec)
(1) Brno, Czechoslovakia	341,948	4.81
(2) Prague, Czechoslovakia	1,092,759	5.88
(3) Corte, Corsica	5,491	3.31
(4) Bastia, France	49,375	4.90
(5) Munich, Germany	1,340,000	5.62
(6) Psychro, Crete	365	2.76
(7) Itea, Greece	2,500	2.27
(8) Iraklion, Greece	78,200	3.85
(9) Athens, Greece	867,023	5.21
(10) Safed, Israel	14,000	3.70
(11) Dimona, Israel	23,700	3.27
(12) Netanya, Israel	70,700	4.31
(13) Jerusalem, Israel	304,500	4.42
(14) New Haven, U.S.A.	138,000	4.39
(15) Brooklyn, U.S.A.	2,602,000	5.05

*Bornstein data

3. Using the data, a calculator, and the model you determined for V (Problem 1f), complete the following table:

TABLE 6.6 Observed mean velocity for 15 locations

Location*	Observed velocity V	Predicted velocities
(1)	4.81	
(2)	5.88	
(3)	3.31	
(4)	4.90	
(5)	5.62	
(6)	2.76	
(7)	2.27	
(8)	3.85	
(9)	5.21	
(10)	3.70	
(11)	3.27	
(12)	4.31	
(13)	4.42	
(14)	4.39	
(15)	5.05	

*For location names, see Table 6.5.

4. From the data in Table 6.6, calculate the mean (i.e., the average) of the Bornstein errors $|V_{observed} - V_{predicted}|$. What do the results suggest about the merit of the model?

5. Do Problems 1–4 with the model $V = m(\log P) + b$. Compare the errors with those computed in Problem 4. Compare the two models. Which is better?

6. Table 6.7 and Figure 6.9 present data representing the commercial harvesting of oysters in Chesapeake Bay. Fit a simple, one-term model to the data. How well does the best one-term model you find fit the data? What is the largest error? The average error?

TABLE 6.7 Oysters in the bay

Year	Oysters harvested (bushels)
1940	3,750,000
1945	3,250,000
1950	2,800,000
1955	2,550,000
1960	2,650,000
1965	1,850,000
1970	1,500,000
1975	1,000,000
1980	1,100,000
1985	750,000
1990	330,000

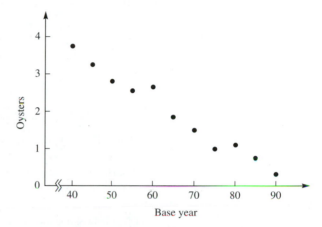

FIGURE 6.9   Oysters (mil. of lbs) versus base year

7. In Table 6.8, X is the Fahrenheit temperature and Y is the number of times a cricket chirps in 1 min. Fit a model to these data. Analyze how well it fits.

TABLE 6.8 Temperature and chirps per minute for 20 crickets

Observation number	X	Y	Observation number	X	Y
(1)	46	40	(11)	61	96
(2)	49	50	(12)	62	88
(3)	51	55	(13)	63	99
(4)	52	63	(14)	64	110
(5)	54	72	(15)	66	113
(6)	56	70	(16)	67	120
(7)	57	77	(17)	68	127
(8)	58	73	(18)	71	137
(9)	59	90	(19)	72	132
(10)	60	93	(20)	71	137

Data inferred from a scatterplot in Frederick E. Croxton, Dudley J. Cowden, and Sidney Klein, *Applied General Statistics*, 3rd. ed. (Englewood Cliffs, NJ: Prentice-Hall, 1967), p. 390.

8. Fit a model to Table 6.9. Do you recognize the data? What relationship can be inferred from them?

TABLE 6.9

Observation number	X	Y
(1)	35.97	0.241
(2)	67.21	0.615
(3)	92.96	1.000
(4)	141.70	1.881
(5)	483.70	11.860
(6)	886.70	29.460
(7)	1783.00	84.020
(8)	2794.00	164.800
(9)	3666.00	248.400

9. The following data measure two characteristics of a ponderosa pine. The variable X is the diameter of the tree, in inches, measured at breast height; Y is a measure of volume—number of board feet divided by 10. Fit a model to the data. Then express Y in terms of X.

Diameter and volume for 20 ponderosa pine trees

Observation number	X	Y	Observation number	X	Y
(1)	36	192	(11)	31	141
(2)	28	113	(12)	20	32
(3)	28	88	(13)	25	86
(4)	41	294	(14)	19	21
(5)	19	28	(15)	39	231
(6)	32	123	(16)	33	187
(7)	22	51	(17)	17	22
(8)	38	252	(18)	37	205
(9)	25	56	(19)	23	57
(10)	17	16	(20)	39	265

Data reported in Croxton, Cowden, and Klein, *Applied General Statistics*, p. 421.

10. The following data represent length and weight of a set of fish (bass). Model weight as a function of the length of the fish.

Length (in.)	12.5	12.625	14.125	14.5	17.25	17.75
Weight (oz)	17	16.5	23	26.5	41	49

11. The following data give the population of the United States from 1800 to 1990. Model the population (in thousands) as a function of the year. How well does your model fit? Is a one-term model appropriate for these data? Why?

Year	1800	1820	1840	1860	1880	1900
Population (thousands)	5308	9638	17,069	31,443	50,156	75,995

Year	1920	1940	1960	1980	1990
Population (thousands)	105,711	131,669	179,323	226,505	248,710

6.1 PROJECTS

1. Complete the requirements of UMAP 551, "The Pace of Life, An Introduction to Empirical Model Fitting," by Bruce King. Prepare a short summary for classroom discussion.

6.1 Further Reading

BORNSTEIN, Marc H. & Helen G. Bornstein, "The Pace of Life." *Nature* 259 (19 February 1976): 557–559.

CROXTON, Fredrick E. Dudley, J. Crowden, & Sidney Klein. *Applied General Statistics*, 3rd ed. Englewood Cliffs, NJ: Prentice-Hall, 1967.

NETER, John & William Wassermann, *Applied Linear Statistical Models*. Homewood, IL: Irwin, 1974.

VELLMAN, Paul F. & David C. Hoaglin. *Applications, Basics, and Computing of Exploratory Data Analysis*. Boston: Duxbury Press, 1981.

YULE, G. Udny. "Why Do We Sometimes Get Nonsense-Correlations between Time Series?—A Study in Sampling and the Nature of Time Series." *Journal of the Royal Statistical Society* 89 (1926): 1–69.

6.2 HIGH-ORDER POLYNOMIAL MODELS

In Section 6.1, we investigated the possibility of finding a simple, one-term model that captures the trend of the collected data. Because of their inherent simplicity, one-term models facilitate model analysis, including sensitivity analysis, optimization, estimation of rates of change and area under the curve. Because of their mathematical simplicity, however, one-term models are limited in their ability to capture the trend of any collection of data. In some cases models with more than one term must be considered. The remainder of this chapter considers one type of multiterm model—namely, the polynomial. Because polynomials are easy to integrate and to differentiate, they are especially popular to use. Yet polynomials also have their disadvantages. For example, it is far more appropriate to approximate a data set having a vertical asymptote using a quotient of polynomials $p(x)/q(x)$ than a single polynomial.

Let's begin by studying polynomials that pass through each point in a data set that includes only one observation for each value of the independent variable.

Consider the data in Figure 6.10a. Through the two given data points, a unique line $y = a_0 + a_1 x$ can be passed. The constants a_0 and a_1 are determined by the conditions that the line pass through the points (x_1, y_1) and (x_2, y_2). Thus,

$$y_1 = a_0 + a_1 x_1$$

and

$$y_2 = a_0 + a_1 x_2$$

In a similar manner a *unique* polynomial function of (at most) degree 2 $y = a_0 + a_1 x + a_2 x^2$ can be passed through three distinct points, as shown in Figure 6.10b. The constants a_0, a_1, and a_2 are determined by solving the following system of linear equations:

$$y_1 = a_0 + a_1 x_1 + a_2 x_1^2$$
$$y_2 = a_0 + a_1 x_2 + a_2 x_2^2$$
$$y_3 = a_0 + a_1 x_3 + a_2 x_3^2$$

Let's explain why the qualifier "at most" is needed in the next-to-last sentence of the preceding paragraph. Note that if the three points in Figure 6.10b just happen to lie along a straight line, then the unique polynomial function of at most degree 2 passing through the points would necessarily be a straight line (a polynomial of degree 1) rather than a quadratic function, as generally would be expected. The descriptor *unique* is also important. There are an infinite number of polynomials of degree greater than 2 and passing through the three points depicted in Figure 6.10b. (Convince yourself of this fact before proceeding by using Figure 6.10c.) There is, however, only one polynomial of degree 2 or less. Although this fact may not be obvious, we later state a theorem in its support. For now remember from high school geometry that a unique circle, which is also represented by an algebraic equation

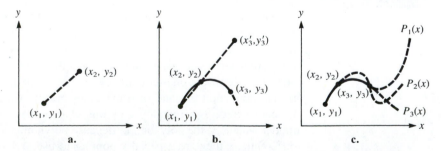

FIGURE 6.10 A unique polynomial of at most degree 2 can be passed through three data points (**a.** and **b.**), but an infinite number of polynomials of degree greater than 2 can be passed through three data points (**c.**)

of degree 2, is determined by three points in a plane. Next we illustrate these ideas in an applied problem; we then discuss the advantages and disadvantages of the procedure.

Example 1   Elapsed Time of a Tape Recorder

We collected data relating the counter on a particular tape recorder with its elapsed playing time. Suppose we are unable to build an explicative model of this system but are still interested in predicting what may occur. How can this difficulty be resolved? As an example, let's construct an empirical model to predict the amount of elapsed time of a tape recorder as a function of its counter reading.

Thus let c_i represent the counter reading and t_i(sec) the corresponding amount of elapsed time. Consider the following data:

c_i	100	200	300	400	500	600	700	800
t_i(sec)	205	430	677	945	1233	1542	1872	2224

One empirical model is a polynomial that passes through each of the data points. Because we have eight data points, a unique polynomial of at most degree 7 is expected. Denote the polynomial symbolically by

$$P_7(c) = a_0 + a_1c + a_2c^2 + a_3c^3 + a_4c^4 + a_5c^5 + a_6c^6 + a_7c^7$$

The eight data points require that the constants a_i satisfy the system of linear algebraic equations:

$$205 = a_0 + 1a_1 + 1^2a_2 + 1^3a_3 + 1^4a_4 + 1^5a_5 + 1^6a_6 + 1^7a_7$$
$$430 = a_0 + 2a_1 + 2^2a_2 + 2^3a_3 + 2^4a_4 + 2^5a_5 + 2^6a_6 + 2^7a_7$$
$$\vdots$$
$$2224 = a_0 + 8a_1 + 8^2a_2 + 8^3a_3 + 8^4a_4 + 8^5a_5 + 8^6a_6 + 8^7a_7$$

Large systems of linear equations can be difficult to solve with great numerical precision. In the preceding illustration, we divided each counter reading by 100 to lessen the numerical difficulties. Because the counter data values are being raised to the seventh power, it is easy to generate numbers differing by several orders of magnitude. It is important to have as much accuracy as possible in the coefficients a_i because each is being multiplied by a number raised to a power as high as 7. For instance, a small a_7 may become significant as c becomes large. This observation suggests why there may be dangers using even good polynomial functions that capture the trend of the data when we are beyond the range of the observations. The following solution to this system was obtained with the aid of a handheld calculator program:

$$a_0 = -13.9999923 \qquad a_4 = -5.354166491$$
$$a_1 = 232.9119031 \qquad a_5 = 0.8013888621$$
$$a_2 = -29.08333188 \qquad a_6 = -0.0624999978$$
$$a_3 = 19.78472156 \qquad a_7 = 0.0019841269$$

Let's see how well the empirical model fits the data. Denoting the polynomial predictions by $P_7(c_i)$, we find

c_i	100	200	300	400	500	600	700	800
t_i	205	430	677	945	1233	1542	1872	2224
$P_7(c_i)$	205	430	677	945	1233	1542	1872	2224

Rounding the predictions for $P_7(c_i)$ to four decimal places gives complete agreement with the observed data (as would be expected) and results in zero absolute deviations. Now we can see the folly of applying any of the criteria of best-fit studied in Chapter 5 as the sole judge for the best model. Can we really consider this model to be better than other models we could propose?

Let's see how well this new model $P_7(c_i)$ captures the trend of the data. The model is graphed in Figure 6.11.

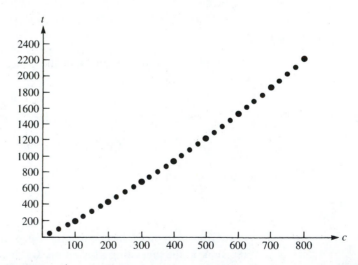

FIGURE 6.11 An empirical model for predicting the elapsed time of a tape recorder

Lagrangian Form of the Polynomial

From the preceding discussion we might expect that given $(n + 1)$ distinct data points, there is a unique polynomial of at most degree n that passes through all the data points. Because there are the same number of coefficients in the polynomial as there are data points, intuitively we would think that only one such polynomial exists. This hypothesis is indeed the case, although we will not prove that fact here. Rather, we present the Lagrangian form of the cubic polynomial followed by a brief discussion of how the coefficients may be found for higher-order polynomials.

Suppose the following data have been collected:

x	x_1	x_2	x_3	x_4
y	y_1	y_2	y_3	y_4

Consider the following cubic polynomial:

$$
\begin{aligned}
P_3(x) = {} & \frac{(x - x_2)(x - x_3)(x - x_4)}{(x_1 - x_2)(x_1 - x_3)(x_1 - x_4)} y_1 \\
& + \frac{(x - x_1)(x - x_3)(x - x_4)}{(x_2 - x_1)(x_2 - x_3)(x_2 - x_4)} y_2 \\
& + \frac{(x - x_1)(x - x_2)(x - x_4)}{(x_3 - x_1)(x_3 - x_2)(x_3 - x_4)} y_3 \\
& + \frac{(x - x_1)(x - x_2)(x - x_3)}{(x_4 - x_1)(x_4 - x_2)(x_4 - x_3)} y_4
\end{aligned}
$$

Convince yourself that the polynomial is indeed cubic and agrees with the value y_i when $x = x_i$. Notice that the x_i values must all be different to avoid division by zero. Observe the pattern for forming the numerator and the denominator for the coefficient of each y_i. This same pattern is followed when forming polynomials of any desired degree. The procedure is justified by the following result.

THEOREM   If $x_0, x_1, \ldots, x_n$ are $(n + 1)$ distinct points and $y_0, y_1, \ldots, y_n$ are corresponding observations at these points, then there exists a unique polynomial $P(x)$, of at most degree n, with the property that

$$
y_k = P(x_k) \quad \text{for each } k = 0, 1, \ldots, n
$$

This polynomial is given by

$$
P(x) = y_0 L_0(x) + \cdots + y_n L_n(x) \tag{6.3}
$$

where

$$
L_k(x) = \frac{(x - x_0)(x - x_1) \cdots (x - x_{k-1})(x - x_{k+1}) \cdots (x - x_n)}{(x_k - x_0)(x_k - x_1) \cdots (x_k - x_{k-1})(x_k - x_{k+1}) \cdots (x_k - x_n)}
$$

Because polynomial (6.3) passes through each of the data points, the resultant sum of absolute deviations is zero. Considering the various criteria of best-fit presented in Chapter 5, we are tempted to use high-order polynomials to fit larger sets of data. After all, the fit is precise. Let's examine both the advantages and disadvantages of using high-order polynomials.

Advantages and Disadvantages of High-Order Polynomials

As we have seen in previous chapters, it may be of interest to determine the area under the curve representing our model or its rate of change at a particular point. Polynomial functions have the distinct advantage of being easily integrated and differentiated. If a polynomial can be found that reasonably represents the underlying behavior, it would be easy to approximate the integral and the derivative of the unknown true model as well. Now consider some of the disadvantages of higher-order polynomials. For the 17 data points presented in Table 6.10, it is clear that the trend of the data is $y = 0$ for all x over the interval $-8 \le x \le 8$.

TABLE 6.10

x_i	-8	-7	-6	-5	-4	-3	-2	-1	0	1	2	3	4	5	6	7	8
y_i	0	0	0	0	0	0	0	0	0	0	0	0	0	0	0	0	0

Suppose Equation (6.3) is used to determine a polynomial that passes through the points. Because there are 17 distinct data points, it is possible to pass a unique polynomial of at most degree 16 through the given points. The graph of a polynomial passing through the data points is depicted in Figure 6.12.

Note that although the polynomial does pass through the data points (within tolerances of computer round-off error), there is severe oscillation of the polynomial near each end of the interval. Thus there would be gross error in estimating y between

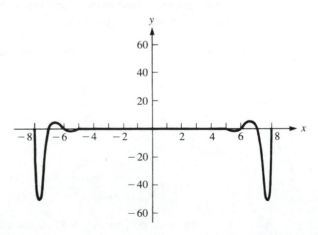

FIGURE 6.12 Fitting a higher-order polynomial through the data points in Table 6.10

the data points near $+8$ or -8. Likewise, consider the error in using the derivative of the polynomial to estimate the rate of change of the data or the area under the polynomial to estimate the area trapped by the data. This tendency of high-order polynomials to oscillate severely near the end points of the interval is a serious disadvantage to using them.

Let's illustrate another disadvantage of high-order polynomials. Consider the three sets of data presented in Table 6.11.[3]

TABLE 6.11

x_i	0.2	0.3	0.4	0.6	0.9
Case 1: y_i	2.7536	3.2411	3.8016	5.1536	7.8671
Case 2: y_i	2.754	3.241	3.802	5.154	7.867
Case 3: y_i	2.7536	3.2411	3.8916	5.1536	7.8671

Let's discuss the three cases. Case 2 is merely Case 1 with less precise data: There is one less significant digit for each of the observed data points. Case 3 is Case 1 with an error introduced in the observation corresponding to $x = 0.4$. Note that the error occurs in the third significant digit (3.8916 versus 3.8016). Intuitively, we would think that the three interpolating polynomials would be similar because the trends of the data are all similar. Let's determine the interpolating polynomials and see if that is really the situation.

Because there are five distinct data points in each case, a unique polynomial of at most degree 4 can be passed through each set of data. Denote the fourth-degree polynomial symbolically as follows:

$$P_4(x) = a_0 + a_1 x + a_2 x^2 + a_3 x^3 + a_4 x^4$$

Table 6.12 tabulates the coefficients a_0, a_1, a_2, a_3, a_4 (to four decimal places) determined by fitting the data points in each of the three cases. Note how sensitive the values of the coefficients are to the data. Nevertheless, the *graphs* of the polynomials are nearly the same over the interval of observations $(0.2, 0.9)$.

TABLE 6.12

	a_0	a_1	a_2	a_3	a_4
Case 1	2	3	4	-1	1
Case 2	2.0123	2.8781	4.4159	-1.5714	1.2698
Case 3	3.4580	-13.2000	64.7500	-91.0000	46.0000

[3]This example was furnished by Professor George Morris, Naval Postgraduate School, Monterey, California.

The graphs of the three fourth-degree polynomials representing each case are presented in Figure 6.13. This example illustrates the sensitivity of the coefficients of high-order polynomials to small changes in the data. Because we do expect measurement error to occur, the tendency of high-order polynomials to oscillate, as well as the sensitivity of their coefficients to small changes in the data, is a disadvantage that restricts their usefulness in modeling. In the next two sections we consider techniques that address the deficiencies noted in this section.

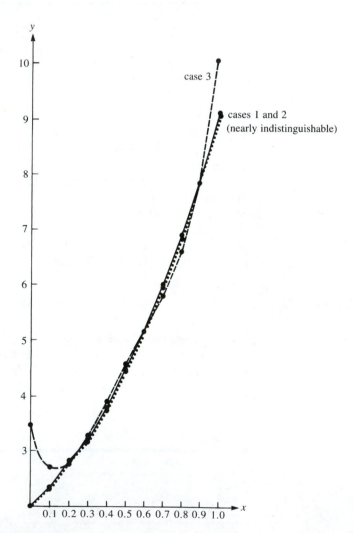

FIGURE 6.13 Small measurement errors can cause huge differences in the coefficients of the higher-order polynomials that result; note that the polynomials diverge outside the range of observations

6.2 PROBLEMS

1. For the tape recorder problem in this section, give a system of equations determining the coefficients of a polynomial that passes through each of the data points. If a computer is available, determine and sketch the polynomial. Does it represent the trend of the data?

2. Consider the pace of life data from Problem 1 in Section 6.1. Consider fitting a 14th-order polynomial to the data. Discuss the disadvantages of using the polynomial to make predictions. If a computer is available, determine and graph the polynomial.

3. In the following data, X is the Fahrenheit temperature and Y is the number of times a cricket chirps in 1 min (see Problem 7 in Section 6.1). Make a scatterplot of the data and discuss the appropriateness of using an 18th-degree polynomial that passes through the data points as an empirical model. If you have a computer available, fit a polynomial to the data and plot the results.

X	46	49	51	52	54	56	57	58	59	60
Y	40	50	55	63	72	70	77	73	90	93
X	61	62	63	64	66	67	68	71	72	
Y	96	88	99	110	113	120	127	137	132	

4. In the following data, X represents the diameter of a ponderosa pine measured at breast height and Y is a measure of volume—number of board feet divided by 10 (see Problem 9 in Section 6.1). Make a scatterplot of the data. Discuss the appropriateness of using a 13th-degree polynomial that passes through the data points as an empirical model. If you have a computer available, fit a polynomial to the data and graph the results.

X	17	19	20	22	23	25	31	32	33	36	37	38	39	41
Y	19	25	32	51	57	71	141	123	187	192	205	252	248	294

6.3   SMOOTHING: LOW-ORDER POLYNOMIAL MODELS

We seek methods that retain many of the conveniences found in high-order polynomials without incorporating their disadvantages. One popular technique is to choose a low-order polynomial regardless of the number of data points. This choice normally results in a situation in which the number of data points exceeds the number of constants necessary to determine the polynomial. Because there are fewer constants to determine than there are data points, the low-order polynomial generally will not pass through all the data points. For example, suppose it is decided to fit a quadratic to a set of 10 data points. Because it is generally impossible to force a quadratic to pass through 10 data points, it must be decided which quadratic best fits the data (according to some criterion, as discussed in Chapter 5). This process, which is **smoothing** called **smoothing**, is illustrated in Figure 6.14. The combination of using low-order polynomials while not requiring that it pass through each data point reduces both the tendency of the polynomial to oscillate and its sensitivity to small changes in the data. This quadratic function smooths the data because it is not required to pass through all the data points.

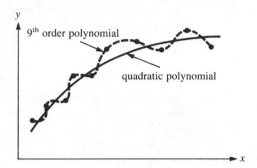

FIGURE 6.14 The quadratic function smooths the data because it is not required to pass through all the data points

The process of smoothing requires two decisions. First, the order of the interpolating polynomial must be selected. Second, the coefficients of the polynomial must be determined according to some criterion for the best-fitting polynomial. The problem that results is an optimization problem of the form addressed in Chapter 5. For example, it may be decided to fit a quadratic model to 10 data points using the least-squares best-fitting criterion. We will review the process of fitting a polynomial to a set of data points using the least-squares criterion and later return to the more difficult question of how to best choose the order of the interpolating polynomial.

Example 1   *Elapsed Time of a Tape Recorder Revisited*

Consider again the tape recorder problem modeled in the previous section. For a particular cassette deck or tape recorder equipped with a counter, relate the counter to the amount of playing time that has elapsed. If we are interested in predicting the elapsed time but are unable to construct an explicative model, it may be possible to construct an empirical model instead. Let's fit a second-order polynomial of the following form to the data:

$$P_2(c) = a + bc + dc^2$$

where c is the counter reading, $P_2(c)$ is the elapsed time, and a, b, and d are constants to be determined. Consider the collected data for the tape recorder problem in the previous section, shown in Table 6.13:

TABLE 6.13 Data collected for the tape recorder problem

c_i	100	200	300	400	500	600	700	800
t_i(sec)	205	430	677	945	1233	1542	1872	2224

Our problem is to determine the constants, a, b, and d so that the resultant quadratic model best fits the data. Although other criteria might be used, we will find the quadratic that minimizes the sum of the squared deviations. Mathematically, the problem is to

$$\text{Minimize } S = \sum_{i=1}^{m} [t_i - (a + bc_i + dc_i^2)]^2$$

The necessary conditions for a minimum to exist ($\partial S/\partial a = \partial S/\partial b = \partial S/\partial d = 0$) yield these equations:

$$ma + \left(\sum c_i\right) b + \left(\sum c_i^2\right) d = \sum t_i$$

$$\left(\sum c_i\right) a + \left(\sum c_i^2\right) b + \left(\sum c_i^3\right) d = \sum c_i t_i$$

$$\left(\sum c_i^2\right) a + \left(\sum c_i^3\right) b + \left(\sum c_i^4\right) d = \sum c_i^2 t_i$$

For the data in Table 6.13, the preceding system of equations becomes

$$8a + 3600b + 2{,}040{,}000d = 9128$$

$$3600a + 2{,}040{,}000b + 1{,}296{,}000{,}000d = 5{,}318{,}900$$

$$2{,}040{,}000a + 1{,}296{,}000{,}000b + 8.772 \times 10^{11}d = 3{,}435{,}390{,}000$$

Solution of the preceding system yields the values $a = 0.14286$, $b = 1.94226$, and $d = 0.00105$, giving the quadratic

$$P_2(c) = 0.14286 + 1.94226c + 0.00105c^2$$

We can compute the deviation between the observations and the predictions made by the model $P_2(c)$:

c_i	100	200	300	400	500	600	700	800
t_i	205	430	677	945	1233	1542	1872	2224
$t_i - P_2(c_i)$	.167	−0.452	0.000	0.524	0.119	−0.214	−0.476	0.333

Note that the deviations are small compared with the order of magnitude of the times.

When we are considering the use of a low-order polynomial for smoothing, two issues come to mind:

1. Should a polynomial be used?
2. If so, what order of polynomial would be appropriate?

The derivative concept can help in answering these two questions.

Divided Differences

Notice that a quadratic function is characterized by the properties that its second derivative is constant and its third derivative is zero. That is, given

$$P(x) = a + bx + cx^2$$

we have

$$P'(x) = b + 2cx$$
$$P''(x) = 2c$$
$$P'''(x) = 0$$

However, the only information available is a set of discrete data points. How can these points be used to estimate the various derivatives? Refer to Figure 6.15 and recall the definition of the derivative:

$$\frac{dy}{dx} = \lim_{\Delta x \to 0} \frac{\Delta y}{\Delta x}$$

Because dy/dx at $x = x_1$ can be interpreted geometrically as the slope of the line tangent to the curve there, we see from Figure 6.15 that unless Δx is small, the ratio $\Delta y/\Delta x$ is probably not a good estimate of dy/dx. Nevertheless, if dy/dx is to be zero, then Δy must go to zero. Thus we can compute the *differences* $y_{i+1} - y_i = \Delta y$ between successive function values in our tabled data to gain insight into what the first derivative is doing.

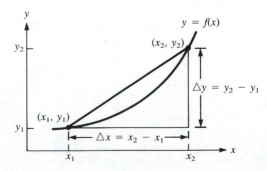

FIGURE 6.15   The derivative of $y = f(x)$ at $x = x_1$ is the limit of the slope of the secant line

Likewise, because the first derivative is itself a function, the process can be repeated to estimate the second derivative. That is, the differences between successive estimates of the first derivative can be computed to approximate the second derivative. Before describing the entire process, we illustrate this idea with an example.

We know that the curve $y = x^2$ passes through the points $(0, 0), (2, 4), (4, 16),$ $(6, 36),$ and $(8, 64)$. Suppose the data displayed in Table 6.14 have been collected.

TABLE 6.14 Hypothetical set of collected data

x_i	0	2	4	6	8
y_i	0	4	16	36	64

Using the data in Table 6.14, we can construct a *difference table* as shown in Table 6.15.

TABLE 6.15 A difference table for the data of Table 6.14

Data		Differences			
x_i	y_i	Δ	Δ^2	Δ^3	Δ^4
0	0				
		4			
2	4		8		
		12		0	
4	16		8		0
		20		0	
6	36		8		
		28			
8	64				

The first differences, denoted by Δ, are constructed by computing $y_{i+1} - y_i$ for $i = 1, 2, 3, 4$. The second differences, denoted by Δ^2, are computed by finding the difference between successive first differences from the Δ column. The process can be continued, column by column, until Δ^{n-1} is computed for n data points. Note from Table 6.15 that the second differences in our example are constant and the third differences are zero. These results are consistent with the fact that a quadratic function has a constant second derivative and a zero third derivative.

Even if the data are essentially quadratic in nature, we would not expect the differences to go to zero precisely because of the various errors present in the modeling and data collection processes. We might, however, expect the data to become small. Our judgment of the significance of small can be improved by **divided differences** computing **divided differences**. Notice that the differences computed in Table 6.15 are estimates of the numerator of each of the various order derivatives. These estimates can be improved by dividing the numerator by the corresponding estimate of the denominator.

Consider the three data points and corresponding estimates of the first and second derivative, called the first and second divided differences, respectively, in Table 6.16. The first divided difference follows immediately from the ratio of

TABLE 6.16 The first and second divided differences estimate the first and second derivatives, respectively

Data		First divided difference	Second divided difference
x_1	y_1		
		$\dfrac{y_2 - y_1}{x_2 - x_1}$	
x_2	y_2		$\dfrac{\dfrac{y_3 - y_2}{x_3 - x_2} - \dfrac{y_2 - y_1}{x_2 - x_1}}{x_3 - x_1}$
		$\dfrac{y_3 - y_2}{x_3 - x_2}$	
x_3	y_3		

$\Delta y / \Delta x$. Because the second derivative represents the rate of change of the first derivative, we can estimate how much the first derivative changes between x_1 and x_3. That is, we can compute the differences between the adjacent first divided differences and divide by the length of the interval over which that change takes place ($x_3 - x_1$ in this case). Refer to Figure 6.16 for a geometrical interpretation of the second divided difference.

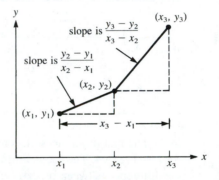

FIGURE 6.16 The second divided difference may be interpreted as the difference between the adjacent slopes (first divided differences) divided by the length of the interval over which the change has taken place

In practice it is easy to construct a divided difference table. We generate the next-higher-order divided difference by taking differences between adjacent current order divided differences and dividing them by the length of the interval over which the change has taken place. Using Δ^n to denote the nth divided difference, a divided difference table for the data of Table 6.14 is displayed in Table 6.17.

It is easy to remember what the numerator should be in each divided difference of the table. To remember what the denominator should be for a given divided difference, we can construct diagonal lines back to y_i of the original data entries and compute the differences in the corresponding x_i. This is illustrated for a third-order divided difference in Table 6.17. This construction becomes more critical when the x_i's are unequally spaced.

TABLE 6.17 A divided difference table for the data of Table 6.14

Data		Divided differences		
x_i	y_i	Δ	Δ^2	Δ^3
0	0			
2	−4	4/2 = 2		
4	16	12/2 = 6	4/4 = 1	0/6 = 0
6	36	20/2 = 10	4/4 = 1	0/6 = 0
8	−64	28/2 = 14	4/4 = 1	

$\Delta x = 6$

Example 2   Elapsed Time of a Tape Recorder Revisited Again

Returning now to our construction of an empirical model for the elapsed time for a tape recorder, how might the order of the smoothing polynomial be chosen? Let's begin by constructing the divided difference table for the given data from Table 6.13. The divided differences are displayed in Table 6.18.

TABLE 6.18 A divided difference table for the tape recorder data

Data		Divided differences			
x_i	y_i	Δ	Δ^2	Δ^3	Δ^4
100	205				
200	430	2.2500	0.0011		
300	677	2.4700	0.0011	0.0000	0.0000
400	945	2.6800	0.0010	0.0000	0.0000
500	1233	2.8800	0.0011	0.0000	0.0000
600	1542	3.0900	0.0011	0.0000	0.0000
700	1872	3.3000	0.0011	0.0000	
800	2224	3.5200			

Note that from Table 6.18 the second divided differences are essentially constant and the third divided differences equal zero to four decimal places. The table suggests the data are essentially quadratic, which supports the use of a quadratic polynomial as an empirical model. The modeler may now want to reinvestigate the assumptions to determine whether a quadratic relationship seems reasonable.

Observations on Difference Tables

Several observations about divided difference tables are in order. First, the x_i's must be distinct and listed in increasing order. It is important to be sensitive to x_i's that are close together because division by a small number can cause numerical difficulties. The scales used to measure both the x_i and y_i must also be considered. For example, suppose the x_i represent distances and are currently measured in miles. If the units are changed to feet, the denominators become much larger, resulting in divided differences that are much smaller. Thus judgment on what is small is relative and qualitative. Remember, however, that we are trying to decide whether a low-order polynomial is worth further investigation. Before accepting the model, we would want to graph it and analyze the goodness of fit.

When using a divided difference table, we must be sensitive to errors and irregularities that occur in the data. Measurement errors can propagate themselves throughout the table and even magnify themselves. For example, consider the following difference table:

Δ^{n-1}	Δ^n	Δ^{n+1}
6.01		
	−.01	
6.00		.02
	.01	
6.01		

Let's suppose that the Δ^{n-1} column is actually constant except for the presence of a relatively small measurement error. Note how the measurement error gives rise to the negative sign in the Δ^n column. Note also that the magnitudes of the numbers in the Δ^{n+1} column are larger than those in the Δ^n column even though the Δ^{n-1} column is essentially constant. The effect of these errors is present not only in subsequent columns but in other rows as well. Because errors and irregularities are normally present in collected data, it is important to be sensitive to the effects they can cause in difference tables.

Historically, divided difference tables were used to determine various forms of interpolating polynomials that passed through a chosen subset of the data points. Today other interpolating techniques, such as smoothing and cubic splines, are more popular. Nevertheless, difference tables are easily constructed on a computer and, like the derivatives they approximate, provide an inexpensive source of useful information about the data.

Example 3   *Vehicular Stopping Distance*

The following problem was presented in Section 2.2: Predict a vehicle's total stopping distance as a function of its speed. In previous chapters, explicative models were constructed describing the vehicle's behavior. Those models will be reviewed in the next section. For the time being, suppose there is no explicative model but only the data displayed in Table 6.19.

TABLE 6.19 Data relating total stopping distance and speed

Speed v (mph)	20	25	30	35	40	45	50	55	60	65	70	75	80
Distance d (ft)	42	56	73.5	91.5	116	142.5	173	209.5	248	292.5	343	401	464

If the modeler is interested in constructing an empirical model by smoothing the data, a divided difference table can be constructed as displayed in Table 6.20.

TABLE 6.20 A divided difference table for the data relating total vehicular stopping distance and speed

| Data | | Divided differences | | | |
v_i	d_i	Δ	Δ^2	Δ^3	Δ^4
20	42				
25	56	2.2800	0.0700		
30	73.5	3.5000	0.0100	-0.0040	0.0006
35	91.5	3.6000	0.1300	0.0080	-0.0007
40	116	4.9000	0.0400	-0.0060	0.0004
45	142.5	5.3000	0.0800	0.0027	0.0000
50	173	6.1000	0.1200	0.0027	-0.0004
55	209.5	7.3000	0.0400	-0.0053	0.0005
60	248	7.7000	0.1200	0.0053	-0.0003
65	292.5	8.9000	0.1200	0.0000	0.0001
70	343	10.1000	0.1500	0.0020	-0.0003
75	401	11.6000	0.1000	-0.0033	
80	464	12.6000			

An examination of the table reveals that the third divided differences are small in magnitude compared with the data and that negative signs have begun to appear. As previously discussed, the negative signs may indicate the presence of measurement error or variations in the data that will not be captured with low-order polynomials. The negative signs will also have a detrimental effect on the differences in the remaining columns. Here we may just decide to use a quadratic model, reasoning that higher-order terms will not reduce the deviations sufficiently to justify their inclusion; but our judgment is qualitative. The cubic term will probably account for some of the deviations not accounted for by the best quadratic model (otherwise the optimal value of the coefficient of the cubic term would be zero, causing the quadratic and cubic polynomials to coincide), but the addition of higher-order terms increases the complexity of the model, its susceptibility to oscillation, and its sensitivity to data errors. These considerations are studied in statistics.

In the following model, v is the speed of the vehicle, $P(v)$ is the stopping distance, and a, b, and c are constants to be determined:

$$P(v) = a + bv + cv^2$$

Our problem is to determine the constants a, b, and c so the resultant quadratic model best fits the data. Although other criteria might be used, we will find the quadratic that minimizes the sum of the squared deviations. Mathematically, the problem is to

$$\text{Minimize } S = \sum_{i=1}^{m} [d_i - (a + bv_i + cv_i^2)]^2$$

The necessary conditions for a minimum to exist ($\partial S/\partial a = \partial S/\partial b = \partial S/\partial c = 0$) yield these equations:

$$ma + \left(\sum v_i\right) b + \left(\sum v_i^2\right) c = \sum d$$

$$\left(\sum v_i\right) a + \left(\sum v_i^2\right) b + \left(\sum v_i^3\right) c = \sum v_i d_i$$

$$\left(\sum v_i^2\right) a + \left(\sum v_i^3\right) b + \left(\sum v_i^4\right) c = \sum v_i^2 d_i$$

Substitution from the data in Table 6.19 gives the following system:

$$13a + 650b + 37{,}050c = 2652.5$$
$$650a + 37{,}050b + 2{,}307{,}500c = 163{,}970$$
$$37{,}050a + 2{,}307{,}500b + 152{,}343{,}750c = 10{,}804{,}975$$

which leads to the solution $a = 50.0594$, $b = -1.9701$, and $c = 0.0886$ (rounded to four decimal places). Therefore, the empirical quadratic model is given by

$$P(v) = 50.0594 - 1.9701v + 0.0886v^2$$

Finally, the fit of $P(v)$ is analyzed in Table 6.21. This empirical model fits better than the model

$$d = 1.104v + 0.0542v^2$$

determined in Section 5.4, because there is an extra parameter (the constant a in this case) that absorbs some of the error. Note, however, that the empirical model predicts a stopping distance of approximately 50 ft when the velocity is zero.

TABLE 6.21 Smoothing the stopping distance using a quadratic polynomial

v_i	20	25	30	35	40	45	50
d_i	42	56	73.5	91.5	116	142.5	173
$d_i - P(v_i)$	-4.097	-0.182	2.804	1.859	2.985	1.680	-0.054
v_i	55	60	65	70	75	80	
d_i	209.5	248	292.5	343	401	464	
$d_i - P(v_i)$	-0.719	-2.813	-3.838	-3.292	0.323	4.509	

Example 4 Growth of a Yeast Culture

In this example we consider a collection of data points for which a divided difference table can help in deciding whether a low-order polynomial will provide a satisfactory empirical model. The data represent the population of yeast cells in a culture measured over time (in hours). A divided difference table for the population data is given by Table 6.22.

TABLE 6.22 A divided difference table for the growth of yeast in a culture

Data		Divided differences			
t_i	P_i	Δ	Δ^2	Δ^3	Δ^4
0	9.60				
1	18.30	8.70			
2	29.00	10.70	1.00		
3	47.20	18.20	3.75	0.92	
4	71.10	23.90	2.85	-0.30	-0.31
5	119.10	48.00	12.05	3.07	0.84
6	174.60	55.50	3.75	-2.77	-1.46
7	257.30	82.70	13.60	3.28	1.51
8	350.70	93.40	5.35	-2.75	-1.51
9	441.00	90.30	-1.55	-2.30	0.11
10	513.30	72.30	-9.00	-2.48	-0.05
11	559.70	46.40	-12.95	-1.32	0.29
12	594.80	35.10	-5.65	2.43	0.94
13	629.40	34.60	-0.25	1.80	-0.16
14	640.80	11.40	-11.60	-3.78	-1.40
15	651.10	10.30	-0.55	3.68	1.87
16	655.90	4.80	-2.75	-0.73	-1.10
17	659.60	3.70	-0.55	0.73	0.37
18	661.80	2.20	-0.75	-0.07	-0.20

Note that the first divided differences Δ are increasing until $t = 8$ hr, when they begin to decrease. This characteristic is reflected in the column Δ^2 with the appearance of a consecutive string of negative signs, indicating a change in concavity. Thus we cannot hope to capture the trend of these data with a quadratic function that has only a single concavity. In the Δ^3 column additional negative signs appear sporadically although the magnitude of the numbers is relatively large.

A scatterplot of the data is given in Figure 6.17. Although the divided difference table suggests that a quadratic function would not be a good model, for illustrative purposes suppose we try to fit a quadratic anyway. Using the least-squares criterion and the equations developed previously for the tape recorder example, we determine the following quadratic model for the data in Table 6.22:

$$P = -93.82 + 65.70t - 1.12t^2$$

The model and the data points are plotted in Figure 6.18. The model fits poorly as we expected and fails to capture the trend of the data. In the problem set you will be asked to fit the cubic model and check its reasonableness. In the next section we will construct a cubic spline model that fits the data much better.

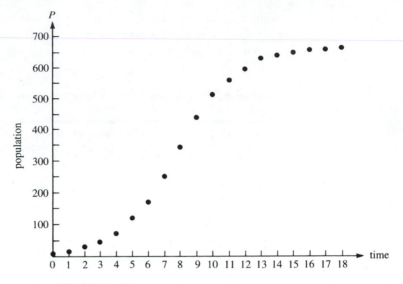

FIGURE 6.17   A scatterplot of the yeast in a culture data

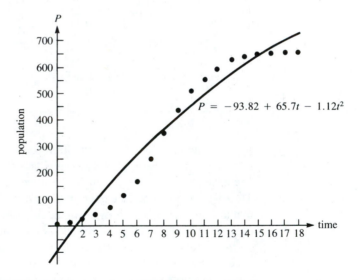

$$P = -93.82 + 65.7t - 1.12t^2$$

FIGURE 6.18   The best-fitting quadratic model fails to capture the trend of the data; note the magnitude of the deviations $P_i - P(t_i)$

6.3 PROBLEMS

For the data sets in Problems 1–4, construct a divided difference table. What conclusions can you make about the data? Would you use a low-order polynomial as an empirical model? If so, what order?

1.

x	0	1	2	3	4	5	6	7
y	2	8	24	56	110	192	308	464

2.

x	0	1	2	3	4	5	6	7
y	23	48	73	98	123	148	173	198

3.

x	0	1	2	3	4	5	6	7
y	7	15	33	61	99	147	205	273

4.

x	0	1	2	3	4	5	6	7
y	1	4.5	20	90	403	1808	8103	36316

5. Construct a scatterplot for the yeast growth in a culture data. Do the data seem reasonable? Construct a divided difference table. Try smoothing with a low-order polynomial using an appropriate criterion. Analyze the fit and compare your model with the quadratic we developed in this section. Graph your model, the data points, and the deviations.

In Problems 6–12, construct a scatterplot of the given data. Is there a trend in the data? Are any of the data points outliers? Construct a divided difference table. Is smoothing with a low-order polynomial appropriate? If so, choose an appropriate polynomial and fit it using the least-squares criterion of best fit. Analyze the goodness of fit by examining appropriate indicators and graphing the model, the data points, and the deviations.

6. In the following data, X is the Fahrenheit temperature and Y is the number of times a cricket chirps in 1 min (see Problem 7 in Section 6.1).

X	46	49	51	52	54	56	57	58	59	60
Y	40	50	55	63	72	70	77	73	90	93
X	61	62	63	64	66	67	68	71	72	
Y	96	88	99	110	113	120	127	137	132	

7. In the following data, X represents the diameter of a ponderosa pine measured at breast height and Y is a measure of volume—number of board feet divided by 10 (see Problem 9 in Section 6.1).

X	17	19	20	22	23	25	31	32	33	36	37	38	39	41
Y	19	25	32	51	57	71	141	123	187	192	205	252	248	294

8. The following data represent the population of the United States from 1790 to 1990.

Year	Observed population
1790	3,929,000
1800	5,308,000
1810	7,240,000
1820	9,638,000
1830	12,866,000
1840	17,069,000
1850	23,192,000
1860	31,443,000
1870	38,558,000
1880	50,156,000
1890	62,948,000
1900	75,995,000
1910	91,972,000
1920	105,711,000
1930	122,755,000
1940	131,669,000
1950	150,697,000
1960	179,323,000
1970	203,212,000
1980	226,505,000
1990	248,710,000

9. The following data were obtained for the growth of a sheep population introduced into a new environment on the island of Tasmania. (Adapted from J. Davidson, "On the Growth of the Sheep Population in Tasmania," *Trans. Roy. Soc. S. Australia 62* (1938): 342–346.)

t (year)	1814	1824	1834	1844	1854	1864
$P(t)$	125	275	830	1200	1750	1650

10. The following represent the pace of life data (see Problem 1 in Section 6.1). P is the population and V is the mean velocity in feet per second over a 50-ft course.

P	365	2500	5491	14000	23700	49375	70700	78200
V	2.76	2.27	3.31	3.70	3.27	4.90	4.31	3.85
P	138000	304500	341948	867023	1092759	1340000	2602000	
V	4.39	4.42	4.81	5.21	5.88	5.62	5.05	

11. The following data represent the length and weight of a bass fish.

length (in.)	12.5	12.625	14.125	14.5	17.25	17.75
weight (oz)	17.0	16.50	23.00	26.5	41.00	49.00

12. The following data represent the weight-lifting results from the 1976 Olympics.

Bodyweight class (lb)		Total winning lifts (lb)		
	Max. weight	Snatch	Jerk	Total weight
Flyweight	114.5	231.5	303.1	534.6
Bantamweight	123.5	259.0	319.7	578.7
Featherweight	132.5	275.6	352.7	628.3
Lightweight	149.0	297.6	380.3	677.9
Middleweight	165.5	319.7	418.9	738.5
Light-heavyweight	182.0	358.3	446.4	804.7
Middle-heavyweight	198.5	374.8	468.5	843.3
Heavyweight	242.5	385.8	496.0	881.8

6.4   CUBIC SPLINE MODELS

The use of polynomials in constructing empirical models that capture the trend of the data is appealing because polynomials are so easy to integrate and differentiate. High-order polynomials, however, tend to oscillate near the endpoints of the data interval, and the coefficients can be sensitive to small changes in the data. Unless the data are essentially quadratic or cubic in nature, smoothing with a low-order polynomial may yield a relatively poor fit somewhere over the range of the data. For instance, the quadratic model that was fit to the data collected for the vehicular braking distance problem in Section 6.3 did not fit well at high velocities.

cubic spline
interpolation   In this section a popular modern technique called **cubic spline interpolation** is introduced. By using different cubic polynomials between successive pairs of data points, we can capture the trend of the data regardless of the nature of the underlying relationship, while simultaneously reducing the tendency toward oscillation and the sensitivity to changes in the data.

Consider the data in Figure 6.19a. What would a draftsperson do if asked to draw a smooth curve connecting the points? One solution would be to use a drawing instrument called a French curve (see Figure 6.19b), which actually contains many

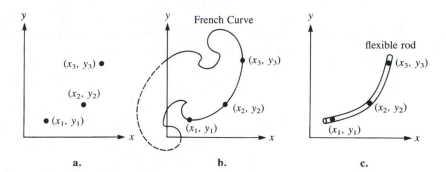

a.   b.   c.

FIGURE 6.19   A draftsperson might attempt to draw a smooth curve through the data points using a French curve or a thin flexible rod called a *spline*

different curves. By manipulating the French curve, we can draw a curve that works reasonably well between two data points and transitions smoothly to another curve for the next pair of data points. Another alternative is to take a thin flexible rod **spline** (called a **spline**) and tack it down at each data point. Cubic spline interpolation is essentially the same idea, except that distinct cubic polynomials are used between successive pairs of data points in a smooth way.

Linear Splines

Probably all of us at one time or another have referred to a table of values (e.g., square root, trigonometric table, or logarithmic table) without locating the value we were seeking. We found, instead, two values that bracket the desired value and made a proportionality adjustment.

For example, consider Table 6.23. Suppose we desire an estimate of the value of y at $x = 1.67$. Probably we would compute $y(1.67) \approx 5 + (2/3)(8 - 5) = 7$. That is, we implicitly assume that the variation in y, between $x = 1$ and $x = 2$, occurs **linear interpolation** linearly. Similarly, $y(2.33) \approx 13\frac{2}{3}$. This procedure is called **linear interpolation**, and for many applications it yields reasonable results, especially when the data are closely spaced.

TABLE 6.23 Linear interpolation

x_i	$y(x_i)$
1	5
2	8
3	25

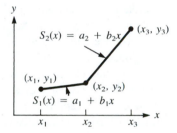

FIGURE 6.20   A linear spline model is a continuous function consisting of line segments

Figure 6.20 is helpful in interpreting the process of linear interpolation geometrically in a manner that mimics what is done with cubic spline interpolation. **linear spline** When x is in the interval $x_1 \leq x < x_2$, the model used is the **linear spline** $S_1(x)$ passing through the data points (x_1, y_1) and (x_2, y_2):

$$S_1(x) = a_1 + b_1 x \quad \text{for } x \text{ in } [x_1, x_2)$$

Similarly, when $x_2 \leq x < x_3$, the linear spline $S_2(x)$ passing through (x_2, y_2) and (x_3, y_3) is used:

$$S_2(x) = a_2 + b_2 x \quad \text{for } x \text{ in } [x_2, x_3]$$

Note that both spline segments meet at the point (x_2, y_2).

Let's determine the constants for the respective splines for the data in Table 6.23. The spline $S_1(x)$ must pass through the points $(1, 5)$ and $(2, 8)$. Mathematically this implies

$$a_1 + 1b_1 = 5$$
$$a_1 + 2b_1 = 8$$

Similarly, the spline $S_2(x)$ must pass through the points $(2, 8)$ and $(3, 25)$, yielding

$$a_2 + 2b_2 = 8$$
$$a_2 + 3b_2 = 25$$

Solution of these linear equations yields $a_1 = 2$, $b_1 = 3$, $a_2 = -26$, and $b_2 = 17$. The linear spline model for the data of Table 6.23 is summarized in Table 6.24.

TABLE 6.24 A linear spline model for the data of Table 6.23

Interval	Spline model
$1 \leq x < 2$	$S_1(x) = 2 + 3x$
$2 \leq x \leq 3$	$S_2(x) = -26 + 17x$

To illustrate how the linear spline model is used, let's predict $y(1.67)$ and $y(2.33)$. Because $1 \leq 1.67 < 2$, $S_1(x)$ is selected to compute $S_1(1.67) \approx 7.01$. Likewise, $2 \leq 2.33 \leq 3$ gives rise to the prediction $S_2(2.33) \approx 13.61$.

Whereas the linear spline method is sufficient for many applications, it fails to capture the trend of the data. Further, if we examine Figure 6.21, we see that the linear spline model does not appear smooth. That is, in the interval $[1, 2)$, $S_1(x)$ has a constant slope of 3, whereas in the interval $[2, 3]$, $S_2(x)$ has a constant slope of 17. Thus at $x = 2$ there is an abrupt change in the slope of the model from 3 to 17 so the first derivatives $S_1'(x)$ and $S_2'(x)$ fail to agree at $x = 2$. As we will discover, the process of cubic spline interpolation incorporates smoothness into the empirical

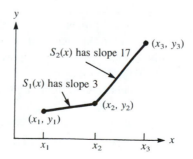

FIGURE 6.21 The linear spline does not appear smooth because the first derivative is not continuous

model by requiring that both the first and second derivatives of adjacent splines agree at each data point.

Cubic Splines

Consider now Figure 6.22. In a manner analogous to linear splines, we define a separate spline function for the intervals $x_1 \leq x < x_2$ and $x_2 \leq x < x_3$ as follows:

$$S_1(x) = a_1 + b_1 x + c_1 x^2 + d_1 x^3 \quad \text{for } x \text{ in } [x_1, x_2)$$
$$S_2(x) = a_2 + b_2 x + c_2 x^2 + d_2 x^3 \quad \text{for } x \text{ in } [x_2, x_3]$$

Because we will want to refer to the first and second derivatives, let's define them as well:

$$S_1'(x) = b_1 + 2c_1 x + 3d_1 x^2 \quad \text{for } x \text{ in } [x_1, x_2)$$
$$S_1''(x) = 2c_1 + 6d_1 x \quad \text{for } x \text{ in } [x_1, x_2)$$
$$S_2'(x) = b_2 + 2c_2 x + 3d_2 x^2 \quad \text{for } x \text{ in } [x_2, x_3]$$
$$S_2''(x) = 2c_2 + 6d_2 x \quad \text{for } x \text{ in } [x_2, x_3]$$

The model is presented geometrically in Figure 6.22.

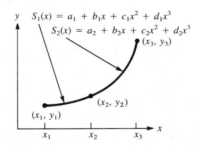

FIGURE 6.22 A cubic spline model is a continuous function with continuous first and second derivatives consisting of cubic polynomial segments

Cubic splines offer the possibility of matching up not only the slopes but also the curvatures at each interior data point. To determine the constants defining each cubic spline segment, we appeal to the requirement that each spline pass through the two data points specified by the interval over which the spline is defined. For the model depicted in Figure 6.22, this requirement yields the equations:

$$y_1 = S_1(x_1) = a_1 + b_1 x_1 + c_1 x_1^2 + d_1 x_1^3$$
$$y_2 = S_1(x_2) = a_1 + b_1 x_2 + c_1 x_2^2 + d_1 x_2^3$$
$$y_2 = S_2(x_2) = a_2 + b_2 x_2 + c_2 x_2^2 + d_2 x_2^3$$
$$y_3 = S_2(x_3) = a_2 + b_2 x_3 + c_2 x_3^2 + d_2 x_3^3$$

Note that there are eight unknowns $(a_1, b_1, c_1, d_1, a_2, b_2, c_2, d_2)$ and only four equations in the preceding system. An additional four independent equations are needed to determine the constants uniquely. Because smoothness of the spline system is also required, adjacent first derivatives must match at the interior data point (in this case when $x = x_2$). This requirement yields the equation

$$S_1'(x_2) = b_1 + 2c_1x_2 + 3d_1x_2^2 = b_2 + 2c_2x_2 + 3d_2x_2^2 = S_2'(x_2)$$

It can also be required that adjacent second derivatives match at each interior point as well:

$$S_1''(x_2) = 2c_1 + 6d_1x_2 = 2c_2 + 6d_2x_2 = S_2''(x_2)$$

To determine unique constants, we still require two additional independent equations. Although conditions on the derivatives at interior data points have been applied, nothing has been said about the derivatives at the exterior endpoints (x_1 and x_3 in Figure 6.22). Two popular conditions may be specified. One is to require that there be no change in the first derivative at the exterior endpoints. Mathematically, because the first derivative is constant, the second derivative must be zero. Application of this condition at x_1 and x_3 yields

$$S_1''(x_1) = 2c_1 + 6d_1x_1 = 0$$
$$S_2''(x_3) = 2c_2 + 6d_2x_3 = 0$$

natural spline   A cubic spline formed in this manner is called a **natural spline**. If we think again of our analog with the thin flexible rod tacked down at the data points, a natural spline allows the rod to be free at the endpoints to assume whatever direction the data points indicate. The natural spline is interpreted geometrically in Figure 6.23a.

Alternatively, if the values of the first derivative at the exterior endpoints are known, the first derivatives of the exterior splines can be required to match the known values. Suppose the derivatives at the exterior endpoints are known and given by $f'(x_1)$ and $f'(x_3)$. Mathematically, this matching requirement yields the equations:

$$S_1'(x_1) = b_1 + 2c_1x_1 + 3d_1x_1^2 = f'(x_1)$$
$$S_2'(x_3) = b_2 + 2c_2x_3 + 3d_2x_3^2 = f'(x_3)$$

clamped spline   A cubic spline formed in this manner is called a **clamped spline**. Again referring to our flexible rod analog, this situation corresponds to clamping the flexible rod in a vise at the exterior endpoints to ensure that the flexible rod has the proper angle. The clamped cubic spline is interpreted geometrically in Figure 6.23b. Unless *precise* information about the first derivative at the endpoints is known, the natural spline is generally used.

Let's illustrate the construction of a natural cubic spline model using the data in Table 6.23. We illustrate the technique with this simple example because the procedure readily extends to problems with more data points.

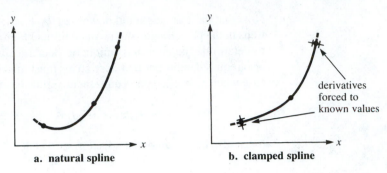

a. natural spline   b. clamped spline

FIGURE 6.23   The conditions for the natural and clamped cubic splines result in first derivatives at the two exterior endpoints that are constant; that constant value for the first derivative is specified in the clamped spline, whereas it is free to assume a natural value in the natural spline

Requiring the spline segment $S_1(x)$ to pass through the two endpoints $(1, 5)$ and $(2, 8)$ of its interval requires that $S_1(1) = 5$ and $S_1(2) = 8$, or

$$a_1 + 1b_1 + 1c_1 + 1d_1 = 5$$
$$a_1 + 2b_1 + 2^2 c_1 + 2^3 d_1 = 8$$

Similarly, $S_2(x)$ must pass through its endpoints of the second interval so that $S_2(2) = 8$ and $S_2(3) = 25$, or

$$a_2 + 2b_2 + 2^2 c_2 + 2^3 d_2 = 8$$
$$a_2 + 3b_2 + 3^2 c_2 + 3^3 d_2 = 25$$

Next, the first derivatives of $S_1(x)$ and $S_2(x)$ are forced to match at the interior data point $x_2 = 2$: $S_1'(2) = S_2'(2)$, or

$$b_1 + 2c_1(2) + 3d_1(2)^2 = b_2 + 2c_2(2) + 3d_2(2)^2$$

Forcing the second derivatives of $S_1(x)$ and $S_2(x)$ to match at $x_2 = 2$ requires $S_1''(2) = S_2''(2)$, or

$$2c_1 + 6d_1(2) = 2c_2 + 6d_2(2)$$

Finally, a natural spline is built by requiring that the second derivatives at the endpoints be zero: $S_1''(1) = S_2''(3) = 0$, or

$$2c_1 + 6d_1(1) = 0$$
$$2c_2 + 6d_2(3) = 0$$

Thus the procedure has yielded a linear algebraic system of eight equations in eight unknowns that can be solved uniquely. The resulting model is summarized in Table 6.25 and graphed in Figure 6.24.

TABLE 6.25 A natural cubic spline model for the data of Table 6.23

Interval	Model
$1 \leq x < 2$	$S_1(x) = 2 + 10x - 10.5x^2 + 3.5x^3$
$2 \leq x \leq 3$	$S_2(x) = 58 - 74x + 31.5x^2 - 3.5x^3$

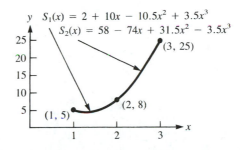

FIGURE 6.24 The natural cubic spline model for the data in Table 6.23 is a smooth curve that is easily integrated and differentiated

Let's illustrate the use of the model by again predicting $y(1.67)$ and $y(2.33)$:

$$S_1(1.67) \approx 5.72$$
$$S_2(2.33) \approx 12.32$$

Compare these values with the values predicted by the linear spline. In which values do you have the most confidence? Why?

The construction of cubic splines for more data points proceeds in the same manner. That is, each spline is forced to pass through the endpoints of the interval over which it is defined, the first and second derivatives of adjacent splines are forced to match at the interior data points, and either the clamped or natural conditions are applied at the two exterior data points. For computational reasons it would be necessary to implement the procedure on a computer. The procedure we described here does not give rise to a computationally or numerically efficient computer algorithm. Our approach was designed to facilitate our understanding of the basic concepts underlying the cubic spline interpolation.[4]

It is revealing to view how the graphs of the different cubic splines fit together to form a single composite interpolating curve between the data points. Consider the following data (from Problem 4 in Section 5.3):

x	7	14	21	28	35	42
y	8	41	133	250	280	297

[4]For a computationally efficient algorithm, see R. L. Burden and J. D. Faires, *Numerical Analysis*, 6th ed. (Pacific Grove, CA: Brooks/Cole, 1997).

Because there are six data points, five distinct cubic polynomials S_1 through S_5, are calculated to form the composite natural cubic spline. Each of these cubics is graphed and overlaid on the same graph to obtain Figure 6.25. Between any two consecutive data points only one of the five cubic polynomials is active, giving the smooth composite cubic spline shown in Figure 6.26.

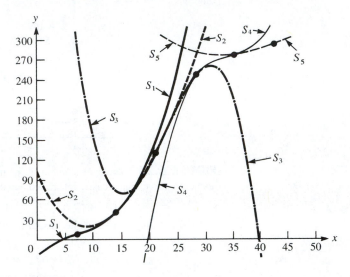

FIGURE 6.25 Between any two consecutive data points only one cubic spline polynomial is active (Graphics by Jim McNulty and Bob Hatton)

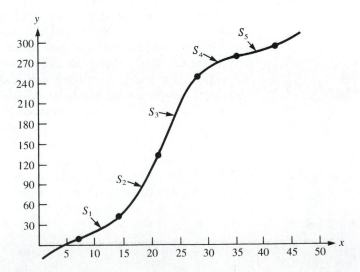

FIGURE 6.26 The smooth composite cubic spline from the cubic polynomials in Figure 6.25

You should be concerned with whether the procedure just described results in a unique solution. Also, you may be wondering why we jumped from a linear spline to a cubic spline without discussing quadratic splines. Intuitively, you would think that first derivatives could be matched with a quadratic spline. These and related issues are discussed in most numerical analysis texts (e.g., see Burden and Faires cited earlier).

Example 1   *Vehicular Stopping Distance Revisited*

Consider again the problem posed in Section 2.2: Predict a vehicle's total stopping distance as a function of its speed. In Section 4.2, we reasoned that the model should have the form

$$d = k_1 v + k_2 v^2$$

where d is the total stopping distance, v the velocity, and k_1 and k_2 are constants of proportionality resulting from the submodels for reaction distance and mechanical braking distance, respectively. We found reasonable agreement between the data furnished for the submodels and the graphically estimated k_1 and k_2 to obtain the model

$$d = 1.1v + 0.054v^2$$

In Section 5.4 we estimated k_1 and k_2 using the least-squares criterion and obtained the model

$$d = 1.104v + 0.0542v^2$$

The fit of the preceding two models is analyzed in Table 5.7. Note in particular that both models break down at high speeds, where they increasingly underestimate the stopping distances. In Section 6.3 we constructed an empirical model by smoothing the data with a quadratic polynomial and analyzed the fit.

Now suppose we are not satisfied with the predictions made by our analytical models or are unable to construct an analytical model, but we find it necessary to make predictions. If we are reasonably satisfied with the collected data, we might consider constructing a cubic spline model for the data presented in Table 6.26.

TABLE 6.26 Data relating total stopping distance and speed

Speed, v (mph)	20	25	30	35	40	45	50
Distance, d (ft)	42	56	73.5	91.5	116	142.5	173
Speed, v (mph)	55	60	65	70	75	80	
Distance, d (ft)	209.5	248	292.5	343	401	464	

Using a computer code, we obtained the cubic spline model summarized in Table 6.27. The first three spline segments are plotted in Figure 6.27. Note how each segment passes through the data points at either end of its interval, and note the smoothness of the transition across adjacent segments.

TABLE 6.27 A cubic spline model for vehicular breaking distance

Interval	Model
$20 \leq v < 25$	$S_1(v) = 42 + 2.596(v - 20) + 0.008(v - 20)^3$
$25 \leq v < 30$	$S_2(v) = 56 + 3.208(v - 25) + 0.122(v - 25)^2 - 0.013(v - 25)^3$
$30 \leq v < 35$	$S_3(v) = 73.5 + 3.472(v - 30) - 0.070(v - 30)^2 + 0.019(v - 30)^3$
$35 \leq v < 40$	$S_4(v) = 91.5 + 4.204(v - 35) + 0.216(v - 35)^2 - 0.015(v - 35)^3$
$40 \leq v < 45$	$S_5(v) = 116 + 5.211(v - 40) - 0.015(v - 40)^2 + 0.006(v - 40)^3$
$45 \leq v < 50$	$S_6(v) = 142.5 + 5.550(v - 45) + 0.082(v - 45)^2 + 0.005(v - 45)^3$
$50 \leq v < 55$	$S_7(v) = 173 + 6.787(v - 50) + 0.165(v - 50)^2 - 0.012(v - 50)^3$
$55 \leq v < 60$	$S_8(v) = 209.5 + 7.503(v - 55) - 0.022(v - 55)^2 + 0.012(v - 55)^3$
$60 \leq v < 65$	$S_9(v) = 248 + 8.202(v - 60) + 0.161(v - 60)^2 - 0.004(v - 60)^3$
$65 \leq v < 70$	$S_{10}(v) = 292.5 + 9.489(v - 65) + 0.096(v - 65)^2 + 0.005(v - 65)^3$
$70 \leq v < 75$	$S_{11}(v) = 343 + 10.841(v - 70) + 0.174(v - 70)^2 - 0.005(v - 70)^3$
$75 \leq v < 80$	$S_{12}(v) = 401 + 12.245(v - 75) + 0.106(v - 75)^2 - 0.007(v - 75)^3$

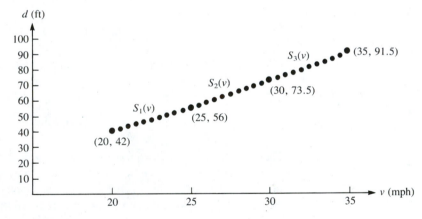

FIGURE 6.27 A plot of the cubic spline model for vehicular braking distance for $20 \leq r \leq 35$

Summary: Constructing Empirical Models

We conclude this chapter by presenting a summary and suggesting a procedure for constructing empirical models using the techniques presented. The procedure begins by examining the data in search of suspect data points that need to be examined more closely and perhaps discarded or obtained anew. Simultaneously, we look to see if a trend exists in the data. Normally, these questions are best considered by constructing a scatterplot, possibly with the aid of a computer if a large amount of data is being considered. If a trend does appear to exist, simple, one-term models are investigated first to see if one adequately captures the trend of the data. A one-term model can be identified using a transformation that converts the data into a straight line (approximately). The transformation may be found among the Ladder of Transformations or among the transformations discussed in Chapter 5. A graphical plot of the transformed data is often useful to determine whether the one-term model linearizes the data. If a model that appears adequate is found, the chosen model can be fit graphically or analytically using one of the criteria discussed in Chapter 5. Next, a careful analysis to determine how well the model fits the data points is obtained by examining indicators such as the sum of the absolute deviations, the largest absolute deviation, or the sum of squared deviations. A plot of the deviations as a function of the independent variable(s) may be useful for determining where the model does not fit well. If the fit proves unsatisfactory, other one-term models may be considered.

If it is determined that one-term models are inadequate, polynomials can be used. If there is a small number of data points, an $n - 1$st-order polynomial through the n data points can be tried. Be sure to note any oscillations, especially near the endpoints of the interval. A careful plot of the polynomial will help reveal this feature. If there is a large number of data points, a low-order polynomial to smooth the data may be considered. A divided difference table is a qualitative aid for determining whether a low-order polynomial is appropriate and in choosing the order of that polynomial. After the order of the polynomial is chosen, the polynomial may be fit and analyzed according to the techniques discussed in Chapter 5. If smoothing with a low-order polynomial proves inadequate, a cubic (or linear) spline that is satisfactory for many data sets can be used. See Figure 6.28 for a flowchart of this discussion.

6.4 PROBLEMS

1. For each of the following data sets, write a system of equations to determine the coefficients of the natural cubic splines passing through the given points. If a computer program is available, solve the system of equations and graph the splines.

   a.

x	2	4	7
y	2	8	12

   b.

x	3	4	6
y	10	15	35

   c.

x	0	1	2
y	0	10	30

   d.

x	0	2	4
y	5	10	40

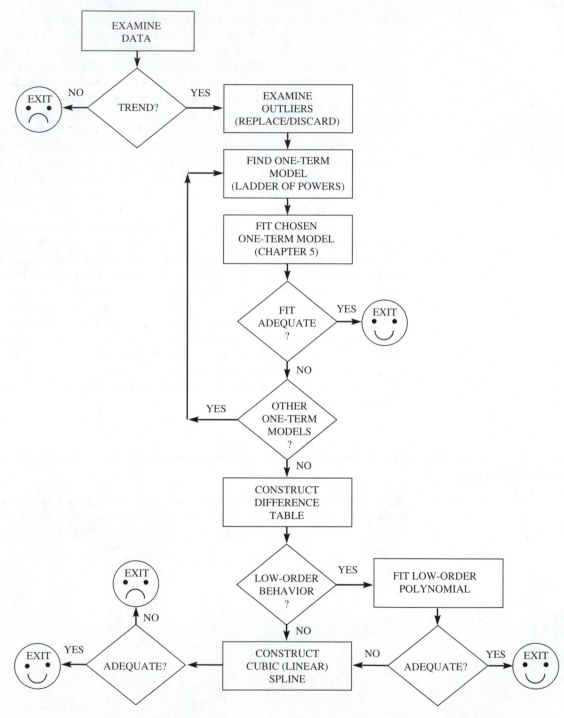

FIGURE 6.28 A flowchart for empirical model building

For Problems 2 and 3, find the natural cubic splines that pass through the given data points. Use the splines to answer the requirements.

2.

x	3.0	3.1	3.2	3.3	3.4	3.5	3.6	3.7	3.8	3.9
y	20.08	22.20	24.53	27.12	29.96	33.11	36.60	40.45	44.70	49.40

a. Estimate the derivative evaluated at $x = 3.45$. Compare your estimate with the derivative of e^x evaluated at $x = 3.45$.

b. Estimate the area under the curve from 3.3 to 3.6. Compare with

$$\int_{3.3}^{3.6} e^x \, dx$$

3.

x	0	$\pi/6$	$\pi/3$	$\pi/2$	$2\pi/3$	$5\pi/6$	π
y	0.00	0.50	0.87	1.00	0.87	0.50	0.00

4. For the data collected in the tape recorder problem relating the elapsed time with the counter reading (Sections 6.2 and 6.3), construct the natural splines that pass through the data points. Compare this model with previous models you have constructed. Which model makes the best predictions?

5. *The Cost of a Postage Stamp.* Consider the following data. Use the procedures in this chapter to capture the trend of the data if one exists. Would you eliminate any data points? Why? Would you be willing to use your model to predict the price of a postage stamp on January 1, 2010? What do the various models you construct predict about the price on January 1, 2010? When will the price reach $1? You might enjoy reading the article upon which this problem is based: Donald R. Byrkit and Robert E. Lee, "The Cost of a Postage Stamp, or Up, Up, and Away," *Mathematics and Computer Education* 17, no. 3 (Summer 1983): 184–190.

Date	First-class stamp	
1885–1917	$0.02	
1917–1919	0.03	(Wartime increase)
1919	0.02	(Restored by Congress)
July 6, 1932	0.03	
August 1, 1958	0.04	
January 7, 1963	0.05	
January 7, 1968	0.06	
May 16, 1971	0.08	
March 2, 1974	0.10	
December 31, 1975	0.13	(Temporary)
July 18, 1976	0.13	
May 1978	0.15	
March 22, 1981	0.18	
November 1, 1981	0.20	

6.4 PROJECTS

1. Construct a computer code for determining the coefficients of the natural splines that pass through a given set of data points. See Burden and Faires cited earlier in this chapter for an efficient algorithm.

For Projects 2–7, use the software you developed in Project 1 to find the splines that pass through the given data points. Use graphics software, if available, to sketch the resulting splines.

2. The following data represent the growth of yeast in a culture (data from R. Pearl, "The Growth of Population," *Quart. Rev. Biol.* 2 (1927): 532–548.

t (hours)	0	1	2	3	4	5	6
Observed yeast biomass	9.6	18.3	29.0	47.2	71.1	119.1	174.6

t (hours)	7	8	9	10	11	12	13
Observed yeast biomass	257.3	350.7	441.0	513.3	559.7	594.8	629.4

t (hours)	14	15	16	17	18
Observed yeast biomass	640.8	651.1	655.9	659.6	661.8

3. The following data represent the population of the United States from 1790 to 1990.

Year	Observed population
1790	3,929,000
1800	5,308,000
1810	7,240,000
1820	9,638,000
1830	12,866,000
1840	17,069,000
1850	23,192,000
1860	31,443,000
1870	38,558,000
1880	50,156,000
1890	62,948,000
1900	75,995,000
1910	91,972,000
1920	105,711,000
1930	122,755,000
1940	131,669,000
1950	150,697,000
1960	179,323,000
1970	203,212,000
1980	226,505,000
1990	248,710,000

4. The following data were obtained for the growth of a sheep population introduced into a new environment on the island of Tasmania. (Adapted from J. Davidson, "On the Growth of the Sheep Population in Tasmania," *Trans. Roy. Soc. S. Australia* 62 (1938): 342–346.)

t (year)	1814	1824	1834	1844	1854	1864
$P(t)$	125	275	830	1200	1750	1650

5. The following data represent the pace of life study (see Problem 1 in Section 6.1). P is the population and V is the mean velocity in feet per second over a 50-ft course.

P	365	2500	5491	14000	23700	49375	70700	78200
V	2.76	2.27	3.31	3.70	3.27	4.90	4.31	3.85
P	138000	304500	341948	867023	1092759	1340000	2602000	
V	4.39	4.42	4.81	5.21	5.88	5.62	5.05	

6. The following data represent the length and weight of a fish (bass).

length (in.)	12.5	12.625	14.125	14.5	17.25	17.75
weight (oz)	17.0	16.50	23.00	26.5	41.00	49.00

7. The following data represent the weight-lifting results from the 1976 Olympics.

Bodyweight class (lb)		Total winning lifts (lb)		
	Max. weight	Snatch	Jerk	Total weight
Flyweight	114.5	231.5	303.1	534.6
Bantamweight	123.5	259.0	319.7	578.7
Featherweight	132.5	275.6	352.7	628.3
Lightweight	149.0	297.6	380.3	677.9
Middleweight	165.5	319.7	418.9	738.5
Light-heavyweight	182.0	358.3	446.4	804.7
Middle-heavyweight	198.5	374.8	468.5	843.3
Heavyweight	242.5	385.8	496.0	881.8

8. You can use the cubic spline software you developed, coupled with some graphics, to draw smooth curves to represent a figure you wish to draw on the computer. Overlay a piece of graph paper on a picture or drawing you wish to produce on a computer. Record enough data to gain ultimately smooth curves. Take more data points where there are abrupt changes (see Figure 6.29).

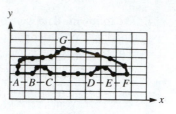

FIGURE 6.29

Now take the data points and determine the splines that pass through them. Note that if natural discontinuities in the derivative occur in the data such as at points A, B, C, D, E, F, and G in Figure 6.29, you will want to terminate one set of spline functions and begin another. You can then graph the spline functions using graphics software. In essence, you are using the computer to connect the dots with smooth curves. Select a figure of interest to you, such as your school mascot, and draw it on the computer.

Chapter Seven

SIMULATION MODELING

INTRODUCTION

In many situations a modeler is unable to construct an analytic (symbolic) model adequately explaining the behavior being observed because of its complexity or the intractability of the proposed explicative model. Yet if it is necessary to make predictions about the behavior, the modeler may conduct experiments (or otherwise gather data) to investigate the relationship between the dependent variable(s) and selected values of the independent variable(s) within some range. We constructed empirical models based on collected data in Chapter 6. To collect the data, the modeler may observe the behavior directly. In other instances, the modeler might duplicate the behavior (possibly in a scaled-down version) under controlled conditions, as will be done when predicting the size of craters in Section 13.4.

In some circumstances, it may not be feasible to observe the behavior directly or to conduct experiments. For instance, consider the service provided by a system of elevators during morning rush hour, as presented in Example 4 of Section 2.2. After identifying an appropriate problem and defining what is meant by good service, we suggested some alternative delivery schemes, such as assigning elevators to even and odd floors or using express elevators. Theoretically, each alternative could be tested for some period of time to determine which one provides the best service for particular arrival and destination patterns of the customers. However, such a procedure would probably be disruptive because it would be necessary to harass the customers constantly as the required statistics are collected. Moreover, the customers would become confused because the elevator delivery system would keep changing. Another problem concerns testing alternative schemes for controlling automobile traffic in a large city. It would be impractical to change directions of the one-way streets and the distribution of traffic signals constantly to conduct tests.

In still other situations, the system for which alternative procedures need to be tested *may not even exist yet*. An example is the situation of several proposed communications networks with the problem of determining which is best for a given office building. Still another example is the problem of determining locations of machines in a new industrial plant. The *cost* of conducting experiments may be prohibitive. This is the case when trying to predict the effects of various alternatives for protecting and evacuating the population in case of failure of a nuclear power plant.

In instances in which the behavior cannot be explained analytically or data collected directly, the modeler might *simulate* the behavior indirectly in some manner and then test the various alternatives being considered to estimate how each affects the behavior. Data can then be collected to determine which alternative is best. An example is to determine the drag force on a proposed submarine. Because it is infeasible to build a prototype, we can build a scaled model to simulate the behavior of the actual submarine. Another example of this type of simulation is using a scaled model of a jet airplane in a wind tunnel to estimate the effects of very high speeds for various designs of the aircraft. There is yet another type of simulation, which you will study in this chapter. It is the method for simulating

Monte Carlo simulation

behavior, called **Monte Carlo simulation**, and is typically accomplished with the aid of a computer.

Suppose we are investigating the service provided by a system of elevators at morning rush hour. In Monte Carlo simulation the arrival of customers at the elevators during the hour and the destination floors they select need to be replicated. That is, the distribution of arrival times and the distribution of floors desired on the simulated trial must portray a possible rush hour. Moreover, after we simulated many trials, the daily distribution of arrivals and destinations that occur must mimic the real-world distributions in proper proportions. When we are satisfied that the behavior is adequately duplicated, we can investigate various alternative strategies for operating the elevators. Using a large number of trials, we can gather appropriate statistics, such as the average total delivery time of a customer or the length of the longest queue. These statistics can help determine the best strategy for operating the elevator system.

This chapter provides a brief introduction to Monte Carlo simulation. Additional studies in probability and statistics are required to delve into the deeper intricacies of computer simulation and understand its appropriate uses. Nevertheless you will gain some appreciation of this powerful component of mathematical modeling. Keep in mind that there is a danger in placing too much confidence in the predictions resulting from a simulation, especially if the assumptions inherent in the simulation are not clearly stated. Moreover, the appearance of using large amounts of data and huge amounts of computer time, coupled with the fact that laymen can understand a simulation model and computer output with relative ease, often leads to overconfidence in the results.

When any Monte Carlo simulation is performed, random numbers are used. We discuss how to generate random numbers in Section 7.2. Loosely speaking, a sequence of random numbers uniformly distributed in an interval m to n is a set of numbers with no apparent pattern, where each number between m and n can appear with equal likelihood. For example, if you toss a six-sided die 100 times and write down the number showing on the die each time, you will have written down a sequence of 100 random integers approximately uniformly distributed over the interval 1 to 6. Now suppose random numbers consisting of six digits can be generated. The tossing of a coin can be duplicated by generating a random number and assigning it a head if the random number is even and a tail if the random number is odd. If this trial is replicated a large number of times, you would expect heads to occur about 50% of the time. However, there is an element of chance involved. It is possible that a run of 100 trials could produce 51 heads and that the next 10 trials (although not very likely) would produce all heads. Thus the estimate with 110 trials is actually worse than the estimate with 100 trials. Processes with an element of chance involved are called **probabilistic**, as opposed to **deterministic**, processes. Monte Carlo simulation is therefore a probabilistic model.

probabilistic
deterministic

The behavior being modeled can be either deterministic or probabilistic. For instance, the area under a curve is deterministic (even though it may be impossible to find it precisely). On the other hand, the time between arrivals of customers at the elevator on a particular morning is probabilistic behavior. Referring to Figure 7.1 you can see that a deterministic model can be used to approximate either a deterministic or probabilistic behavior and, similarly, a Monte Carlo simulation can be used to approximate a deterministic behavior (as you will see with a Monte Carlo approximation to an area under a curve) or a probabilistic one. But, as we would expect, the real power of Monte Carlo simulation lies in modeling a probabilistic behavior.

FIGURE 7.1   The behavior and the model can be either deterministic or probabilistic

A principal advantage of Monte Carlo simulation is the relative ease with which it can sometimes be used to approximate complex probabilistic systems. Additionally, Monte Carlo simulation provides performance estimation over a wide range of conditions rather than a restricted range as often required by an analytical model. Furthermore, because a particular submodel can be changed rather easily in a Monte Carlo simulation (such as the arrival and destination patterns of customers at the elevators), there is the potential of conducting a sensitivity analysis. Still another advantage is that the modeler has control over the level of detail in a simulation. For example, a long time frame can be compressed or a small time frame expanded, giving a great advantage over experimental models. Finally, there are powerful, high-level simulation languages (such as GPSS, GASP, PROLOG, SIMAN, SLAM, and DYNAMO) that eliminate much of the tedious labor in constructing a simulation model.

On the negative side, simulation models are typically expensive to develop and operate. They may require many work hours to construct, as well as large amounts of computer time and memory when being run. Another disadvantage is that the probabilistic nature of the simulation model limits the conclusions that can be drawn from a particular run unless a sensitivity analysis is conducted. Such an analysis often requires many more runs just to consider a small number of combinations of conditions that can occur in the various submodels. This limitation forces the modeler to estimate which combinations might occur for a particular set of conditions.

A simulation model can also be difficult to verify because of its many components and its probabilistic nature. The modeler must ensure that the random number generator is operating correctly. Typically, it is difficult to validate a simulation model using real-world data. (This point is particularly important when well-known probability distributions are assumed for various submodels. Each distribution needs to be checked for its appropriateness to the situation.) Nevertheless, the modeler may be forced to develop a simulation model precisely because it is impossible to obtain data (making model validation impossible anyway). Finally, even though a simulation correctly estimates which of the various alternatives seems best, it still cannot provide an optimal solution (as we will study in the next chapter) because all the possible alternatives have not been considered. So considerable judgment is required to determine which alternatives to simulate.

7.1

SIMULATING DETERMINISTIC BEHAVIOR: AREA UNDER A CURVE

In this section we illustrate the use of Monte Carlo simulation to model a deterministic behavior—the area under a curve. We begin by finding an approximate value to the area under a nonnegative curve. Specifically, suppose $y = f(x)$ is some given continuous function satisfying $0 \le f(x) \le M$ over the closed interval $a \le x \le b$. Here the number M is simply some constant that *bounds* the function. This situation is depicted in Figure 7.2. Notice that the area we seek is wholly contained within the rectangular region of height M and length $b - a$ (the length of the interval over which f is defined).

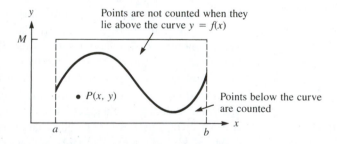

FIGURE 7.2 The area under the nonnegative curve $y = f(x)$ over $a \le x \le b$ is contained within the rectangle of height M and base length $b - a$

Now we select a point $P(x, y)$ at random from within the rectangular region. We will do so by generating two random numbers, x and y, satisfying $a \le x \le b$ and $0 \le y \le M$ and interpreting them as a point P with coordinates x and y. Once $P(x, y)$ is selected, we ask whether it lies within the region below the curve; that is, does the y-coordinate satisfy $0 \leq y \leq f(x)$? If the answer is yes, we count the point P by adding 1 to some counter. Two counters will be necessary: one to count the total points generated and a second to count those points that lie below the curve (see Figure 7.2). We can then calculate an approximate value for the area under the curve by the following formula:

$$\frac{\text{area under curve}}{\text{area of rectangle}} \approx \frac{\text{number of points counted below curve}}{\text{total number of random points}}$$

As discussed in the introduction, the Monte Carlo technique is probabilistic and typically requires a large number of trials before the deviation between the predicted and true values becomes small. A discussion of the number of trials needed to ensure a predetermined level of confidence in the final estimate requires a background in statistics. As a general rule, however, to double the accuracy of the result (i.e., cut the expected error in half), about four times as many experiments are necessary.

The following algorithm gives the sequence of calculations needed for a general computer simulation of this Monte Carlo technique for finding the area under a curve.

Monte Carlo Area Algorithm

Input   Total number n of random points to be generated in the simulation.

Output   AREA = approximate area under the specified curve $y = f(x)$ over the given interval $a \leq x \leq b$, where $0 \leq f(x) < M$.

Step 1   Initialize: COUNTER = 0.

Step 2   For $i = 1, 2, \ldots, n$, do Steps 3–5.

Step 3   Calculate random coordinates x_i and y_i, satisfying $a \leq x_i \leq b$ and $0 \leq y_i < M$.

Step 4   Calculate $f(x_i)$ for the random x_i coordinate.

Step 5   If $y_i \leq f(x_i)$, then increment the COUNTER by 1. Otherwise, leave COUNTER as is.

Step 6   Calculate AREA = $M(b - a)$ COUNTER$/n$.

Step 7   OUTPUT (AREA)
STOP

Table 7.1 gives the results of several different simulations to obtain the area beneath the curve $y = \cos x$ over the interval $-\pi/2 \leq x \leq \pi/2$, where $0 \leq \cos x < 2$.

The actual area under the curve $y = \cos x$ over the given interval is 2 square units. Note that even with the relatively large number of points generated, the error is significant. For functions of one variable, the Monte Carlo technique is generally not competitive with quadrature techniques you will learn in numerical analysis. The lack of an error bound and the difficulty in finding an upper bound M are disadvantages as well. Nevertheless, the Monte Carlo technique can be extended to functions of several variables and becomes more practical in that situation.

TABLE 7.1 Monte Carlo approximation to the area under the curve $y = \cos x$ over the interval $-\pi/2 \le x \le \pi/2$

Number of points	Approximation to area	Number of points	Approximation to area
100	2.07345	2000	1.94465
200	2.13628	3000	1.97711
300	2.01064	4000	1.99962
400	2.12058	5000	2.01439
500	2.04832	6000	2.02319
600	2.09440	8000	2.00669
700	2.02857	10000	2.00873
800	1.99491	15000	2.00978
900	1.99666	20000	2.01093
1000	1.96664	30000	2.01186

Volume Under a Surface

Let's consider finding that part of the volume of the sphere

$$x^2 + y^2 + z^2 \le 1$$

that lies in the first octant, $x > 0, y > 0, z > 0$ (see Figure 7.3).

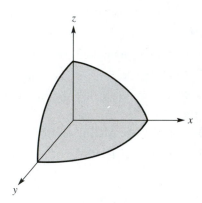

FIGURE 7.3 Volume of a sphere $x^2 + y^2 + z^2 \le 1$ that lies in the first octant, $x > 0, y > 0, z > 0$

The methodology to approximate the volume is similar to that of finding the area under a curve. However, now we will use an approximation for the volume under the surface by the following rule:

$$\frac{\text{volume under surface}}{\text{volume of box}} \approx \frac{\text{number of points counted below surface in 1st octant}}{\text{total number of points}}$$

The following algorithm gives the sequence of calculations required to use Monte Carlo techniques to find the approximate volume of the region.

Monte Carlo Volume Algorithm

Input   Total number n of random points to be generated in the simulation.

Output   VOLUME = approximate volume enclosed by the specified function, $z = f(x, y)$ in the first octant, $x > 0, y > 0, z > 0$.

Step 1   Initialize: COUNTER = 0.

Step 2   For $i = 1, 2, \ldots, n$, do Steps 3–5.

Step 3   Calculate random coordinates x_i, y_i, z_i that satisfy: $0 \leq x_i \leq 1, 0 \leq y_i \leq 1, 0 \leq z_i \leq 1$.
(In general, $a \leq x_i \leq b, c \leq y_i \leq d, 0 \leq z_i \leq M$.)

Step 4   Calculate $f(x_i, y_i)$ for the random coordinate (x_i, y_i).

Step 5   If random $z_i \leq f(x_i, y_i)$, then increment the COUNTER by 1. Otherwise, leave COUNTER as is.

Step 6   Calculate VOLUME = $M(d - c)(b - a)$COUNTER$/n$.

Step 7   OUTPUT (VOLUME)
STOP

Table 7.2 gives the results of several Monte Carlo runs to obtain the approximate volume of

$$x^2 + y^2 + z^2 \leq 1$$

that lies in the first octant, $x > 0, y > 0, z > 0$.

TABLE 7.2 Monte Carlo approximation to the volume in the first octant under the surface $x^2 + y^2 + z^2 \leq 1$

Number of points	Approximate volume
100	0.4700
200	0.5950
300	0.5030
500	0.5140
1000	0.5180
2000	0.5120
5000	0.5180
10000	0.5234
20000	0.5242

The actual volume in the first octant is known to be approximately 0.5236 cubic units ($\pi/6$ with $R = 1$). Generally, although not uniformly, the error becomes smaller as the number of points generated increases.

7.1 PROBLEMS

1. Each ticket in a lottery contains a single hidden number according to the following scheme: 55% of the tickets contain a 1, 35% contain a 2, and 10% contain a 3. A participant in the lottery wins a prize by obtaining all three numbers 1, 2, and 3. Describe an experiment that could be used to determine how many tickets you would expect to buy to win a prize.

2. Two record companies, A and B, produce classical music recordings. Label A is a budget label and 5% of A's new compact discs exhibit a significant degree of warpage. Label B is manufactured under tighter quality control (and consequently is more expensive) than A, so only 2% of its compact discs are warped. You purchase one label A and one label B recording at your local store on a regular basis. Describe an experiment that could be used to determine how many times you would expect to make such a purchase before buying two warped compact discs for a given sale.

3. Using Monte Carlo simulation, write an algorithm to calculate an approximation to π by considering the number of random points selected inside the quarter circle:

$$Q: x^2 + y^2 = 1, \quad x \geq 0, \quad y \geq 0$$

where the quarter circle is taken to be inside the square:

$$S: 0 \leq x \leq 1 \quad \text{and} \quad 0 \leq y \leq 1$$

Use the equation $\pi/4 =$ area Q/area S.

4. Using Monte Carlo simulation, write an algorithm to calculate that part of the volume of an ellipsoid:

$$\frac{x^2}{2} + \frac{y^2}{4} + \frac{z^2}{8} \leq 16$$

that lies in the first octant, $x > 0$, $y > 0$, $z > 0$.

5. Using Monte Carlo simulation, write an algorithm to calculate the volume trapped between the two paraboloids:

$$z = 8 - x^2 - y^2 \quad \text{and} \quad z = x^2 + 3y^2$$

Note that the two paraboloids intersect on the elliptic cylinder:

$$x^2 + 2y^2 = 4$$

7.2 GENERATING RANDOM NUMBERS

In the previous section, we developed algorithms for Monte Carlo simulations to find areas and volumes. A key ingredient common to each algorithm is the need for random numbers. Random numbers have a variety of applications, such as gambling problems, finding area and volume, and modeling larger complex systems such as large-scale combat operations or air traffic control situations.

In some sense a computer does not really generate random numbers because computers use deterministic algorithms. But we can generate sequences of pseudorandom numbers that, for all practical purposes, may be considered random. There is no single best random number generator or best test to ensure randomness.

There are complete courses of study for random numbers and simulations that cover in depth the methods and tests for pseudorandom number generators. Our purpose here is to introduce a few random number methods we can use to generate sequences of numbers that are nearly random.

Many programming languages, such as Pascal, Basic, and other software (Minitab, Quattro Pro, and Excel, just to name a few) have built-in random number generators for convenience to the user.

Middle-Square Method

The middle-square method was developed in 1946 by John von Neumann, S. Ulm, and N. Metropolis at Los Alamos Laboratories to simulate neutron collisions as part of the Manhattan Project. Their middle-square method works as follows:

1. Start with a four-digit number x_0, called the *seed*.
2. Square it to obtain an eight-digit number (add a leading zero if necessary).
3. Take the middle four digits as the next random number.

Continuing in this manner, we obtain a sequence that appears to be random over the integers from 0 to 9999. These integers can then be scaled to any interval a to b. For example, if we wanted numbers from 0 to 1 we would divide the four-digit numbers by 10,000. Let's illustrate the middle-square method.

Pick a seed, say $x_0 = 2041$, and square it (adding a leading zero) to get 04165681. The middle four digits give the next random number, 1656. Generating nine random numbers in this way yields

n	0	1	2	3	4	5	6	7	8
x_n	2041	1656	7423	1009	0180	1049	1004	80	64

We can use more than four digits if we wish, but we always take the middle number of digits equal to the number of digits in the seed. For example, if $x_0 = 653217$ (6 digits), its square $426,692,449,089$ has 12 digits. Thus take the middle 6 digits as the random number, namely, 692449.

The middle-square method is reasonable, but it has a major drawback in its tendency to degenerate to zero (where it will stay forever). With the seed 2041, the random sequence does seem to be approaching zero. How many numbers can be generated until we are almost at zero?

Linear Congruence

The linear congruence method was introduced by D. H. Lehmer in 1951, and a great majority of pseudorandom numbers used today are based on this method. One advantage it has over other methods is that seeds can be selected that generate patterns that eventually cycle (we illustrate this concept with an example). However, the length of the cycle is so large that the pattern does not repeat itself on large computers for most applications. The method requires the choice of three integers: a, b, and c. Given some initial seed, say x_0, we generate a sequence by the rule

$$x_{n+1} = (a \times x_n + b) \bmod(c)$$

where c is the modulus, a is the multiplier, and b is the increment. The qualifier $\bmod(c)$ in the equation means to obtain the remainder after dividing the quantity $(a \times x_n + b)$ by c. For example, with $a = 1$, $b = 7$, and $c = 10$,

$$x_{n+1} = (1 \times x_n + 7) \bmod(10)$$

means x_{n+1} is the integer remainder upon dividing $x_n + 7$ by 10. Thus if $x_n = 115$, $x_{n+1} = remainder\left(\frac{122}{10}\right) = 2$.

Before investigating the linear congruence methodology, we need to discuss **cycling**, which is a major problem that occurs with random numbers. Cycling means the sequence repeats itself, and although undesirable, it is unavoidable. At some point, all pseudorandom number generators begin to cycle. Let's illustrate cycling with an example.

If we set our seed at $x_0 = 7$, we find $x_1 = (1 \times 7 + 7) \bmod(10)$ or $14 \bmod(10)$, which is 4. Repeating this same procedure, we obtain the following sequence:

$$7, 4, 1, 8, 5, 2, 9, 6, 3, 0, 7, 4, \ldots$$

and the original sequence repeats again and again. Note that there is cycling after 10 numbers. The methodology produces a sequence of integers between 0 and $c - 1$ (which includes the possible remainders after dividing the integers by c). Cycling is guaranteed with at most c numbers in the random number sequence. Nevertheless, c can be chosen to be very large and a and b chosen in such a way to obtain a full set of c numbers before cycling begins to occur. Many computers use $c = 2^{31}$ for the large value of c. Again, we can scale the random numbers to obtain a sequence between any limits a and b, as required.

A second problem that can occur with the linear congruence method is lack of statistical independence among the members in the list of random numbers. Any correlations between the nearest neighbors, the next-nearest neighbors, the third-nearest neighbors, and so forth, are generally unacceptable. (Because we live in a three-dimensional world, third-nearest neighbor correlations can be particularly damaging in physical applications.) Pseudorandom number sequences can never be completely statistically independent because they are generated by a mathematical formula or algorithm. Nevertheless, the sequence will appear (for practical purposes) independent when it is subjected to certain statistical tests. These concerns are best addressed in a course in statistics.

7.2 PROBLEMS

1. Use the middle-square method to generate
   a. 10 random numbers using $x_0 = 1009$.
   b. 20 random numbers using $x_0 = 653217$.
   c. 15 random numbers using $x_0 = 3043$.
   d. Comment about the results of each sequence. Was there cycling? Did each sequence degenerate rapidly?

2. Use the linear congruence method to generate
   a. 10 random numbers using $a = 5$, $b = 1$, and $c = 8$.
   b. 15 random numbers using $a = 1$, $b = 7$, and $c = 10$.
   c. 20 random numbers using $a = 5$, $b = 3$, and $c = 16$.
   d. Comment about the results of each sequence. Was there cycling? If so, when did it occur?

7.2 PROJECTS

1. Complete the requirement for UMAP module 269, "Monte Carlo: The Use of Random Digits to Simulate Experiments," by Dale T. Hoffman. The Monte Carlo technique is presented, explained, and used to find approximate solutions to several realistic problems. Simple experiments are included for student practice.

2. "Random Numbers," by Mark D. Myerson, UMAP 590. This module discusses methods for generating random numbers in more depth than our presentation and includes tests for determining the randomness of a string of numbers. Complete this module and prepare a short report on testing for randomness.

3. Write a computer program to generate uniformly distributed random integers in the interval $m < x < n$, where m and n are integers, according to the following algorithm:

Step 1   Let $d = 2^{31}$ and choose N (the number of random numbers to generate).

Step 2   Choose any seed integer Y such that
$$999999 > Y > 100000$$

Step 3   Let $i = 1$.

Step 4   Let $Y = (15625\, Y + 22221)\, \text{mod}(d)$.

Step 5   Let $X_i = m + \text{floor}[(n - m + 1)Y/d]$.

Step 6 Increment i by 1: $i = i + 1$.

Step 7 Go to Step 4 unless $i = N + 1$.

Here floor (p) means the largest integer not exceeding p.

For most choices of Y the numbers $X_1, X_2, \ldots$ form a sequence of (pseudo) random integers as desired. One possible recommended choice is $Y = 568731$.

To generate random numbers (not just integers) in an interval a to b with $a < b$, use the preceding algorithm, replacing the formula in Step 5 by

$$\text{Let } X_i = a + \frac{Y(b - a)}{d - 1}$$

4. Write a program to generate 1000 integers between 1 and 5 in a random fashion so that 1 occurs 22% of the time, 2 occurs 15% of the time, 3 occurs 31% of the time, 4 occurs 26% of the time, and 5 occurs 6% of the time. Over what interval would you generate the random numbers? How do you decide which of the integers from 1 to 5 has been generated according to its specified chance of selection?

5. Write a program or use a spreadsheet to find the approximate area or volumes in Problems 3–5 in Section 7.1.

7.3 SIMULATING PROBABILISTIC BEHAVIOR

One of the keys to good Monte Carlo simulation practices is an understanding of the axioms of probability. The term *probability* refers to the study of both randomness and uncertainty, as well as the quantifying of the likelihoods associated with various outcomes. Probability can be seen as a long-term average. For example, if the probability of an event occurring is 1 out of 5, then in the long run, the chance of the event happening is 1/5. Over the long haul, the probability of an event can be thought of as the ratio of

$$\frac{\text{number of favorable events}}{\text{total number of events}}$$

Our goal in this section is to show how to model simple probabilistic behavior to build intuition and understanding before developing submodels of probabilistic processes to incorporate in simulations (Sections 7.4 and 7.5).

We will examine three simple probabilistic models:

1. Flip of a fair coin
2. Roll of a fair die or pair of dice
3. Roll of an unfair die or pair of unfair dice

A Fair Coin

Most people realize that the chance of obtaining a head or a tail on a coin is 1/2. What happens if we actually start flipping a coin? Will one out of every two flips be a head? Probably not. Again, probability is a long-term average. Thus, in the long run, the ratio of heads to the number of flips approaches 0.5. Let's define $f(x)$ as follows, where x is a random number between [0, 1]:

$$f(x) = \begin{cases} \text{Head,} & 0 \le x \le 0.5 \\ \text{Tail,} & 0.5 < x \le 1 \end{cases}$$

Note that $f(x)$ assigns the outcome head or tail to a number between [0, 1]. We want to take advantage of the cumulative nature of this function as we make random assignments to numbers between [0, 1]. In the long run we expect to find the following percent occurrences:

Random number interval	Percent occurrences	Cumulative occurrences
$x < 0$	0.00	0
$0 < x < 0.5$	0.50	0.5
$0.5 < x < 1.0$	0.50	1

Let's illustrate using the following algorithm.

Monte Carlo Fair Coin Algorithm

Input   Total number n of random flips of a fair coin to be generated in the simulation.

Output   Probability of getting a head when we flip a fair coin.

Step 1   Initialize: COUNTER = 0.

Step 2   For $i = 1, 2, \ldots, n$, do Steps 3 and 4.

Step 3   Obtain a random number x_i between 0 and 1.

Step 4   If $0 \le x_i \le 0.5$, then COUNTER = COUNTER + 1. Otherwise leave COUNTER as is.

Step 5   Calculate $P(\text{head}) = \text{COUNTER}/n$.

Step 6   OUTPUT Probability of heads, $P(\text{head})$.
STOP

Table 7.3 illustrates our results for various choices n of the number of random x_i generated. Note that as n gets large, the probability of heads occurring is about 0.5, or half the time.

Roll of a Fair Die

Rolling a fair die adds a new twist to the process. In the flip of a coin, only one event is assigned. Now we must devise a method to assign six events because a die consists of the numbers $\{1, 2, 3, 4, 5, 6\}$. The probability of each event occurring is 1/6 because each number is equally likely to occur. As before, this probability of a

TABLE 7.3 Results from flipping a fair coin

Number of flips	Number of heads	Percent heads
100	49	0.49
200	102	0.51
500	252	0.504
1000	492	0.492
5000	2469	0.4930
10,000	4993	0.4993

particular number occurring is defined to be

$$\frac{\text{number of occurrences of the particular number } \{1, 2, 3, 4, 5, 6\}}{\text{total number of trials}}$$

We can use the following algorithm to generate our experiment for a roll of a fair die.

Monte Carlo Roll of a Fair Die Algorithm

Input   Total number n of random rolls of a die in the simulation.

Output   The percentage or probability for rolls $\{1, 2, 3, 4, 5, 6\}$.

Step 1   Initialize COUNTER 1 through COUNTER 6 to zero.

Step 2   For $i = 1, 2, \ldots, n$, do Steps 3 and 4.

Step 3   Obtain a random number satisfying $0 \le x_i \le 1$.

Step 4   If x_i belongs to these intervals, then increment the appropriate COUNTER.

$$0 \le x_i \le \tfrac{1}{6} \qquad \text{COUNTER 1} = \text{COUNTER 1} + 1$$
$$\tfrac{1}{6} < x_i \le \tfrac{2}{6} \qquad \text{COUNTER 2} = \text{COUNTER 2} + 1$$
$$\tfrac{2}{6} < x_i \le \tfrac{3}{6} \qquad \text{COUNTER 3} = \text{COUNTER 3} + 1$$
$$\tfrac{3}{6} < x_i \le \tfrac{4}{6} \qquad \text{COUNTER 4} = \text{COUNTER 4} + 1$$
$$\tfrac{4}{6} < x_i \le \tfrac{5}{6} \qquad \text{COUNTER 5} = \text{COUNTER 5} + 1$$
$$\tfrac{5}{6} < x_i \le 1 \qquad \text{COUNTER 6} = \text{COUNTER 6} + 1$$

Step 5   Calculate probability of each roll $j = \{1, 2, 3, 4, 5, 6\}$ by COUNTER$(j)/n$.

Step 6   OUTPUT probabilities.
STOP

TABLE 7.4 Results from a roll of a fair die (n = number of trials)

Die value	10	100	1000	10000	100000	Expected results
1	0.300	0.190	0.152	0.1703	0.1652	0.1667
2	0.00	0.150	0.152	0.1652	0.1657	0.1667
3	0.100	0.090	0.157	0.1639	0.1685	0.1667
4	0.00	0.160	0.180	0.1653	0.1685	0.1667
5	0.400	0.150	0.174	0.1738	0.1676	0.1667
6	0.200	0.160	0.185	0.1615	0.1652	0.1667

We see that with 100,000 runs we are close (for these trials) to the expected results.

Roll of an Unfair Die

Let's consider a probability model where each event is not equally likely. Assume the die is biased according to the following empirical distribution:

Roll value	P(roll)
1	0.1
2	0.1
3	0.2
4	0.3
5	0.2
6	0.1

The cumulative occurrences for the function to be used in our algorithm would be

Value of x_i	Assignment
$[0, 0.1]$	ONE
$(0.1, 0.2]$	TWO
$(0.2, 0.4]$	THREE
$(0.4, 0.7]$	FOUR
$(0.7, 0.9]$	FIVE
$(0.9, 1.0]$	SIX

We model the roll of an unfair die using the following algorithm:

Monte Carlo Roll of an Unfair Die Algorithm

Input   Total number n of random rolls of a die in the simulation.

Output   The percentage or probability for rolls $\{1, 2, 3, 4, 5, 6\}$.

Step 1   Initialize COUNTER 1 through COUNTER 6 to zero.

Step 2   For $i = 1, 2, \ldots, n$, do Steps 3 and 4.

Step 3   Obtain a random number satisfying $0 \leq x_i \leq 1$.

Step 4   If x_i belongs to these intervals, then increment the appropriate COUNTER.

$$0 \leq x_i \leq 0.1 \qquad \text{COUNTER } 1 = \text{COUNTER } 1 + 1$$

$$0.1 < x_i \leq 0.2 \qquad \text{COUNTER } 2 = \text{COUNTER } 2 + 1$$

$$0.2 < x_i \leq 0.4 \qquad \text{COUNTER } 3 = \text{COUNTER } 3 + 1$$

$$0.4 < x_i \leq 0.7 \qquad \text{COUNTER } 4 = \text{COUNTER } 4 + 1$$

$$0.7 < x_i \leq 0.9 \qquad \text{COUNTER } 5 = \text{COUNTER } 5 + 1$$

$$0.9 < x_i \leq 1.0 \qquad \text{COUNTER } 6 = \text{COUNTER } 6 + 1$$

Step 5   Calculate probability of each roll $j = \{1, 2, 3, 4, 5, 6\}$ by COUNTER(j)/n.

Step 6   OUTPUT probabilities.
STOP

The results are shown in Table 7.5. Note that a large number of trials is required for the model to approach the expected results.

TABLE 7.5 Results from a roll of an unfair die

Die value	100	1000	5000	10000	40000	Expected results
1	0.130	0.980	0.094	0.0948	0.0948	0.1
2	0.08	0.099	0.099	0.0992	0.0992	0.1
3	0.210	0.199	0.192	0.1962	0.1962	0.2
4	0.35	0.320	0.308	0.3082	0.3081	0.3
5	0.280	0.184	0.201	0.2012	0.2011	0.2
6	0.150	0.120	0.104	0.1044	0.1045	0.1

In the next section we will see how to use these ideas to simulate a real-world probabilistic situation.

7.3 PROBLEMS

1. You arrive at the beach for a vacation and are dismayed to learn that the local weather station is predicting a 50% chance of rain every day. Using Monte Carlo simulation, predict the chance that it rains three consecutive days during your vacation.

2. Use Monte Carlo simulation to approximate the probability of three heads occurring when five fair coins are flipped.

3. Use Monte Carlo simulation to simulate the sum of 100 consecutive rolls of a fair die.

4. Given loaded dice according to the distribution given in the following table, use Monte Carlo simulation to simulate the sum of 300 rolls of two unfair dice.

Roll	Die 1	Die 2
1	0.1	0.3
2	0.1	0.1
3	0.2	0.2
4	0.3	0.1
5	0.2	0.05
6	0.1	0.25

5. Make up a game that uses a flip of a fair coin, and then use Monte Carlo simulation to predict the results of the game.

7.3 PROJECTS

1. *Blackjack*—Construct and perform a Monte Carlo simulation of Blackjack (also called Twenty-one). The rules of Blackjack are as follows:

   Most casinos use six or eight decks of cards when playing this game to inhibit card counters. You will use two decks of cards in your simulation (104 cards total). There are only two players, you and the dealer. Each player receives two cards to begin play. The cards are worth their face value in points for 2–10, 10 points for face cards (Jack, Queen, and King), and either 1 or 11 points for aces. The object of the game is to obtain a total as close to 21 as possible without going over (called busted), with your total more than the dealer's.

   If the first two cards total 21 (ace-10 or ace-face card), this is called blackjack and is an automatic winner (unless both you and the dealer have blackjack, in which case it is a tie, or push, and your bet remains on the table). Winning via blackjack pays you 3 to 2, or 1.5 to 1 (a $1 bet reaps $1.50 and you don't lose the $1 you bet).

   If neither you nor the dealer has blackjack, you (the player) can take as many cards as you want, one at a time, to try to get as close to 21 as possible. If you go over 21, you lose and the game ends. When you are satisfied with your score, you stand. The dealer then draws cards according to the following rules:

   The dealer stands on 17, 18, 19, 20, or 21. The dealer must draw a card if the total is 16 or less. The dealer always counts aces as 11 unless it causes him or her to bust, in which case it is counted as a 1. For example, an ace-6 combination for the dealer is 17, not 7 (the dealer has no option), and the dealer must stand on 17. However, if the dealer has an ace-4 (for 15) and draws a King, then the new total is 15, because the ace reverts to its value of 1 (so as not to go over 21). The dealer would then draw another card.

   If the dealer goes over 21, you win (including your bet money; you gain $1 for every $1 you bet). If the dealer's total exceeds your total, you lose all the money you bet. If the dealer's total equals your total, it is a push (no money exchanges hands; you don't lose your bet, but neither do you gain any money).

   What makes the game exciting in a casino is that the dealer's original two cards are one up, one down, so you do not know the dealer's total and must play the odds based on the one card showing. You do not need to incorporate this twist into your simulation for this project. Here's what you are required to do:

   Run through 12 sets of 2 decks playing the game. You have an unlimited bankroll (don't you wish!) and bet $2 on each hand. Each time the 2 decks run out, the hand in play continues with 2 fresh decks (104 cards). At that point, record your standing (plus or minus X dollars). Then start again at 0 for the next deck. So your output will be the 12 results from playing each of the 12 sets of decks, which you can then average or total to determine your overall performance.

What about *your* strategy? That's up to you! But here's the catch—you will assume that you can see neither of the dealer's cards (so you have no idea what cards the dealer has). Choose a strategy to play, and then play it throughout the entire simulation. (Blackjack enthusiasts can consider implementing doubling down and splitting pairs into their simulation, but this is not necessary.)

Provide your instructor with the simulation algorithm, computer code, and output results from each of the 12 sets of decks.

2. *Darts*—Construct and perform a Monte Carlo simulation of a darts game. The rules are

Dart board area	Points
Bull's-eye	50
Yellow ring	25
Blue ring	15
Red ring	10
White ring	5

From the origin (the center of the bull's-eye) the radius of each ring is as follows:

Ring	Thickness (in.)	Distance to outer ring edge from the origin (in.)
Bull's-eye	1.0	1.0
Yellow	1.5	2.5
Blue	2.5	5.0
Red	3.0	8.0
White	4.0	12.0

The round board has a radius of 1 ft (12 in.).

Make an assumption about the distribution of how the darts hit on the board. Write an algorithm and code it in the computer language of your choice. Run 1000 simulations to determine the mean score for throwing five darts. Also determine which ring has the highest expected value (point value times the probability of hitting that ring).

3. *Craps*—Construct and perform a Monte Carlo simulation of the popular casino game of craps. The rules are as follows:

There are two basic bets in craps, pass and don't pass. In the *pass* bet you wager that the shooter (the person throwing the dice) will win; in the *don't pass* bet, you wager that the shooter will lose. We will play by the rule that on an initial roll of 12 (boxcars), both pass and don't pass bets are losers. Both are even-money bets.

Conduct of the game:

Roll a 7 or 11 on the first roll: Shooter wins (pass bets win and don't pass bets lose).

Roll a 12 on the first roll: Shooter loses (boxcars, pass and don't pass bets lose).

Roll a 2 or 3 on the first roll: Shooter loses (pass bets lose, don't pass bets win).

Roll 4, 5, 6, 8, 9, 10 on the first roll: This becomes the point. The object then becomes to roll the point again before rolling a 7.

The shooter continues to roll the dice until the point or a 7 appears. Pass bettors win if the shooter rolls the point again before rolling a 7. Don't pass bettors win if the shooter rolls a 7 before rolling the point again.

Write an algorithm and code it in the computer language of your choice. Run the simulation to estimate the probability of winning a pass bet and the probability of winning a don't pass bet. Which is the better bet? As the number of trials increases, to what do the probabilities converge?

4. *Horse Race*—Construct and perform a Monte Carlo simulation of a horse race. You can be creative here and use odds from the newspaper or simulate the Mathematical Derby with the following entries and odds:

Mathematical Derby

Entry's name	Odds
Euler's Folly	7–1
Leapin' Leibniz	5–1
Newton Lobell	9–1
Count Cauchy	12–1
Pumped up Poisson	4–1
Loping L'Hôpital	35–1
Steamin' Stokes	15–1
Dancin' Dantzig	4–1

Construct and perform a Monte Carlo simulation of 1000 horse races. Which horse won the most races? Which horse won the fewest races? Do these results surprise you? Provide the tallies of how many races each horse won with your output.

5. *Roulette*—In American roulette, there are 38 spaces on the wheel; 0, 00, and 1 through 36. Half the spaces numbered 1–36 are red and half are black. The two spaces 0 and 00 are green.

Simulate the playing of 1000 games betting either red or black (which pay even money, 1:1). Bet $1 on each game and keep track of your earnings. What are the earnings per game betting red/black according to your simulation? What was your longest winning streak? Longest losing streak?

Simulate 1000 games betting green (pays 17:1; if you win, you add $17 to your kitty; and if you lose, you lose $1). What are your earnings per game betting green according to your simulation? How does it differ from your earnings betting red/black? What was your longest winning streak betting green? Longest losing streak? Which strategy do you recommend using, and why?

6. *The Price Is Right*—On the popular TV game show "The Price Is Right," at the end of each half-hour, the three winning contestants face off in what is called the Showcase Showdown. The game consists of spinning a large wheel with 20 spaces on which the pointer can land, numbered in 5-cent increments from $0.05 to $1. The contestant who has won the least amount of money at this point in the show spins first, followed by the one who has won the next most, followed by the biggest winner for that half-hour.

    The object of the game is to get as close to $1 as possible without going over that amount with a maximum of two spins. If the first player does not go over, the other two will use one or both spins in an attempt to overtake the leader.

    But what of the person spinning first? How high a value on the first spin does he or she need to not want a second spin? Remember, the person can lose if

    a. Either of the other two players surpasses the player's total or

    b. The player spins again and goes over $1.

7. *Let's Make a Deal*—You are dressed to kill in your favorite costume and the host picks you out of the audience. You are offered the choice of three wallets. Two wallets contain a single $50 bill, and the third contains a $1,000 bill. You choose one of the wallets, 1, 2, or 3. The host, who knows which wallet contains the $1,000, then shows you one of the other two wallets, which has $50 inside. The host does this purposely, because he has at least one wallet with $50 inside. If he also has the $1,000 wallet, he simply shows you the $50 one he holds. Otherwise, he shows you one of his two $50 wallets. The host then asks you if you want to trade your choice for the one he's still holding. Should you trade?

    Develop an algorithm and construct a computer simulation to support your answer.

7.4   *INVENTORY MODEL: GASOLINE AND CONSUMER DEMAND*

In the previous section we began modeling probabilistic behaviors using Monte Carlo simulation. In this section we will learn a method to approximate more probabilistic processes. Additionally, we will check to determine how well the simulation duplicates the process being studied. We begin by considering an inventory control problem.

You have been hired as a consultant by a chain of gasoline stations to determine how often and how much gasoline should be delivered to the various stations. Each time gasoline is delivered, a cost of d dollars is incurred, which is in addition to the cost of gasoline and is independent of the amount delivered. The gasoline stations are near interstate highways, so demand is fairly constant. Other factors determining the costs include the capital tied up in the inventory, the amortization costs of equipment, insurance, taxes, and security measures. We assume that, in the short run, the demand and price of gasoline are constant for each station, yielding a constant total revenue as long as the station does not run out of gasoline. Because total profit is total revenue minus total cost, and total revenue is constant by assumption, total profit can be maximized by minimizing total cost. Thus, we identify the following problem: *Minimize the average daily cost of delivering and storing sufficient gasoline at each station to meet consumer demand.*

After discussing the relative importance of the various factors determining the average daily cost, we develop the following model:

$$\text{average daily cost} = f(\text{storage costs, delivery costs, demand rate})$$

Turning our attention to the various submodels, we argue that although the cost of storage may vary with the amount stored, it is reasonable to assume the cost per unit stored would be constant over the range of values under consideration. Similarly, the delivery cost is assumed constant per delivery, independent of the amount delivered, over the range of values being considered. Plotting the daily demand for gasoline at a particular station is likely to give a graph similar to the one shown in Figure 7.4a. If the frequency of each demand level over a fixed time period (say, 1 yr) is plotted, then a plot similar to that shown in Figure 7.4b might be obtained.

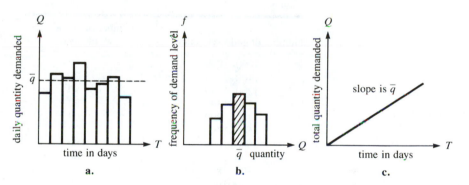

FIGURE 7.4 A constant demand rate

If demands are fairly tightly packed about the most frequently occurring demand, then we would accept the daily demand as being constant. In some cases it may be reasonable to assume a constant demand. Finally, even though the demands occur in discrete time periods, a continuous submodel for demand can be taken for the purposes of simplification. A continuous submodel is depicted in Figure 7.4c, where the slope of the line represents the constant daily demand. Notice the importance of each of the preceding assumptions in producing the linear submodel.

From these assumptions we will construct in Chapter 8 an analytical model for the average daily cost and use it to compute an optimal time between deliveries as well as an optimal delivery quantity:

$$T^* = \sqrt{\frac{2d}{sr}}$$
$$Q^* = rT^*$$

where

$$T^* = \text{optimal time between deliveries in days}$$
$$Q^* = \text{optimal delivery quantity of gasoline in gallons}$$
$$r = \text{demand rate in gallons per day}$$
$$d = \text{delivery cost in dollars per delivery}$$
$$s = \text{storage cost per gallon per day}$$

We will see how the analytical model depends heavily on a set of conditions that, although reasonable in some cases, would never be met precisely in the real world. It is difficult to develop analytical models that take into account the probabilistic nature of the submodels.

Suppose we decide to check our submodel for constant demand rate by inspecting the sales for the last 1000 days at a particular station. Thus, the data displayed in Table 7.6 are collected.

TABLE 7.6 History of demand at a particular gasoline station

Number of gallons demanded	Number of occurrences (in days)
1000–1099	10
1100–1199	20
1200–1299	50
1300–1399	120
1400–1499	200
1500–1599	270
1600–1699	180
1700–1799	80
1800–1899	40
1900–1999	30
	1000

For each of the 10 intervals of demand levels give in Table 7.6, compute the relative frequency of occurrence by dividing the number of occurrences by the total number of days, 1000. This computation results in an estimate of the probability of occurrence for each demand level. These probabilities are displayed in Table 7.7 and plotted in histogram form in Figure 7.5.

If we are satisfied with the assumption of a constant demand rate, we might estimate this rate at 1550 gal per day (from Figure 7.5). Then the analytical model could be used to compute the optimal time between deliveries and the delivery quantities from the delivery and storage costs.

Suppose, however, we are not satisfied with the assumption of constant daily demand. How could we simulate the submodel for the demand suggested by Figure 7.5? First we could build a cumulative histogram by consecutively adding together the probabilities of each individual demand level, as displayed in Figure 7.6. Note in Figure 7.6 that the difference in height between adjacent columns

TABLE 7.7 Probability of the occurrence of each demand level

Number of gallons demanded	Probability of occurrence
1000–1099	0.01
1100–1199	0.02
1200–1299	0.05
1300–1399	0.12
1400–1499	0.20
1500–1599	0.27
1600–1699	0.18
1700–1799	0.08
1800–1899	0.04
1900–1999	0.03
	1.00

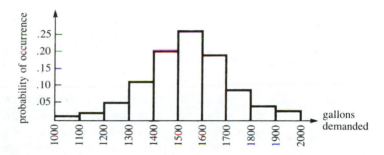

FIGURE 7.5 The relative frequency of each of the demand intervals in Table 7.7

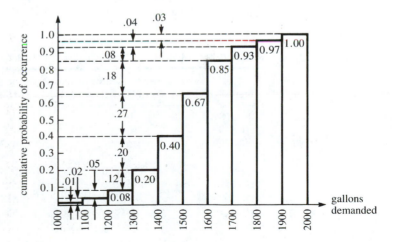

FIGURE 7.6 A cumulative histogram of the demand submodel from the data in Table 7.7

represents the probability of occurrence of the subsequent demand interval. Thus we can construct a correspondence between the numbers in the interval $0 \leq x \leq 1$ and the relative occurrence of the various demand intervals. This correspondence is displayed in Table 7.8.

TABLE 7.8 Using random numbers uniformly distributed over $0 \leq x \leq 1$ to duplicate the occurrence of the various demand intervals

Random number	Corresponding demand	Percent occurrence
$0 \leq x < 0.01$	1000–1099	0.01
$0.01 \leq x < 0.03$	1100–1199	0.02
$0.03 \leq x < 0.08$	1200–1299	0.05
$0.08 \leq x < 0.20$	1300–1399	0.12
$0.20 \leq x < 0.40$	1400–1499	0.20
$0.40 \leq x < 0.67$	1500–1599	0.27
$0.67 \leq x < 0.85$	1600–1699	0.18
$0.85 \leq x < 0.93$	1700–1799	0.08
$0.93 \leq x < 0.97$	1800–1899	0.04
$0.97 \leq x \leq 1.00$	1900–1999	0.03

Thus, if the numbers between 0 and 1 are randomly generated, so that each number has an equal probability of occurring, the histogram of Figure 7.5 can be approximated. Using a random number generator on a handheld programmable calculator, we generated random numbers between 0 and 1, and then used the assignment procedure suggested by Table 7.7 to determine the demand interval corresponding to each random number. The results for 1,000 and 10,000 trials are presented in Table 7.9.

For the gasoline inventory problem, we ultimately want to be able to determine a specific demand, rather than a demand interval, for each day simulated. How

TABLE 7.9 A Monte Carlo approximation of the demand submodel

	Number of occurrences/expected no. of occurrences	
Interval	1,000 trials	10,000 trials
1000–1099	8/10	91/100
1100–1199	16/20	198/200
1200–1299	46/50	487/500
1300–1399	118/120	1205/1200
1400–1499	194/200	2008/2000
1500–1599	275/270	2681/2700
1600–1699	187/180	1812/1800
1700–1799	83/80	857/800
1800–1899	34/40	377/400
1900–1999	39/30	284/300
	1,000/1,000	10,000/10,000

can this be accomplished? There are several alternatives. Consider the plot of the midpoints of each demand interval as displayed in Figure 7.7. Because we want a continuous model capturing the trend of the plotted data, we can use methods discussed in Chapter 6.

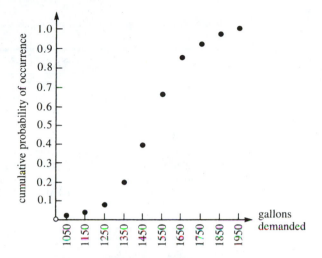

FIGURE 7.7 A cumulative plot of the demand submodel displaying only the center point of each interval

In many instances, especially where the subintervals are small and the data fairly approximate, a linear spline model is suitable. A linear spline model for the data in Figure 7.7 is presented in Figure 7.8, and the individual spline functions are given in Table 7.10. The interior spline functions—$S_2(q)$ through $S_9(q)$—were computed by passing a line through the adjacent data points. $S_1(q)$ was computed by passing a line through $(1000, 0)$ and the first data point $(1050, 0.01)$. $S_{10}(q)$ was

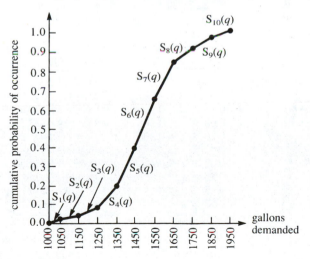

FIGURE 7.8 A linear spline model for the demand submodel

computed by passing a line through the points (1850, 0.97) and (2000, 1.00). Note that if we use the midpoints of the intervals, we have to make a decision on how to construct the two exterior splines. If the intervals are small, it is usually easy to construct a linear spline function that captures the trend of the data.

TABLE 7.10 Linear splines for the empirical demand submodel

Demand interval	Linear spline
$1000 \leq q < 1050$	$S_1(q) = 0.0002q - 0.2$
$1050 \leq q < 1150$	$S_2(q) = 0.0002q - 0.2$
$1150 \leq q < 1250$	$S_3(q) = 0.0005q - 0.545$
$1250 \leq q < 1350$	$S_4(q) = 0.0012q - 1.42$
$1350 \leq q < 1450$	$S_5(q) = 0.002q - 2.5$
$1450 \leq q < 1550$	$S_6(q) = 0.0027q - 3.515$
$1550 \leq q < 1650$	$S_7(q) = 0.0018q - 2.12$
$1650 \leq q < 1750$	$S_8(q) = 0.0008q - 0.47$
$1750 \leq q < 1850$	$S_9(q) = 0.0004q + 0.23$
$1850 \leq q \leq 2000$	$S_{10}(q) = 0.0002q + 0.6$

Now suppose we wish to simulate a daily demand for a given day. To do this, we generate a random number x between 0 and 1 and compute a corresponding demand q. That is, x is the independent variable from which a unique corresponding q is calculated. This calculation is possible because the function depicted in Figure 7.8 is strictly increasing. (Think about whether this situation will always be the case.) Thus, the problem is to find the inverse functions for the splines listed in Table 7.10. For instance, given $x = S_1(q) = 0.0002q - 0.2$, we can solve for $q = (x + 0.2)5000$. In the case of linear splines, it is easy to find the inverse functions summarized in Table 7.11.

TABLE 7.11 Inverse linear splines provide for the daily demand as a function of a random number in [0, 1]

Random number	Inverse linear spline
$0 \leq x < 0.01$	$q = (x + 0.2)5000$
$0.01 \leq x < 0.03$	$q = (x + 0.2)5000$
$0.03 \leq x < 0.08$	$q = (x + 0.545)2000$
$0.08 \leq x < 0.20$	$q = (x + 1.42)833.3\underline{3}$
$0.20 \leq x < 0.40$	$q = (x + 2.5)500$
$0.40 \leq x < 0.67$	$q = (x + 3.515)370.37$
$0.67 \leq x < 0.85$	$q = (x + 2.12)555.5\underline{5}$
$0.85 \leq x < 0.93$	$q = (x + 0.47)1250$
$0.93 \leq x < 0.97$	$q = (x - 0.23)2500$
$0.97 \leq x \leq 1.00$	$q = (x - 0.6)5000$

Let's illustrate how Table 7.11 can be used to represent the daily demand submodel. To simulate a demand for a given day, we generate a random number between 0 and 1, say $x = 0.214$. Because $0.20 \leq 0.214 < 0.40$, the spline

$q = (x + 2.5)500$ is used to compute $q = 1357$. Thus 1357 gal is the simulated demand for that day.

Note that the inverse splines presented in Table 7.11 could have been constructed directly from the data in Figure 7.6 by choosing x as the independent variable instead of q. We will follow this procedure later when computing the cubic spline demand submodel. (The preceding development was presented to ease your understanding of the process and also because it mimics what you will do after studying probability.) Figure 7.6 is an example of a cumulative distribution function. Many types of behavior approximate well-known probability distributions, which can be used as the basis for Figure 7.6 rather than experimental data. The inverse function must then be found to use as the demand submodel in the simulation, and this may prove to be difficult. In such cases, the inverse function is approximated with an empirical model such as a linear spline or cubic spline. For an excellent introduction to some types of behavior that follow well-known probability distributions, see UMAP 340, "The Poisson Random Process," by Carroll Wilde, listed in the Projects at the end of this section.

If we want a smooth, continuous submodel for demand, we can construct a cubic spline submodel. We will construct the splines directly as a function of the random number x. That is, using a computer program, we calculate the cubic splines for the following data points:

x	0	0.01	0.03	0.08	0.2	0.4	0.67	0.85	0.93	0.97	1.0
q	1000	1050	1150	1250	1350	1450	1550	1650	1750	1850	2000

The splines are presented in Table 7.12. If the random number $x = 0.214$ is generated, the empirical cubic spline model yields the demand $q = 1350 + 715.46(0.014) - 1572.5(0.014)^2 + 2476(0.014)^3 = 1359.7$ gal.

TABLE 7.12 An empirical cubic spline model for demand

Random number	Cubic spline
$0 \leq x < 0.01$	$S_1(x) = 1000 + 4924.92x + 750788.75x^3$
$0.01 \leq x < 0.03$	$S_2(x) = 1050 + 5150.18(x - 0.01) + 22523.66(x - 0.01)^2 - 1501630.8(x - 0.01)^3$
$0.03 \leq x < 0.08$	$S_3(x) = 1150 + 4249.17(x - 0.03) - 67574.14(x - 0.03)^2 + 451815.88(x - 0.03)^3$
$0.08 \leq x < 0.20$	$S_4(x) = 1250 + 880.37(x - 0.08) + 198.24(x - 0.08)^2 - 4918.74(x - 0.08)^3$
$0.20 \leq x < 0.40$	$S_5(x) = 1350 + 715.46(x - 0.20) - 1572.51(x - 0.20)^2 + 2475.98(x - 0.20)^3$
$0.40 \leq x < 0.67$	$S_6(x) = 1450 + 383.58(x - 0.40) - 86.92(x - 0.40)^2 + 140.80(x - 0.40)^3$
$0.67 \leq x < 0.85$	$S_7(x) = 1550 + 367.43(x - 0.67) + 27.12(x - 0.67)^2 + 5655.69(x - 0.67)^3$
$0.85 \leq x < 0.93$	$S_8(x) = 1650 + 926.92(x - 0.85) + 3081.19(x - 0.85)^2 + 11965.43(x - 0.85)^3$
$0.93 \leq x < 0.97$	$S_9(x) = 1750 + 1649.66(x - 0.93) + 5952.90(x - 0.93)^2 + 382645.25(x - 0.93)^3$
$0.97 \leq x \leq 1.00$	$S_{10}(x) = 1850 + 3962.58(x - 0.97) + 51870.29(x - 0.97)^2 - 576334.88(x - 0.97)^3$

An empirical submodel for demand can be constructed in a variety of other ways. For example, rather than using the intervals for gallons demanded as given in Table 7.6, we can use smaller intervals. If the intervals are small enough, the midpoint of an interval could itself be a reasonable approximation to the demand for the entire interval. Thus a cumulative histogram similar to that in Figure 7.6

could serve as a submodel directly. If preferred, a continuous submodel could be constructed readily from the refined data.

The purpose of our discussion has been to demonstrate how a submodel for a probabilistic behavior can be constructed using Monte Carlo simulation and experimental data. Now let's see how the inventory problem can be simulated in general terms.

An inventory strategy consists of specifying a delivery quantity Q and a time T between deliveries, given values for storage cost per gallon per day s and a delivery cost d. If s and d are known, then a specific inventory strategy can be tested using a Monte Carlo simulation algorithm, as follows.

Summary of Monte Carlo Inventory Algorithm Terms

Q	Delivery quantity of gasoline in gallons
T	Time between deliveries in days
I	Current inventory in gallons
d	Delivery cost in dollars per delivery
s	Storage cost per gallon per day
C	Total running cost
c	Average daily cost
N	Number of days to run the simulation
K	Days remaining in the simulation
x_i	A random number in the interval $[0, 1]$
q_i	A daily demand
Flag	An indicator used to terminate the algorithm

Monte Carlo Inventory Algorithm

Input Q, T, d, s, N

Output c

Step 1   Initialize:
$$K = N$$
$$I = 0$$
$$C = 0$$
$$\text{Flag} = 0$$

Step 2   Begin the next inventory cycle with a delivery:
$$I = I + Q$$
$$C = C + d$$

Step 3   Determine if the simulation will terminate during this cycle:
If $T \geq K$, then set $T = K$ and Flag $= 1$

Step 4   Simulate each day in the inventory cycle (or portion remaining):
For $i = 1, 2, \ldots, T$, do Steps 5–9

Step 5   Generate the random number x_i.

Step 6   Compute q_i using the demand submodel.

Step 7   Update the current inventory: $I = I - q_i$.

Step 8   Compute the daily storage cost and total running cost, unless the inventory has been depleted: If $I \le 0$, then set $I = 0$ and GOTO Step 9.
Else $C = C + I * s$.

Step 9   Decrement the number of days remaining in the simulation:

$$K = K - 1$$

Step 10   If Flag $= 0$, then GOTO Step 2. Else GOTO Step 11.

Step 11   Compute the average daily cost: $c = C/N$.

Step 12   Output c.
STOP

Various strategies can now be tested with the algorithm to determine the average daily costs. You probably want to refine the algorithm to keep track of other measures of effectiveness, such as unsatisfied demands and number of days without gasoline, as suggested in the following problem set.

7.4 PROBLEMS

1. Modify the inventory algorithm to keep track of unfilled demands and the total number of days the gasoline station is without gasoline for at least part of the day.

2. Most gasoline stations have a storage capacity Q_{max} that cannot be exceeded. Refine the inventory algorithm to take this consideration into account. Because of the probabilistic nature of the demand submodel at the end of the inventory cycle, there might still be significant amounts of gasoline remaining. If several cycles occur in succession, the excess might build up to Q_{max}. Because there is a financial cost in carrying excess inventory, this situation would be undesirable. What alternatives can you suggest? Modify the inventory algorithm to take your alternatives into account.

3. In many situations the time T between deliveries and the order quantity Q is not fixed. Instead, an order is placed for a specific amount of gasoline. Depending on how many orders are placed in a given time interval, the time to fill an order varies. You have no reason to believe the performance of the delivery operation will change. Therefore, you have examined records for the last 100 deliveries and found the following lag times or extra days required to fill your order:

Lag time (in days)	Number of occurrences
2	10
3	25
4	30
5	20
6	13
7	2
Total:	100

Construct a Monte Carlo simulation for the lag time submodel. If you have a handheld calculator or a computer available, test your submodel by running 1000 trials and comparing the number of occurrences of the various lag times with the historical data.

4. Problem 3 suggests an alternative inventory strategy. When the inventory reaches a certain level (an order point), an order can be placed for an optimal amount of gasoline. Construct an algorithm that simulates this process and incorporates probabilistic submodels for demand and lag times. How could you use this algorithm to search for the optimal order point and the optimal order quantity?

5. In the case that a gasoline station runs out of gas, the customer is simply going to go to another station. In many situations (name a few), however, some customers will place a back order or collect a rain check. If the order is not filled within a time period varying from customer to customer in a probabilistic fashion, the customer will cancel his or her order. Suppose we examine historical data for 100 customers and find the data shown in Table 7.13. That is, 200 customers will not even place an order, and an additional 150 customers will cancel if the order is not filled within 1 day.

TABLE 7.13 Hypothetical data for a backorder submodel

Number of days customer is willing to wait before canceling	Number of occurrences	Cumulative occurrences
0	200	200
1	150	350
2	200	550
3	200	750
4	150	900
5	50	950
6	50	1000
	1000	

a. Construct a Monte Carlo simulation for the back order submodel. If you have a calculator or computer available, test your submodel by running 1000 trials and comparing the number of occurrences of the various cancellations with the historical data.

b. Consider the algorithm you modified in Problem 1. Further modify the algorithm to consider back orders. Do you think back orders should be penalized in some fashion? If so, how would you do it?

7.4 PROJECTS

1. Complete the requirements of UMAP module 340, "The Poisson Random Process," by Carroll O. Wilde. Probability distributions are introduced to obtain practical information on random arrival patterns, interarrival times or gaps between arrivals, waiting line buildup, and service loss rates. The Poisson distribution, the exponential distribution, and Erlang's formulas are used. The module requires an introductory probability course, the ability to use summation notation, and basic concepts of the derivative and the integral from calculus. Prepare a 10-minute summary of the module for a classroom presentation.

2. Assume a storage cost of $0.001 per gallon per day and a delivery charge of $500 per delivery. Construct a computer code of the algorithm you constructed in Problem 4, and compare various order points and order quantity strategies.

7.5   QUEUING MODELS

Example 1   A Harbor System

Consider a small harbor with unloading facilities for ships. Only one ship can be unloaded at any one time. Ships arrive for unloading of cargo at the harbor, and the time between the arrival of successive ships varies from 15 to 145 min. The unloading time required for a ship depends on the type and amount of cargo and varies from 45 to 90 min. We seek answers to the following questions:

1. What is the average and maximum time per ship in the harbor?
2. If the *waiting time* for a ship is the time between its arrival and the start of unloading, what are the average and maximum waiting times per ship?
3. What percent of the time are the unloading facilities idle?

To get some reasonable answers, we can simulate the activity in the harbor using a computer or programmable calculator. We assume the arrival times between successive ships and unloading time per ship are uniformly distributed over their respective time intervals. For instance, the arrival time between ships can be any integer between 15 and 145, and any integer within that interval can appear with equal likelihood. Before giving a general algorithm to simulate the harbor system, let's consider a hypothetical situation with five ships.

We have the following data for each ship:

	Ship 1	Ship 2	Ship 3	Ship 4	Ship 5
Time between successive ships	20	30	15	120	25
Unload time	55	45	60	75	80

Because Ship 1 arrives 20 min after the clock commences at $t = 0$ min, the harbor facilities are idle for 20 min at the start. Ship 1 immediately begins to unload. The unloading takes 55 min, and meanwhile Ship 2 arrives on the scene at $t = 20 + 30 = 50$ min after the clock begins. Ship 2 cannot start to unload until Ship 1 finishes unloading at $t = 20 + 55 = 75$ min. This means that Ship 2 must wait $75 - 50 = 25$ min before unloading begins. The situation is depicted in the following timeline diagram:

Now before Ship 2 starts to unload, Ship 3 arrives at time $t = 50 + 15 = 65$ min. Because the unloading of Ship 2 starts at $t = 75$ min and takes 45 min, unloading of Ship 3 cannot start until $t = 75 + 45 = 120$ min, when Ship 2 is finished. Thus, Ship 3 must wait $120 - 65 = 55$ min. The situation is depicted in the next timeline diagram:

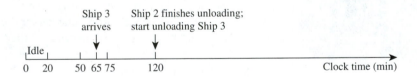

Ship 4 doesn't arrive in the harbor until $t = 65 + 120 = 185$ min. Ship 3 has already finished unloading at $t = 120 + 60 = 180$ min, and the harbor facilities are idle for $185 - 180 = 5$ min. The unloading of Ship 4 commences immediately upon its arrival, as depicted in the next diagram:

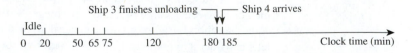

Finally Ship 5 arrives at $t = 185 + 25 = 210$ min, before Ship 4 finishes unloading at $t = 185 + 75 = 260$ min. Thus Ship 5 must wait $260 - 210 = 50$ min before it starts to unload. The simulation is complete when Ship 5 finishes unloading at $t = 260 + 80 = 340$ min. The final situation is shown in the next diagram:

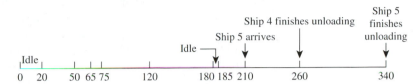

In Figure 7.9, we summarize the waiting and unloading times for each of the five hypothetical ship arrivals. In Table 7.14, we summarize the results of the entire simulation of the five hypothetical ships. Note that the total waiting time spent by all five ships before unloading is 130 min. This waiting time represents a cost to the shipowners and is a source of customer dissatisfaction with the docking facilities. On the other hand, the docking facility has only 25 min of total idle time. It is in use 315 out of the total 340 min in the simulation, or approximately 93% of the time.

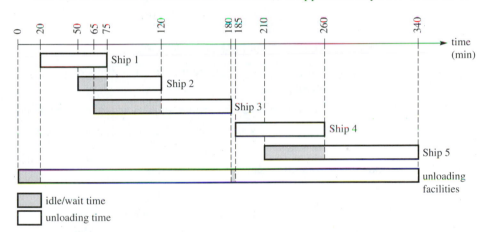

FIGURE 7.9 Idle and unloading times for the ships and docking facilities

TABLE 7.14 Summary of the harbor system simulation

Ship no.	Random time between ship arrivals	Arrival time	Start service	Queue length at arrival	Wait time	Random unload time	Time in harbor	Dock idle time
1	20	20	20	0	0	55	55	20
2	30	50	75	1	25	45	70	0
3	15	65	120	2	55	60	115	0
4	120	185	185	0	0	75	75	5
5	25	210	260	1	50	80	130	0
Totals (if appropriate):					130			25
Averages (if appropriate):					26	63	89	

Note: All times are given in minutes after the start of the clock at time $t = 0$.

Suppose the owners of the docking facilities are concerned with the quality of service they are providing and want various management alternatives to be evaluated to see if improvement in service justifies the added cost. Several statistics can help in evaluating the quality of the service. For example, the maximum time a ship spends in the harbor is 130 min by Ship 5, whereas the average is 89 min. Generally customers are sensitive to the amount of time spent waiting. In this example, the maximum amount of time spent waiting for a facility is 55 min, whereas the average time spent waiting is 26 min. Some customers are apt to take their business elsewhere if queues are too long. In this case, the longest queue is 2. The following Monte Carlo simulation algorithm will compute such statistics to assess various management alternatives.

Summary of Harbor System Algorithm Terms

$between_i$ Time between successive arrivals of Ships i and $i - 1$ (a random integer varying between 15 and 145 min)

$arrive_i$ Time from start of clock at $t = 0$ when Ship i arrives at the harbor for unloading

$unload_i$ Time required to unload Ship i at the dock (a random integer varying between 45 and 90 min)

$start_i$ Time from start of clock at which Ship i commences its unloading

$idle_i$ Time for which dock facilities are idle immediately *before* commencement of unloading Ship i

$wait_i$ Time Ship i waits in the harbor after arrival before unloading commences

$finish_i$ Time from start of clock at which service for Ship i is completed at the unloading facilities

$harbor_i$ Total time Ship i spends in the harbor

HARTIME   Average time per ship in the harbor

MAXHAR   Maximum time of a ship in the harbor

WAITIME   Average waiting time per ship before unloading

MAXWAIT   Maximum waiting time of a ship

IDLETIME   Percent of total simulation time unloading facilities are idle

Harbor System Simulation Algorithm

Input   Total number n of ships for the simulation.

Output   HARTIME, MAXHAR, WAITIME, MAXWAIT, and IDLETIME.

Step 1   Randomly generate $between_1$ and $unload_1$. Then set $arrive_1 = between_1$.

Step 2   Initialize all output values:

$$HARTIME = unload_1, \qquad MAXHAR = unload_1,$$

$$WAITIME = 0, \qquad MAXWAIT = 0, \qquad IDLETIME = arrive_1$$

Step 3   Calculate finish time for unloading of Ship_1:

$$\text{finish}_1 = \text{arrive}_1 + \text{unload}_1$$

Step 4   For $i = 2, 3, \ldots, n$, do Steps 5–16.

Step 5   Generate the random pair of integers between_i and unload_i over their respective time intervals.

Step 6   Assuming the time clock begins at $t = 0$ minutes, calculate the time of arrival for Ship_i:

$$\text{arrive}_i = \text{arrive}_{i-1} + \text{between}_i$$

Step 7   Calculate the time difference between the arrival of Ship_i and the finish time for unloading the previous Ship_{i-1}:

$$\text{timediff} = \text{arrive}_i - \text{finish}_{i-1}$$

Step 8   For nonnegative timediff, the unloading facilities are idle:

$$\text{idle}_i = \text{timediff} \quad \text{and} \quad \text{wait}_i = 0$$

For negative timediff, Ship_i must wait before it can unload:

$$\text{wait}_i = -\text{timediff} \quad \text{and} \quad \text{idle}_i = 0$$

Step 9   Calculate the start time for unloading Ship_i:

$$\text{start}_i = \text{arrive}_i + \text{wait}_i$$

Step 10   Calculate the finish time for unloading Ship_i:

$$\text{finish}_i = \text{start}_i + \text{unload}_i$$

Step 11   Calculate the time in harbor for Ship_i:

$$\text{harbor}_i = \text{wait}_i + \text{unload}_i$$

Step 12   Sum harbor_i into total harbor time HARTIME for averaging.

Step 13   If $\text{harbor}_i >$ MAXHAR, then set MAXHAR $= \text{harbor}_i$. Otherwise leave MAXHAR as is.

Step 14   Sum wait_i into total waiting time WAITIME for averaging.

Step 15   Sum idle_i into total idle time IDLETIME.

Step 16   If $\text{wait}_i >$ MAXWAIT, then set MAXWAIT $= \text{wait}_i$. Otherwise leave MAXWAIT as is.

Step 17   Set HARTIME $=$ HARTIME$/n$, WAITIME $=$ WAITIME$/n$, and IDLETIME $=$ IDLETIME$/\text{finish}_n$.

Step 18   OUTPUT (HARTIME, MAXHAR, WAITIME, MAXWAIT, IDLETIME)
STOP

Table 7.15 gives the results, according to the preceding algorithm, of six independent simulation runs of 100 ships each.

TABLE 7.15 Harbor system simulation results for 100 ships

Average time of a ship in the harbor	106	85	101	116	112	94
Maximum time of a ship in the harbor	287	180	233	280	234	264
Average waiting time of a ship	39	20	35	50	44	27
Maximum waiting time of a ship	213	118	172	203	167	184
Percentage of time dock facilities are idle	0.18	0.17	0.15	0.20	0.14	0.21

Note: All times are given in minutes. Time between successive ships is 15–145 min. Unloading time per ship varies from 45 to 90 min.

Now suppose you are a consultant for the owners of the docking facilities. What would be the effect of hiring additional labor or acquiring better equipment for unloading cargo so the unloading time interval is reduced to between 35 and 75 min per ship? Table 7.16 gives the results based on our simulation algorithm.

TABLE 7.16 Harbor system simulation results for 100 ships

Average time of a ship in the harbor	74	62	64	67	67	73
Maximum time of a ship in the harbor	161	116	167	178	173	190
Average waiting time of a ship	19	6	10	12	12	16
Maximum waiting time of a ship	102	58	102	110	104	131
Percentage of time dock facilities are idle	0.25	0.33	0.32	0.30	0.31	0.27

Note: All times are given in minutes. Time between successive ships is 15–145 min. Unloading time per ship varies from 35 to 75 min.

You can see from Table 7.16 that a reduction of the unloading time per ship by 10–15 min decreases the time ships spend in the harbor, especially the waiting times. However, the percent of the total time during which the dock facilities are idle nearly doubles. The situation is favorable for shipowners because it increases the availability of each ship for hauling cargo over the long run. Thus the traffic coming into the harbor is likely to increase. If the traffic increases to the extent that the time between successive ships is reduced to between 10 and 120 min, the simulated results are shown in Table 7.17. We can see from this table that the ships again spend more time in the harbor with the increased traffic, but now harbor facilities are idle much less of the time. Moreover, both the shipowners and the dock owners are benefiting from the increased business.

TABLE 7.17 Harbor system simulation results for 100 ships

Average time of a ship in the harbor	114	79	96	88	126	115
Maximum time of a ship in the harbor	248	224	205	171	371	223
Average waiting time of a ship	57	24	41	35	71	61
Maximum waiting time of a ship	175	152	155	122	309	173
Percentage of time dock facilities are idle	0.15	0.19	0.12	0.14	0.17	0.06

Note: All times are given in minutes. Time between successive ships is 10–120 min. Unloading time per ship varies from 35 to 75 min.

Suppose that we are not satisfied with the assumption that the arrival time between ships (i.e., their interarrival times) and the unloading time per ship are uniformly distributed over the time intervals $15 \leq between_i \leq 145$ and $45 \leq unload_i \leq 90$, respectively. So it is decided to collect experimental data for the harbor system and incorporate the results in our model, as discussed for the demand submodel in the previous section. We observe (hypothetically) 1200 ships using the harbor to unload their cargoes and collect the data displayed in Table 7.18.

TABLE 7.18 Data collected for 1200 ships using the harbor facilities

Time between arrivals	Number of occurrences	Probability of occurrence	Unloading time	Number of occurrences	Probability of occurrence
15–24	11	0.009			
25–34	35	0.029			
35–44	42	0.035	45–49	20	0.017
45–54	61	0.051	50–54	54	0.045
55–64	108	0.090	55–59	114	0.095
65–74	193	0.161	60–64	103	0.086
75–84	240	0.200	65–69	156	0.130
85–94	207	0.172	70–74	223	0.185
95–104	150	0.125	75–79	250	0.208
105–114	85	0.071	80–84	171	0.143
115–124	44	0.037	85–90	109	0.091
125–134	21	0.017		1200	1.000
135–145	3	0.003			
	1200	1.000			

Note: All times are given in minutes.

Following the procedures outlined in Section 7.4, we consecutively add together the probabilities of each individual time between arrivals interval, as well as probabilities of each individual unloading time interval. These computations result in cumulative histograms depicted in Figure 7.10.

Next we use random numbers uniformly distributed over the interval $0 \leq x \leq 1$ to duplicate the various interarrival times and unloading times based on the cumulative histograms. We then use the midpoints of each interval and construct linear splines through adjacent data points. (We ask you to complete this construction in Problem 1.) Because it is easy to calculate the inverse splines directly, we do so and summarize the results in Tables 7.19 and 7.20.

Finally, we incorporate our linear spline submodels into the simulation model for the harbor system by generating $between_i$ and $unload_i$ for $i = 1, 2, \ldots, n$ in Steps 1 and 5 of our algorithm, according to the rules displayed in Tables 7.19 and 7.20. Using these submodels, Table 7.21 gives the results of six independent simulation runs of 100 ships each.

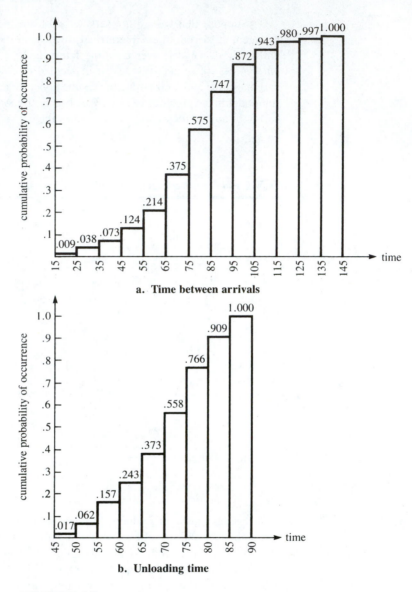

a. Time between arrivals

b. Unloading time

FIGURE 7.10 Cumulative histograms of the time between ship arrivals and the unloading times, from the data in Table 7.18

TABLE 7.19 The inverse linear splines provide for the time between arrivals of successive ships as a function of a random number in the interval [0, 1]

Random number interval	Corresponding arrival time	Inverse linear spline
$0 \leq x < 0.009$	$15 \leq b < 20$	$b = 555.6x + 15.0000$
$0.009 \leq x < 0.038$	$20 \leq b < 30$	$b = 344.8x + 16.8966$
$0.038 \leq x < 0.073$	$30 \leq b < 40$	$b = 285.7x + 19.1429$
$0.073 \leq x < 0.124$	$40 \leq b < 50$	$b = 196.1x + 25.6863$
$0.124 \leq x < 0.214$	$50 \leq b < 60$	$b = 111.1x + 36.2222$
$0.214 \leq x < 0.375$	$60 \leq b < 70$	$b = 62.1x + 46.7080$
$0.375 \leq x < 0.575$	$70 \leq b < 80$	$b = 50.0x + 51.2500$
$0.575 \leq x < 0.747$	$80 \leq b < 90$	$b = 58.1x + 46.5698$
$0.747 \leq x < 0.872$	$90 \leq b < 100$	$b = 80.0x + 30.2400$
$0.872 \leq x < 0.943$	$100 \leq b < 110$	$b = 140.8x - 22.8169$
$0.943 \leq x < 0.980$	$110 \leq b < 120$	$b = 270.3x - 144.8649$
$0.980 \leq x < 0.997$	$120 \leq b < 130$	$b = 588.2x - 456.4706$
$0.997 \leq x \leq 1.000$	$130 \leq b \leq 145$	$b = 5000.0x - 4855$

TABLE 7.20 The inverse linear splines provide for the unloading time of a ship as a functon of a random number in the interval [0, 1]

Random number interval	Corresponding unloading time	Inverse linear spline
$0 \leq x < 0.017$	$45 \leq u < 47.5$	$u = 147x + 45.000$
$0.017 \leq x < 0.062$	$47.5 \leq u < 52.5$	$u = 111x + 45.611$
$0.062 \leq x < 0.157$	$52.5 \leq u < 57.5$	$u = 53x + 49.237$
$0.157 \leq x < 0.243$	$57.5 \leq u < 62.5$	$u = 58x + 48.372$
$0.243 \leq x < 0.373$	$62.5 \leq u < 67.5$	$u = 38.46x + 53.154$
$0.373 \leq x < 0.558$	$67.5 \leq u < 72.5$	$u = 27x + 57.419$
$0.558 \leq x < 0.766$	$72.5 \leq u < 77.5$	$u = 24x + 59.087$
$0.766 \leq x < 0.909$	$77.5 \leq u < 82.5$	$u = 35x + 50.717$
$0.909 \leq x \leq 1.000$	$82.5 \leq u \leq 90$	$u = 82.41x + 7.582$

TABLE 7.21 Harbor system simulation results for 100 ships

Average time of a ship in the harbor	108	95	125	78	123	101
Maximum time of a ship in the harbor	237	188	218	133	250	191
Average waiting time of a ship	38	25	54	9	53	31
Maximum waiting time of a ship	156	118	137	65	167	124
Percentage of time dock facilities are idle	0.09	0.09	0.08	0.12	0.06	0.10

Note: Based on the data exhibited in Table 7.18. All times are given in minutes.

Example 2   *Morning Rush Hour*

In the previous example, we initially considered a harbor system with a single facility for unloading ships. Such problems are often called *single-server queues*. In this example, we consider a system with four elevators, illustrating *multiple-server queues*. We discuss the problem and present the algorithm in Appendix B.

Consider an office building with 12 floors in a metropolitan area of some city. During the morning rush hour, from 7:50 A.M. to 9:10 A.M., workers enter the lobby of the building and take an elevator to their floor. There are four elevators servicing the building. The time between arrivals of the customers at the building varies in a probabilistic manner every 0–30 sec, and on arrival each customer selects the first available elevator (numbered 1–4). When a person enters an elevator and selects a floor, the elevator waits 15 sec before closing its doors. If another person arrives within the 15-sec interval, the waiting cycle is repeated. If no person arrives within the 15-sec interval, the elevator departs to deliver all of its passengers. We assume no other passengers are picked up along the way. After delivering its last passenger, the elevator returns to the main floor, picking up no passengers on the way down. The maximum occupancy of an elevator is 12 passengers. As people arrive in the lobby and no elevator is available (because all four elevators are transporting their loads of passengers), a queue begins to form in the lobby.

The management of the building wants to provide good elevator service to its customers and is interested in exactly what service it is now giving. Some customers claim they have to wait too long in the lobby before an elevator returns. Others complain that they spend too much time riding the elevator, and still others say there is considerable congestion in the lobby during the morning rush hour. What is the real situation? Can the management resolve the complaints by a more effective means of scheduling or using the elevators? You may wish to reread the discussion in Example 4 in Section 2.2 to review some of the formulations of the elevator problem.

We wish to simulate the elevator system using an algorithm for computer implementation that will give answers to the following questions:

1. How many customers are actually being serviced in a typical morning rush hour?
2. If the *waiting time* of a person is the time the person stands in a queue—the time from arrival at the lobby until entry into an available elevator—what are the average and maximum times a person waits in a queue?
3. What is the length of the longest queue? (The answer to this question will provide the management with information about congestion in the lobby.)
4. If the *delivery time* is the time it takes a customer to reach his or her floor after arrival in the lobby, including any waiting time for an elevator to become available, what are the average and maximum delivery times?
5. What are the average and maximum times a customer actually spends in the elevator?
6. How many stops are made by each elevator? What percent of the total morning rush hour time is each elevator actually in use?

An algorithm is presented in Appendix B.

7.5 PROBLEMS

1. Using the data from Table 7.18 and the cumulative histograms of Figure 7.10, construct cumulative plots of the time between arrivals and unloading time submodels (as in Figure 7.7). Calculate equations for the linear splines over each random number interval. Compare your results with the inverse splines given in Tables 7.19 and 7.20.

2. Use a smooth polynomial to fit the data in Table 7.18 to obtain arrivals and unloading times. Compare results with those in Tables 7.19 and 7.20.

3. Modify the ship harbor system algorithm so it keeps track of the number of ships waiting in the queue.

4. Most small harbors have a maximum number of ships N_{max} that can be accommodated in the harbor area while they wait to be unloaded. If a ship cannot get into the harbor, assume it goes elsewhere to unload its cargo. Refine the ship harbor algorithm to take these considerations into account.

5. Suppose the owners of the docking facilities decide to construct a second facility to accommodate unloading of more ships. When a ship enters the harbor, it goes to the next available facility, which is facility 1 if both facilities are available. Using the same assumption for interarrival times between successive ships and unloading times as in the text example, modify the algorithm for a system with two facilities.

6. Construct a Monte Carlo simulation of a baseball game. Use individual batting statistics to simulate the probability of a single, double, triple, home run, or an out. In a more refined model, how would you handle walks, hit batters, steals, and double plays?

7.5 PROJECTS

1. Write a computer simulation to implement the ship harbor algorithm.

2. Write a computer simulation to implement a baseball game between your two favorite teams (see Problem 6).

3. Pick a traffic intersection with a traffic light. Collect data on vehicle arrival times and clearing times. Build a Monte Carlo simulation to model traffic flow at this intersection.

Chapter Eight

CONTINUOUS OPTIMIZATION MODELS

INTRODUCTION

In Chapter 5 we discussed the problem of fitting tested and accepted models to collected data. Several curve-fitting criteria were identified, which led to distinct mathematical models that we attempted to solve. In particular, the criteria of minimizing the largest absolute deviation and minimizing the sum of the absolute deviations led to mathematical models that were tedious to formulate and generally quite difficult to solve without a computer. Partly because of the difficulty in solving those mathematical models, the alternative criterion of minimizing the sum of the squared deviations was studied. Minimizing the sum of squared deviations, least squares, is an example of unconstrained optimization. In Chapter 9 we will revisit the criteria of minimizing the largest absolute deviation and minimizing the sum of the absolute deviations.

There are other situations in mathematical modeling that require us to determine the best or optimal solution. It may be the problem of determining the *maximum* profit a firm can make (as in Chapter 1) or of finding the *minimum* sum of squared deviations between a fitted model and a set of data points (as in Chapter 5). **optimization** The process of finding the best solution to such problems is known as **optimization**. As we see in this chapter, the solution of a model requiring optimization can be difficult to obtain. The study of optimization constitutes a large and interesting field of mathematics in which extensive research is being conducted. Whereas many optimization problems can be solved by a direct application of elementary single and multivariable calculus, others require the application of specialized mathematics, which is best studied in a separate course or even a sequence of courses. In this chapter we first present an overview of the field of continuous optimization. Later we present scenarios that naturally give rise to models requiring optimization.

In Sections 8.1 and 8.2 we provide a general classification of optimization problems and address situations that lead to models illustrating many types of optimization problems. In both sections the emphasis is on *model formulation*, which will allow you additional practice on the first several steps of the modeling process while simultaneously providing a preview of the kinds of problems you will learn to solve in advanced mathematics courses.

In Section 8.3 we address a special class of problems that can be solved using only elementary calculus. In the illustrative problem in that section we develop a model for determining an optimal inventory strategy. That problem is concerned with deciding in what quantities and how often should goods be ordered to minimize the total cost of carrying an inventory. The restrictions on the submodels are severe, so the sensitivity of the solutions to the assumptions is examined. (A simulation model with relaxed assumptions was constructed in Section 7.4.) The emphasis in Section 8.3 is on *model solution* and *model sensitivity*.

Section 8.4 is an optional section, where we address *equality constrained* optimization problems. In the illustrative example we develop a model for transferring oil from the constrained space of a storage tank. The emphasis is on *model solution* and *model sensitivity* of this class of optimization problems.

Section 8.5, in which *graphical optimization* is presented, is also an optional section. The illustrative problem addresses the management of a fishing industry and is concerned with whether a free market can lead to a satisfactory solution for fishers, consumers, and ecologists, or whether some type of government intervention is necessary. The graphical analysis in the example provides a qualitative preview for the type of analytical models we develop in Chapters 10 and 11 using differential equations.

The projects in this chapter allow for a more detailed study of the optimization topics we discuss. For instance, you can study Lagrange multipliers or elementary ideas in the calculus of variations using UMAP modules referenced in the projects.

In the next chapter we discuss the Simplex Method and solving linear optimization problems. We also discuss discrete optimization models for which we examine pattern searches to approximate solutions to continuous models.

8.1 CLASSIFYING OPTIMIZATION PROBLEMS

To provide a framework for discussing a class of optimization problems, we offer a basic model for such problems. The problems are classified according to the characteristics of the basic model possessed by the problems. We discuss, too, variations from the basic model itself. The basic model is

$$\text{Optimize } f_j(\mathbf{X}) \text{ for } j \text{ in } J \tag{8.1}$$

Subject to

$$g_i(\mathbf{X}) \begin{Bmatrix} \geq \\ = \\ \leq \end{Bmatrix} b_i \text{ for all } i \text{ in } I$$

Now let's explain the notation. To optimize means to maximize or minimize. The subscript j indicates there may be one or more functions to optimize. The functions are distinguished by the integer subscripts that belong to the finite set J. We seek the vector $\mathbf{X}_0$ giving the optimal value for the set of functions $f_j(\mathbf{X})$. The

decision variables
objective functions

components of the vector $\mathbf{X}$ are called the **decision variables** of the model, and the functions $f_j(\mathbf{X})$ are called the **objective functions**. By *subject to*, we connote there may be certain side conditions that must be met. For example, if the objective is to minimize costs of producing a particular product, it might be specified that all contractual obligations for the product be met as side conditions. Side condi-

constraints

tions are typically called **constraints**. The integer subscript i indicates there may be one or more constraint relationships that must be satisfied. A constraint may be an equality (such as precisely meeting the demand for a product) or inequality (such as not exceeding budgetary limitations or providing the minimal nutritional requirements in a diet problem). Finally, each constant b_i represents the level the associated constraint function $g_i(\mathbf{X})$ must achieve and, because of the way opti-

right-hand side

mization problems are typically written, is often called the **right-hand side** in the model. Thus the solution vector $\mathbf{X}_0$ must optimize each of the objective functions $f_j(\mathbf{X})$ and simultaneously satisfy each constraint relationship. We now consider a problem illustrating these basic ideas.

Example 1   *Determining a Production Schedule*

A carpenter makes tables and bookcases. He is trying to determine how many of each type of furniture he should make each week. The carpenter wishes to determine a weekly production schedule for tables and bookcases that maximizes his profits. It costs \$5 and \$7 to produce tables and bookcases, respectively. The revenues are estimated by the expressions:

$$50x_1 - 0.2x_1^2, \text{ where } x_1 \text{ is the number of tables produced per week}$$

and

$$65x_2 - 0.3x_2^2, \text{ where } x_2 \text{ is the number of bookcases produced per week}$$

In this example, the problem is to decide how many tables and bookcases to make every week. Consequently the decision variables are the quantities of tables and bookcases to be made per week. We assume this is a schedule so noninteger values of tables and bookcases make sense. The objective function is a nonlinear expression representing the net weekly profit to be realized from selling the tables and bookcases. Profit is revenue minus costs (see Chapter 1). The profit function is

$$f(x_1, x_2) = 50x_1 - 0.2x_1^2 + 65x_2 - 0.3x_2^2 - 5x_1 - 7x_2$$

There are no constraints in this problem.

Let's consider a variation to this example. The carpenter realizes a net unit profit of \$25 per table and \$30 per bookcase. He is trying to determine how many of each piece of furniture he should make each week. He has up to 600 board-feet of lumber to devote weekly to the project and up to 40 hr of labor. He can use lumber and labor productively elsewhere if they are not used in the production of tables and bookcases. He estimates that it requires 20 board-feet of lumber and 5 hr of labor to complete a table and 30 board-feet of lumber and 4 hr of labor for a bookcase. Moreover, he has signed contracts to deliver four tables and two bookcases every week. The carpenter wishes to determine a weekly production schedule for tables and bookcases that maximizes his profits. The formulation yields

$$\text{Maximize } 25x_1 + 30x_2$$

Subject to

$$20x_1 + 30x_2 \le 600 \quad \text{(lumber)}$$
$$5x_1 + 4x_2 \le 40 \quad \text{(labor)}$$
$$x_1 \ge 4 \quad \text{(contract)}$$
$$x_2 \ge 2 \quad \text{(contract)}$$

This type of linear programming problem will be solved in Chapter 9.

There are various ways of classifying optimization problems. These classifications are not meant to be mutually exclusive but to describe mathematical characteristics possessed by the problem under investigation. We now describe several of these classifications.

unconstrained
constrained
An optimization problem is said to be **unconstrained** if there are no constraints and **constrained** if one or more side conditions are present. The production schedule problems just described illustrate an unconstrained and constrained problem.

linear program
An optimization problem is said to be a **linear program** (to be studied in the next chapter) if it satisfies the following properties:[1]

1. There is a unique objective function.
2. Whenever a decision variable appears in either the objective function or one of the constraint functions, it must appear only as a power term with an exponent of 1, possibly multiplied by a constant.
3. No term in the objective function or in any of the constraints can contain products of the decision variables.
4. The coefficients of the decision variables in the objective function and each constraint are constants.
5. The decision variables are permitted to assume fractional as well as integer values.

These properties ensure, among other things, that the effect of any decision variable is *proportional* to its value. Let's examine each property more closely.

multiobjective
goal programs
Property 1 limits the problem to a single objective function. Problems with more than one objective function are called **multiobjective**, or **goal programs**. (We

---

[1]Although a program satisfying properties 1–5 is linear even if it contains variables that are allowed to assume negative values, the Simplex Method requires using nonnegative variables only. A simple substitution allows this (see Problem 5 in Section 5.2). Many computer codes will make this substitution for you.

give an example of a multiobjective problem in the next section.) Properties 2 and 3 are self-explanatory, and any optimization problem that fails to satisfy either one of **nonlinear**   them is said to be **nonlinear**. The first production schedule objective function had both decision variables as squared terms and thus violated Property 2. Property 4 is restrictive for many scenarios you might wish to model. Consider examining the amount of board-feet and labor required to make tables and bookcases. It might be possible to know exactly the number of board-feet and labor required to produce each item and incorporate these into constraints. But often it is impossible to predict precisely the required values in advance (consider trying to predict the market price of corn), or the coefficients represent average values with large deviations from the actual values occurring in practice. The coefficients may be time dependent as well. **dynamic programs**   Time-dependent problems in a certain class are called **dynamic programs** (we give an example in the next section). If the coefficients are not constant but instead are **stochastic program**   probabilistic in nature, the problem is classified as a **stochastic program**. Finally, if one or more of the decision variables is restricted to integer values (hence violating **integer program**   Property 5), the resulting problem is called an **integer program** (or a **mixed-integer mixed-integer program**   program** if the integer restriction applies to only a subset of the decision variables). In the production scheduling problem, it makes sense to allow fractional numbers of tables and bookcases in determining a weekly schedule. (Why?)

8.1 Projects

Complete the requirements in the referenced UMAP module or monograph.

1. "Geometric Programming," by Robert E. D. Woolsey, UMAP 737. This unit provides alternative optimization formulations, including geometric programming. Familiarity with basic differential calculus is required.

2. "Municipal Recycling: Location and Optimality," by Jannett Highfill and Michael McAsey. *The UMAP Journal*, Vol. 15(1), 1994. This article considers optimization in municipal recycling. Read the article and prepare a 10-minute classroom presentation.

8.2   Formulation of Optimization Models

In this section we illustrate classes of optimization models other than the linear program. First let's review the criteria for fitting models to data, studied in Chapter 5. The criterion of minimizing the largest absolute deviation from the fitted model to any corresponding data point led to the optimization problem formulated in Section 5.1:

$$\text{Minimize } \{\text{Maximum } |y_i - f(x_i)|\}$$

Verify that the optimization problem represented by a polynomial curve fit is a linear program (and hence can be solved by the Simplex Method). Minimizing the largest absolute deviation may also arise in a variety of physical problems. For instance, consider the problem of blending various types of foods containing distinct levels of certain nutrients to obtain a diet that meets specified nutrient levels. It may be just as important not to exceed the specified levels as not to fall short of them. In such a case it may be desirable to find a combination of foods available that minimizes the largest absolute deviation from any specified level, perhaps weighting the deviations to reflect the relative importance of the various nutrients. Another physical problem occurs when a factory discharges water into a river. It would do no good to satisfy all but one of the environmental standards, especially if excessive deviation in either direction from any established standard threatens wildlife. A better approach might be to blend the components of the discharge in such a way that the largest deviation from any standard is minimized.

Unconstrained Optimization Problems

Another criterion considered for fitting a model to data points is minimizing the sum of absolute deviations. For the model $y = f(x)$, if $y(x_i)$ represents the function evaluated at $x = x_i$ and (x_i, y_i) denotes the corresponding data point for $i = 1, 2, \ldots, m$ points, then this criterion can be formulated as follows: Find the parameters of the model $y = f(x)$ to

$$\text{Minimize } \sum_{i=1}^{m} |y_i - y(x_i)| \tag{8.2}$$

This last condition illustrates an unconstrained optimization problem. Because the derivative of the function being minimized fails to be continuous (because of the presence of the absolute value), it is impossible to solve this problem with a straightforward application of the elementary calculus. A discrete numerical solution based on pattern search is presented in Section 9.5.

Finally, the criterion of minimizing the sum of squared deviations (least squares) was considered. Using the same notation as earlier, the problem is to find the parameters of the model $y = f(x)$ to

$$\text{Minimize } \sum_{i=1}^{m} [y_i - y(x_i)]^2 \tag{8.3}$$

As we saw in Chapter 5, this unconstrained optimization problem can be solved by direct application of the calculus. For many problems, however, the computations may be difficult, or even impossible, to carry out. In such cases, numerical solution techniques (Section 9.5) can be used.

Example 1 Locating a Plant's Headquarters

Consider an unconstrained optimization problem that arises in a physical scenario. Suppose we are attempting to locate a plant's headquarters in such a way as to minimize the distance traveled to subsidiary plants. Assume that all plants are visited with the same frequency and neglect vertical distances, giving a two-dimensional model. These assumptions might be reasonable where travel is by private aircraft and each plant and the headquarters have an airstrip or when the travel is by automobile and the terrain is flat. Let (a, b) denote the headquarters' location and (x_i, y_i) the coordinates of the ith plant. The problem is illustrated for four plants in Figure 8.1.

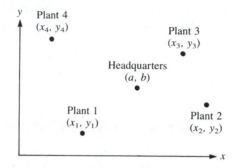

FIGURE 8.1 Locate the headquarters so distance traveled to visit the plants is minimized

The problem is to minimize the sum of the distances from the headquarters to the plants, because each plant is visited with the same frequency. Mathematically, this requirement means to

$$\text{Minimize } D = \sum_{i=1}^{m} \sqrt{(a - x_i)^2 + (b - y_i)^2} \qquad \textbf{(8.4)}$$

Note the similarities and the differences between this problem and the unconstrained optimization problems previously presented. Again distance is being minimized, as in Equation (8.2), but the distance concept has changed because there are two dimensions instead of one. Because the distance D is a function of the two variables a and b, the necessary condition for a relative minimum to exist is that the two partial derivatives $\partial D / \partial a$ and $\partial D / \partial b$ both equal zero (vanish). Solving the system of equations that results from applying this necessary condition presents a difficult task even for a small number of plants. Unfortunately, such is often the case in optimization problems. Thus, even a conceptually simple problem can lead to computational difficulties.

We conclude this discussion with brief remarks on approximating solutions to the optimization problems—such as those represented by Equations (8.2), (8.3), and (8.4)—that may be difficult to solve analytically. One set of numerical techniques **gradient method**   is known as the **gradient method**. Such techniques attempt to locate a relative minimum by repeatedly stepping in the direction of the gradient vector, which is the direction of steepest descent. A critical question is how large a step to take.

search techniques

Moreover, it may be difficult to find and evaluate the gradient function itself. Other methods, known as **search techniques**, avoid this difficulty by using the principle that $f(x_0 + \Delta x) - f(x_0) > 0$ in the vicinity of the relative minimum: If $f(x_0 + \Delta x) - f(x_0) < 0$, then the point $(x_0 + \Delta x)$ represents a more optimal location than the current point x_0. Because the method finds only a relative minimum, knowledge of the general location of the desired global minimum is essential and is usually available for applied problems. Good starting points assist both the gradient and search techniques immensely in most situations. We study search techniques in Section 9.5.

Integer Optimization Programs

The requirements specified for a problem may restrict one or more of its decision variables to integer values. For example, in seeking the right mix of various sized cars, vans, and trucks for a company's transportation fleet to minimize cost under some set of conditions, it would not make sense to determine a fractional part of a vehicle. Integer optimization problems also arise in coded problems, where binary (0 and 1) variables represent specific states such as yes/no or on/off. The following is an example.

Example 2   **Space Shuttle Cargo**

There are various items to be taken on a space shuttle. Unfortunately, there are restrictions on the allowable weight and volume capacities. Suppose there are m different items, each given some numerical value c_j and having weight w_j and volume v_j. (How might you determine c_j in an actual problem?) Suppose the goal is to maximize the value of the items that are to be taken without exceeding the weight limitation W or volume limitation V. We can formulate a model as follows:

$$\text{Let } y_j = \begin{cases} 1, & \text{if item } j \text{ is taken (yes)} \\ 0, & \text{if item } j \text{ is not taken (no)} \end{cases}$$

Then the problem is to

$$\text{Maximize } \sum_{j=1}^{m} c_j y_j$$

Subject to

$$\sum_{j=1}^{m} v_j y_j \leq V$$

$$\sum_{j=1}^{m} w_j y_j \leq W$$

The ability to use binary variables such as y_j permits great flexibility in modeling. They can be used to represent yes/no decisions, such as whether or not to finance a given project in a capital budgeting problem, or to restrict variables to being on or off. For example, the variable x can be restricted to the values 0 and a by using the binary variable y as a multiplier:

$$x = ay, \quad \text{where } y = 0 \text{ or } 1$$

Another illustration restricts x to be either in the interval (a, b) or to be zero by using the binary variable y:

$$ay < x < yb, \quad \text{where } y = 0 \text{ or } 1$$

Example 3   *Approximation by a Piecewise Linear Function*

The power of using binary variables to represent intervals can be more fully appreciated when it is desired to approximate a nonlinear function by a piecewise linear one. Specifically, suppose the nonlinear function in Figure 8.2a represents a cost function and we want to find its minimum value over the interval $0 \le x \le a_3$. If the function is particularly complicated, it could be approximated by a piecewise linear function such as that shown in Figure 8.2b. (The piecewise linear function might occur naturally in a problem, such as when different rates are charged for electrical use based on the amount of consumption.)

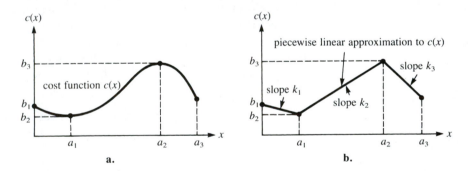

FIGURE 8.2   Using a piecewise linear function to approximate a nonlinear function

Using the approximation suggested in Figure 8.2, our problem is to find the minimum of the following function:

$$c(x) = \begin{cases} b_1 + k_1(x - 0) & \text{if } 0 \le x < a_1 \\ b_2 + k_2(x - a_1) & \text{if } a_1 \le x < a_2 \\ b_3 + k_3(x - a_2) & \text{if } a_2 \le x \le a_3 \end{cases}$$

Define the three new variables $x_1 = (x - 0)$, $x_2 = (x - a_1)$, and $x_3 = (x - a_2)$ for each of the three intervals, and use the binary variables y_1, y_2, and y_3 to restrict the

x_i to the appropriate interval:

$$0 \le x_1 \le y_1 a_1$$
$$0 \le x_2 \le y_2(a_2 - a_1)$$
$$0 \le x_3 \le y_3(a_3 - a_2)$$

where y_1, y_2, and y_3 equal 0 or 1. Because we want exactly one x_i to be active at any time, we impose the following constraint:

$$y_1 + y_2 + y_3 = 1$$

Now that only one of the x_i is active at any one time, our objective function becomes

$$c(x) = y_1(b_1 + k_1 x_1) + y_2(b_2 + k_2 x_2) + y_3(b_3 + k_3 x_3)$$

Observe that whenever $y_i = 0$, the variable $x_i = 0$ as well. Thus, products of the form $x_i y_i$ are redundant, and the objective function can be simplified to give the following model:

$$\text{Minimize } k_1 x_1 + k_2 x_2 + k_3 x_3 + y_1 b_1 + y_2 b_2 + y_3 b_3$$

Subject to

$$0 \le x_1 \le y_1 a_1$$
$$0 \le x_2 \le y_2(a_2 - a_1)$$
$$0 \le x_3 \le y_3(a_3 - a_2)$$
$$y_1 + y_2 + y_3 = 1$$

where y_1, y_2, and y_3 equal 0 or 1.

mixed-integer programming   The preceding model is classified as a **mixed-integer programming** problem because only some decision variables are restricted to integer values. The properties that allow for the linear program and the Simplex Method are lost in integer programming problems, so there is considerable difficulty in solving them. One methodology that has proved successful devises rules for finding good feasible solutions quickly, and then uses tests to show which of the remaining feasible solutions can be discarded. Unfortunately, some tests work well for certain classes of problems but fail miserably for others.

Multiobjective Programming: An Investment Problem

Consider the following problem: An investor has $40,000 to invest. She is considering investments in savings at 7%, municipal bonds at 9%, and stocks that have been consistently averaging 14%. Because there are varying degrees of risk involved in the investments, the investor listed the following goals for her portfolio:

1. A yearly return of at least $5,000.
2. An investment of at least $10,000 in stocks.
3. The investment in stocks should not exceed the combined total in bonds and savings.
4. A liquid savings account between $5,000 and $15,000.
5. The total investment must not exceed $40,000.

We can see from the portfolio that the investor has more than one objective. Unfortunately, as is often the case with real-world problems, not all goals can be achieved simultaneously. If the investment returning the lowest yield is set as low as possible (in this example, $5,000 into savings), the best return possible without violating Goals 2 through 5 is obtained by investing $15,000 in bonds and $20,000 in stocks. However, this portfolio falls short of the desired yearly return of $5,000. How are problems with more than one objective reconciled?

Let's begin by formulating each objective mathematically: Let x denote the investment in savings, y the investment in bonds, and z the investment in stocks. Then the goals are as follows:

Goal 1. $\quad 0.07x + 0.09y + 0.14z \geq 5000$

Goal 2. $\qquad\qquad\qquad\qquad z \geq 10000$

Goal 3. $\qquad\qquad\qquad\qquad z \leq x + y$

Goal 4. $\qquad\qquad\quad 5000 \leq x \leq 15000$

Goal 5. $\qquad\qquad\quad x + y + z \leq 40000$

We have seen that the investor will have to compromise on one or more of her goals to find a feasible solution. Suppose the investor feels she must have a return of $5,000, at least $10,000 in stocks, cannot spend more than $40,000, and is willing to compromise on Goals 3 and 4. She wants to find a solution that minimizes the amounts by which these two goals fail to be achieved. Let's formulate these new requirements mathematically and illustrate a method applicable to similar problems. Thus, let G_3 denote the amount by which Goal 3 and G_4 the amount by which Goal 4 fail to be satisfied. Then the model is to

$$\text{Minimize } G_3 + G_4$$

Subject to

$$0.07x + 0.09y + 0.14z \geq 5000$$
$$z \geq 10000$$
$$z - G_3 \leq x + y$$
$$5000 - G_4 \leq x \leq 15000$$
$$x + y + z \leq 40000$$

where x, y, and z are positive.

This last condition is included to ensure that negative investments do not result. This problem is now a linear program that can be solved by the Simplex Method (see Chapter 9). If the investor feels some goals are more important than others, the objective function can be weighted to emphasize those goals. Furthermore, a sensitivity analysis of the weights in the objective function will identify the breakpoints for the range over which various solutions are optimal. This process generates a number of solutions to be carefully considered by the investor before making her investments. Normally, this is the best that can be accomplished when qualitative decisions are to be made.

Calculus of Variations

We now consider several departures from the basic optimization Model (8.1). Recall that in that model we seek a vector **X** optimizing the value of the objective function(s) while simultaneously satisfying any side conditions that are present. Situations arise, however, when an optimizing *function*, rather than a *vector*, is sought. For example, consider a bead sliding under the force of gravity down a wire from a high point to a low point not directly beneath it. The situation is illustrated in Figure 8.3. The problem is to find the path that minimizes the time of descent from the high point to the low point. That is, from the set of all function curves passing through the two points, find the one that minimizes the travel time. At first we might think the solution is a straight line joining the two points, because that curve does minimize the distance between them. However, the straight line proves to be incorrect and the calculus of variations shows the optimal path is steeper. In fact, the solution curve

cycloid   is a **cycloid**, which is the path traced by a point on the circumference of a wheel rolling along a horizontal straight line without slipping.

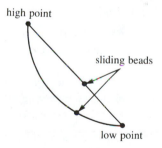

FIGURE 8.3   The path that minimizes the time of descent from the high point to the low point is a cycloid

Calculus of variations problems often arise in various types of control problems. For example, it is of interest to find the function that describes the thrust rate (fuel per unit time) for a rocket so the amount of time it takes the rocket to achieve a certain height is minimized. A related problem is to minimize the amount of fuel it takes to achieve the height while ignoring the time it takes. Problems of this sort

optimal control   are often called **optimal control problems** and, as is typical of most calculus of
problems   variations problems, tend to be difficult to solve.

Dynamic Programming Problems

dynamic programs

Often the model being optimized requires that decisions be made at various time intervals rather than all at once. In the 1950s the American mathematician Richard Bellman developed a technique for optimizing such models in *stages* rather than simultaneously and referred to such problems as **dynamic programs**. Here's a scenario that lends itself to solution by dynamic programming techniques.

A rancher enters the cattle business with an initial herd of k cattle. He intends to retire and sell any remaining cattle after N years. Each year the rancher is faced with the decision of how many cattle to sell and how many to keep. If he sells, he estimates his profit to be p_i in year i. Also, the number of cattle kept in year i is expected to double in year $i + 1$.

Although this scenario neglects many factors an analyst might consider in the real world (name several), we can see that the cattle rancher is faced with a trade-off decision each year: take a profit or build for the future.

The purpose of our discussion has been to provide an overview of the field of optimization. We hope you can appreciate the many applications in which optimization problems arise. The solutions to these problems tend to be difficult, but solution methods are studied in the next chapter and in more advanced courses such as linear and nonlinear optimization.

8.2 PROBLEMS

Use the model-building process described in Chapter 2 to analyze the following scenarios. After identifying the problem to be solved using the process, you may find it helpful to answer the following questions in words before formulating the optimization model:

A. Identify the decision variables: *What decision is to be made?*
B. Formulate the objective function: *How do the decisions to be made affect the objective?*
C. Formulate the constraint set: *What constraints must be satisfied? Be sure to consider whether negative values of the decision variables are allowed by the problem and to ensure they are so constrained if required.*

After constructing the model, check the assumptions for a linear program as well as compare the form of the model to the examples in this section. Try to determine which method of optimization can be applied to obtain a solution.

1. *Resource Allocation*—You have just become the manager of a plant producing plastic products. Although the plant operation involves many products and supplies, you are interested in only three of the products: (1) a vinyl–asbestos floor covering, the output of which is measured in boxed lots, each covering a certain area; (2) a pure vinyl countertop, measured in linear yards; and (3) a vinyl–asbestos wall tile, measured in squares, each covering 100 ft^2.

   Of the many resources needed to produce these plastic products, you have identified four: vinyl, asbestos, labor, and time on a trimming machine. A recent inventory shows that on any given day you have 1500 lb of vinyl and 200 lb of asbestos available for use. Additionally, after talking to your shop foreman and to various labor leaders, you realize that you have 3 person-days of labor available per day and your trimming machine is available for 1 machine-day on any given day. The following table indicates the amount of each of the four resources required to produce a unit of the three desired products, where the units are 1 box of floor cover, 1 yd of countertop, and 1 square of wall tiles. Available resources are also tabulated.

	Vinyl (lb)	Asbestos (lb)	Labor (person-days)	Machine (machine-days)	Profit
Floor cover (per box)	30	3	0.02	0.01	$0.8
Countertop (per yard)	20	0	0.1	0.05	5
Wall tile (per square)	50	5	0.2	0.05	5.5
Available (per day)	1500	200	3	1	—

Formulate a mathematical model to help determine how to allocate resources to maximize profits.

2. *Nutritional Requirements*—A rancher has determined that the minimum weekly nutritional requirements for an average-sized horse include 40 lb of protein, 20 lb of carbohydrates, and 45 lb of roughage. These are obtained from the following sources in varying amounts at the prices indicated:

	Protein (lb)	Carbohydrates (lb)	Roughage (lb)	Cost
Hay (per bale)	0.5	2.0	5.0	$1.80
Oats (per sack)	1.0	4.0	2.0	3.50
Feeding blocks (per block)	2.0	0.5	1.0	0.40
High-protein concentrate (per sack)	6.0	1.0	2.5	1.00
Requirements per horse (per week)	40.0	20.0	45.0	

Formulate a mathematical model to determine how to meet the minimum nutritional requirements at minimum cost.

3. *Scheduling Production*—A manufacturer of an industrial product has to meet the following shipping schedule:

Month	Required shipment (units)
January	10,000
February	40,000
March	20,000

The monthly production capacity is 30,000 units and the production cost per unit is $10. Because the company does not warehouse, the service of a storage company is used whenever needed. The storage company figures its monthly bill by multiplying the number of units in storage on the last day of the month by $3. The manufacturer does not have any beginning inventory on the first day of January, and it doesn't want to have any ending inventory at the end of March. Formulate a mathematical model to assist in minimizing the sum of the production and storage costs for the 3-month period.

How does the formulation change if the production cost is $10x + 10$ dollars, where x is the number of items produced?

4. *Mixing Nuts*—A candy store sells three different assortments of mixed nuts, each assortment containing varying amounts of almonds, pecans, cashews, and walnuts. To preserve the store's reputation for quality, certain maximum and minimum percentages of the various nuts are required for each type of assortment, as shown in the following table:

Nut assortment	Requirements	Selling price per pound
Regular	Not more than 20% cashews Not less than 40% walnuts Not more than 25% pecans No restriction on almonds	$0.89
Deluxe	Not more than 35% cashews Not less than 25% almonds No restriction on walnuts and pecans	1.10
Blue Ribbon	Between 30% and 50% cashews Not less than 30% almonds No restriction on walnuts and pecans	1.80

The following table gives the cost per pound and the maximum quantity of each type of nut available from the store's supplier each week:

Nut type	Cost per pound	Maximum quantity available per week (lb)
Almonds	$0.45	2000
Pecans	0.55	4000
Cashews	0.70	5000
Walnuts	0.50	3000

The store would like to determine the exact amounts of almonds, pecans, cashews, and walnuts that should go into each weekly assortment to maximize its weekly profit. Formulate a mathematical model that will assist the store management in solving the mixing problem. *Hint:* How many decisions need to be made? For example, do we need to distinguish between the cashews in the regular mix and the cashews in the Deluxe mix?

5. *Producing Electronic Equipment*—An electronics firm is producing three lines of products for sale to the government: transistors, micromodules, and circuit assemblies. The firm has four physical processing areas designated as follows: transistor production, circuit printing and assembly, transistor and module quality control, and circuit assembly test and packing.

   The various production requirements are as follows: Production of one transistor requires 0.1 standard hour of transistor production area capacity, 0.5 standard hour of transistor quality control area capacity, and $0.70 in direct costs. Production of micromodules requires 0.4 standard hour of the quality control area capacity, three transistors, and $0.50 in direct costs. Production of one circuit assembly requires 0.1 standard hour of the capacity of the circuit printing area, 0.5 standard hour of the test and packing area, one transistor, three micromodules, and $2 in direct costs.

   Suppose the three products (transistors, micromodules, and circuit assemblies) can be sold in unlimited quantities at prices of $2, $8, and $25 each, respectively. There are 200 hours of production time open in each of the four process areas in the coming month. Formulate a mathematical model to help determine the production that will produce the highest revenue for the firm.

6. *Purchasing Various Trucks*—A truck company allocated $800,000 for the purchase of new vehicles and is considering three types: Vehicle A has a 10-ton payload capacity and is expected to average 45 mph; it costs $26,000. Vehicle B has a 20-ton payload capacity and is expected to average 40 mph; it costs $36,000. Vehicle C is a modified form of Vehicle B and carries sleeping quarters for one driver. This modification reduces the capacity to an 18-ton payload and raises the cost to $42,000, but its operating speed is still expected to average 40 mph.

   Vehicle A requires a crew of one driver and, if driven on three shifts per day, could be operated for an average of 18 hr per day. Vehicles B and C must have crews of two drivers each to meet local legal requirements. Vehicle B could be driven an average of 18 hr per day with three shifts, and Vehicle C could average 21 hr per day with three shifts. The company has 150 drivers available each day to make up crews and will not be able to hire additional trained crews in the near future. The local labor union prohibits any driver from working more than one shift per day. Also, maintenance facilities are such that the total number of vehicles must not exceed 30. Formulate a mathematical model to help determine the number of each type vehicle the company should purchase to maximize its shipping capacity in ton-miles per day.

7. *A Farming Problem*—A farm family owns 100 acres of land and has $25,000 in funds available for investment. Its members can produce a total of 3500 work-hours worth of labor during the winter months (mid-September to mid-May) and 4000 work-hours during the summer. If any of these work-hours are not needed, younger members of the family will use them to work on a neighboring farm for $4.80 per hour during the winter and $5.10 per hour during the summer.

Cash income can be obtained from three crops (soybeans, corn, and oats) and two types of livestock (dairy cows and laying hens). No investment funds are needed for the crops. However, each cow requires an initial investment outlay of $400, and each hen requires $3. Each cow requires 1.5 acres of land and 100 work-hours of work during the winter months and another 50 work-hours during the summer. Each cow produces a net annual cash income of $450 for the family. The corresponding figures for each hen are no acreage, 0.6 work-hours during the winter, 0.3 more work-hours in the summer, and an annual net cash income of $3.50. The chicken house accommodates a maximum of 3000 hens, and the size of the barn limits the cow herd to a maximum of 32 head.

Estimated work-hours and income per acre planted in each of the three crops are as shown in the following table:

Crop	Winter work-hours	Summer work-hours	Net annual cash income (per acre)
Soybean	20	30	$175.00
Corn	35	75	300.00
Oats	10	40	120.00

Formulate a mathematical model to assist in determining how much acreage should be planted in each of the crops and how many cows and hens should be kept to *maximize net cash income*.

8. *Cow to Market*—A cow currently weighs 800 lb and is gaining 35 lb per week. It costs $6.50 a week to maintain the cow. The market price today is $0.95 per pound, but is falling $0.01 per day. Formulate a mathematical model to help determine how long the cow should be kept until it is sold to maximize profits.

9. *Manufacturing VCRs*—A manufacturer of VCRs is planning to introduce two new products, a mini-VCR with four heads (current market selling price is $239) and a compact disc VCR (current market selling price is $421). The cost to the company is $159 per mini-VCR and $279 per compact disc VCR, plus an additional $450,000 in fixed costs. In a competitive market, sales volume affects the selling price. It is assumed by the analysts that the market selling price drops by $0.01 for each additional unit sold. Furthermore, the sales of one type of VCR affects the sales of the other type. It is estimated that the selling price of the

mini-VCR will decrease by $0.03 per compact disc VCR sold, and the compact disc VCR will decrease by $0.04 for each mini-VCR sold.

Additionally, we must be concerned with production capacity. The company has the capacity to produce at most 12,000 units. There is a circuit board that must be used in each unit (according to new government regulations), but it is in short supply. There are at most 8000 boards available for the mini-VCR and 5500 boards available for the compact disc VCR. Formulate a mathematical model to assist in determining how many of each new product should be produced to maximize profits.

8.2 PROJECTS

For Projects 1–3, complete the requirements in the referenced UMAP module or monograph.

1. "Unconstrained Optimization," by Joan R. Hundhausen and Robert A. Walsh, UMAP 522. This unit introduces gradient search procedures with examples and applications. Acquaintance with elementary partial differentiation, chain rules, Taylor series, gradients, and vector dot products is required.

2. "Calculus of Variations with Applications in Mechanics," by Carroll O. Wilde, UMAP 468. This module provides a brief introduction to finding functions that yield the maximum or minimum value of certain definite integral forms, with applications in mechanics. Students learn Euler's equations for some definite integral forms and Hamilton's principle and its application to conservative dynamical systems. The basic physics of kinetic and potential energy, the multivariate chain rules, and ordinary differential equations are required.

3. *The High Cost of Clean Water: Models for Water Quality Management*, by Edward Beltrami, UMAP Expository Monograph. To cope with the severe wastewater disposal problems caused by increases in the nation's population and industrial activity, the U.S. Environmental Protection Agency (EPA) has fostered the development of regional wastewater management plans. This monograph discusses the EPA plan developed for Long Island and formulates a model that allows for the articulation of the trade-offs between cost and water quality. The mathematics involves partial differential equations and mixed-integer linear programming.

8.3   AN INVENTORY PROBLEM: MINIMIZING THE COST OF DELIVERY AND STORAGE

Scenario   You have been hired as a consultant by a chain of gasoline stations to determine how often and how much gasoline should be delivered to the various stations. After some questioning, you determine that each time gasoline is delivered the stations incur a charge of d dollars, which is in addition to the cost of the gasoline and is independent of the amount delivered.

Costs are also incurred when the gasoline is stored. One such cost is capital tied up in inventory—money that is invested in the stored gasoline and that cannot be used elsewhere. The cost is normally computed by multiplying the cost of the gasoline to the company by the current interest rate for the period the gasoline was stored. Other costs include amortization of the tanks and equipment necessary to store the gasoline, insurance, taxes, and security measures.

The gasoline stations are located near interstate highways, where demand is fairly constant throughout the week. Records indicating the gallons sold daily are available for each station.

Problem identification Assume the firm wishes to maximize its profits and that demand and price are constant in the short run. Because total revenue is constant, total profit can be maximized by minimizing the total costs. There are many components of total costs, such as overhead and employee costs. If these costs are affected by the amount and the timing of the deliveries, they should be considered. Let's assume the costs are not so affected and focus our attention on the following problem: *Minimize the average daily cost of delivery and storing sufficient gasoline at each station to meet consumer demand.* Intuitively we expect such a minimum to exist. If the delivery charge is very high and the storage cost very low, we would expect very large orders of gasoline delivered infrequently. On the other hand, if the delivery cost is very low and the storage costs very high, we would expect small orders of gasoline delivered very frequently.

Assumptions In the following presentation, we consider factors important to deciding how large an inventory to maintain. Obvious factors are delivery costs, storage costs, and demand rate for the product. Perishability of the product being stored may also be of paramount concern. In the case of gasoline, the effect of condensation may become more and more appreciable as the gasoline level gets lower and lower in the tank. The market stability of the selling price of the product and the cost of raw materials need to be considered. For example, if the market price of the product is volatile, the seller would be reluctant to store large quantities of the product. On the other hand, an expected large increase in the price of raw materials in the near future argues for large inventories. The stability of the demand for the product by the consumer is another factor to be taken into account. There may be many seasonal fluctuations in the demand for the product, or a technological breakthrough may cause the product to become obsolete. The time horizon being considered can also be extremely important. In the short run, contracts may be signed for warehouse space, some of which will not be needed in the long run. Another consideration is the importance to the owner of an occasional unsatisfied demand (stock-outs). Some owners would opt for a more costly inventory strategy to be assured that they will never run out of stock. From this discussion you can see that the inventory decision is not an easy one, and it is not difficult to build scenarios where any one of the preceding factors may dictate a particular strategy. We restrict our initial model to the following variables:

average daily cost = f(storage costs, delivery costs, demand rate)

The Submodels

Storage costs   We need to consider how the storage cost per unit varies with the number of units being stored. Are we renting space and receiving a discount when storage exceeds certain levels, as suggested in Figure 8.4a? Or do we rent the least expensive storage first (adding more space as needed), as suggested by Figure 8.4b? Do we need to rent an entire warehouse or floor? If so, the per unit price is likely to decrease as the quantity stored increases until another warehouse or floor needs to be rented, as suggested in Figure 8.4c. Does the company own its own storage facilities? If so, what alternative use can be made of them? In our model we take per unit storage as a constant.

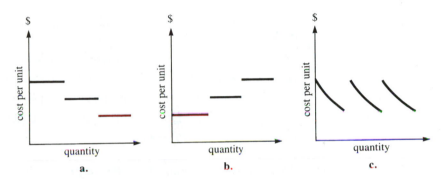

FIGURE 8.4   Submodels for storage costs

Delivery costs   In many cases the delivery charge depends on the amount delivered. For example, if a larger truck or an extra flatcar is needed, an additional charge is made. In our model we consider a constant delivery charge independent of the amount delivered.

Demand   If we plot the daily demand for gasoline at a particular station, we will likely get a graph similar to the one shown in Figure 8.5a. If we plot the frequency

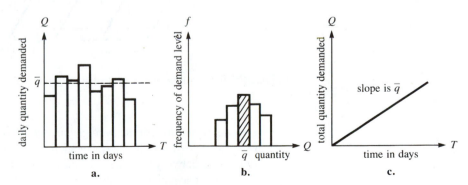

FIGURE 8.5   A constant demand rate

of each demand level over a fixed time period (e.g., 1 yr), we might get a plot similar to that shown in Figure 8.5b. If demands are fairly tightly packed about the most frequently occurring demand, then we accept the daily demand as being constant. We will assume that such is the case for our model. Finally, even though we realize that the demands occur in discrete time periods, for purposes of simplification we take a continuous submodel for demand. This continuous submodel is depicted in Figure 8.5c, where the slope of the line represents the constant demand rate. Notice the importance of our assumptions in producing the linear submodel. Also, as suggested in Figure 8.5b, about half the time demand exceeds its average value. We will return to examine the importance of this assumption in the implementation phase when we consider the possibility of unsatisfied demands.

Model formulation   We use the following notation for constructing our model:

$$s = \text{storage costs per gallon per day}$$
$$d = \text{delivery cost in dollars per delivery}$$
$$r = \text{demand rate in gallons per day}$$
$$Q = \text{quantity of gasoline in gallons}$$
$$T = \text{time in days}$$

Now suppose an amount of gasoline, say $Q = q$, is delivered at time $T = 0$ and the gasoline is used up after $T = t$ days. The same cycle is then repeated, as illustrated in Figure 8.6. The slope of each line segment in Figure 8.6 is $-r$ (the negative of the demand rate). The problem is to determine an order quantity Q^* and a time between orders T^* that minimizes the delivery and storage costs.

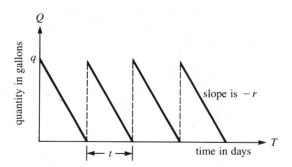

FIGURE 8.6   An inventory cycle consists of an order quantity q consumed in t days

We seek an expression for the average daily cost, so consider the delivery and storage cost for a cycle of length t days. The delivery costs are the constant amount d because only one delivery is made over the single time period. To compute the storage costs, take the average daily inventory $q/2$, multiply by the number of days in storage t, and multiply that by the storage cost per item per day s. In our notation, this gives

$$\text{cost per cycle} = d + s\frac{q}{2}t$$

which on dividing by t yields the average daily cost

$$c = \frac{d}{t} + \frac{sq}{2}$$

Model solution   Apparently the cost function to be minimized has two independent variables, q and t. However, from Figure 8.6 notice that the two variables are related. For a single cyclic period, the amount delivered equals the amount demanded. This translates to $q = rt$. Substitution into the average daily cost equation yields

$$c = \frac{d}{t} + \frac{srt}{2} \qquad (8.5)$$

Equation (8.5) is the sum of a hyperbola and a linear function. The situation is depicted in Figure 8.7.

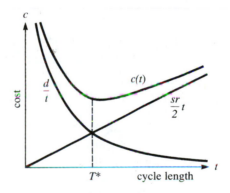

FIGURE 8.7   The average daily cost c is the sum of a hyperbola and a linear function

Let's find the time between orders T^* that minimizes the average daily cost. Differentiating c with respect to t and setting $c' = 0$ yields

$$c' = -\frac{d}{t^2} + \frac{sr}{2} = 0 \qquad (8.6)$$

This last equation gives the following (positive) critical point:

$$T^* = \left(\frac{2d}{sr}\right)^{1/2} \qquad (8.7)$$

This critical point provides a relative minimum for the cost function because the second derivative

$$c'' = \frac{2d}{t^3}$$

is always positive for positive values of t. It is clear from Figure 8.7 that T^* gives a global minimum as well. And note from Equation (8.7) that $d/T^* = (sr/2)T^*$, so that T^* is the point at which the linear function and the hyperbola intersect in Figure 8.7.

Interpretation of the model   Given a (constant) demand rate r, Equation (8.7) suggests a proportionality between the optimal period T^* and $(d/s)^{1/2}$. Intuitively we would expect T^* to increase as the delivery cost d increases and to decrease as storage costs s increase. Thus our model at least makes common sense. However, the nature of the relationship (8.7) is interesting.

For our submodels of demand rate, storage costs, and delivery cost, we assumed simple relationships. To analyze models that have more complex submodels, we need to analyze what we did mathematically. Note that to determine the number of item-days of storage, we computed the area under the curve for one cycle. Thus the storage costs for one cycle could be computed as an integral:

$$s \int_0^t (q - rx)\,dx = s \left(qt - \frac{rt^2}{2} \right) = \frac{sqt}{2}$$

The last equality in this equation follows from the substitution $r = q/t$ and agrees with our previous result for storage costs per cycle. It is important to recognize the underlying mathematical structure to facilitate generalization to other assumptions. As we will see, it is also helpful in analyzing the sensitivity of the model to changes in the assumptions.

One of our assumptions was to neglect the cost of the gasoline in the analysis. But does the cost of gasoline actually affect the optimal order quantity and period? Because the amount purchased in each cycle is rt, if the cost per gallon is p dollars, then the constant amount $p(q/t) = pr$ would have to be added to the average daily cost. Because this amount is constant, it cannot affect T^*. (Why?) Thus we are correct in neglecting the cost of gasoline. In a more refined model, the interest lost in capital invested in the inventory could be considered.

Implementation of the model   Consider again the graph in Figure 8.6. Now the model assumes the entire inventory is used up in each cyclic period, yet all demands are supposed to be satisfied immediately. Note that this assumption is based on an average daily demand of r gallons per day. This assumption means that, over the long run, for roughly half of the time cycles the stations will run out of stock before the end of the period and the next delivery time, and for the other half of the time cycles the stations will still have some gasoline left in the storage tanks when the next delivery arrives. Such a situation probably won't do much for our credibility as consultants! So let's consider recommending a buffer stock to help prevent the stock-outs, as suggested in Figure 8.8. Note in Figure 8.8 that the optimal time period T^* and the optimal order quantity $Q^* = rT^*$ are indicated as labels, because we know those values from our model given in Equation (8.7).

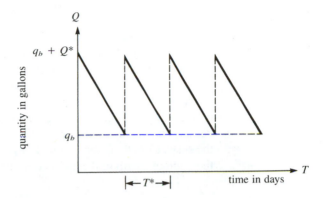

FIGURE 8.8   A buffer stock q_b helps prevent stock-outs

Let's examine the effect of the buffer stock on our inventory strategy. We observed that the storage costs for one cycle are given by the area under the curve over one period multiplied by the constant s. Now the effect of the buffer stock is to add an additional constant area $q_b t$ to the previous area under the curve. Thus the constant amount $q_b s$ is added to the average daily cost and consequently the value of T^* does not change from what it was without the buffer stock. Thus the stations continue to order $Q^* = rT^*$ as before, although the maximum inventory now becomes $Q^* + q_b$. By determining the daily cost of maintaining the buffer stock, the decision maker can determine how large a buffer stock to carry. What other information would be useful to the decision maker in determining how large a buffer stock to carry?

Our mathematical analysis has been straightforward and precise. Yet can we obtain the necessary data to estimate r, s, and d precisely? Probably not. Moreover, how sensitive is the cost to changes in these parameters? These are issues a consultant would want to consider. Note, too, that we would probably round T^* to an integer value. How should it be rounded? Is it better to overestimate or underestimate T^*? Of course, in a given situation, we could substitute various values of T^* for t in Equation (8.5) and see how the cost varies.

Let's use the derivative to determine the shape of the average daily cost curve more generally. We know that the average daily cost curve is minimum at $t = T^*$. The first derivative of the average daily cost represents its rate of change, or the marginal cost, and from Equation (8.6) is $-d/t^2 + sr/2$. The derivative of the marginal cost is $2d/t^3$, which, for positive t, is always positive and decreases as t increases. Note that the derivative of the marginal cost becomes large without bound as t approaches zero and approaches zero as t becomes large. Thus to the left of

T^* the marginal cost is negative and becomes increasingly steeper as t approaches zero. To the right of T^* the marginal cost approaches the constant value $sr/2$. Relate these results to the graph depicted in Figure 8.7 and interpret them economically.

8.3 PROBLEMS

1. Consider an industrial situation in which it is necessary to set up an assembly line. Suppose that each time the line is set up a cost c is incurred. Assume c is in addition to the cost of producing any item and is independent of the amount produced. Suggest submodels for the production rate. Now assume a constant production rate k and a constant demand rate r. What assumptions are implied by the model in Figure 8.9? Next assume a storage cost of s (in dollars per unit per day) and compute the optimal length of the production run P^* to minimize the costs. List all of your assumptions. How sensitive is the average daily cost to the optimal length of the production run?

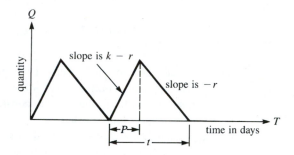

FIGURE 8.9   Determine the optimal length of the production run

2. Consider a company that allows backordering. That is, the company notifies customers that a temporary stock-out exists and that their orders will be filled shortly. What conditions might argue for such a policy? What effect does such a policy have on storage costs? Should costs be assigned to stock-outs? Why? How would you make such an assignment? What assumptions are implied by the model in Figure 8.10? Suppose a loss of goodwill cost of w dollars per unit per day is assigned to each stock-out. Compute the optimal order quantity Q^* and interpret your model.

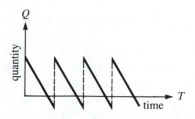

FIGURE 8.10   An inventory strategy that permits stock-outs

3. In the inventory model discussed in the text we assumed a constant delivery cost independent of the amount delivered. Actually, in many cases the cost varies in discrete amounts depending on the size of the truck needed, the number of platform cars required, and so forth. How would you modify the model to take into account these changes? We also assumed a constant cost for raw materials. However, often bulk-order discounts are given. How would you incorporate these discount effects into the model?

4. Discuss the assumptions implicit in the two graphical models depicted in Figure 8.11. Suggest scenarios in which each model might apply. How would you determine the submodel for demand in Figure 8.11b? Discuss how you would compute the optimal order quantity in each case.

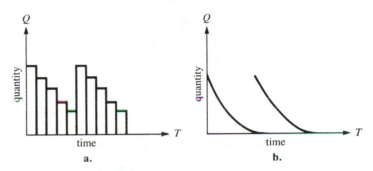

FIGURE 8.11   Two inventory submodels

5. What is the optimal speed and safe following distance that allows the maximum flow rate (cars per unit time)? The solution to the problem would be of use in controlling traffic in tunnels, for roads under repair, or for other congested areas. In the following schematic, l is the length of a typical car and d is the distance between cars:

Justify that the flow rate is given by

$$f = \frac{\text{velocity}}{\text{distance}}$$

Let's assume a car length of 15 ft. In Section 4.2 the safe stopping distance

$$d = 1.1v + 0.054v^2$$

was determined, where d is measured in feet and v in miles per hour. Find the velocity in miles per hour and the corresponding following distance d that maximizes traffic flow. How sensitive is the solution to changes in v? Suggest a practical rule. How would you enforce it?

6. Consider an athlete competing in the shot put. What factors influence the length of his or her throw? Construct a model that predicts the distance thrown as a function of the initial velocity and angle of release. What is the optimal angle of release? If the athlete cannot maximize the initial velocity at the angle of release you propose, should he or she be more concerned with satisfying the angle of release or with generating a high initial velocity? What are the trade-offs?

7. John Smith is responsible for periodically buying new trucks to replace older trucks in his company's fleet of vehicles. He is expected to determine the length of time a truck should be retained to minimize the average cost of owning the truck. Assume the purchase price of a new truck is \$9,000 with trade-in. Also assume the maintenance cost (in dollars) per truck for t years can be expressed analytically by the following empirical model:

$$C(t) = 640 + 180(t + 1)t$$

where t is the time in years the company owns the truck.

   a. Determine $E(t)$, the total cost function for a single truck retained for a period of t years.

   b. Determine $E_A(t)$, the *average* annual cost function for a single truck kept in the fleet for t years.

   c. Graphically depict $E_A(t)$ as a function of t. Justify the shape of your graph.

   d. Analytically determine t^*, the optimal period a truck should be retained in the fleet. Remember that the objective is to minimize the average cost of owning a truck.

   e. Suppose we have to round t^* to the nearest whole year. In general, would it be better to round up or round down? Justify your answer.

8. Find the optimal period to keep the cow of Problem 9 in Section 8.2 to maximize profits.

8.3 PROJECTS

1. "The Human Cough," by Philip M. Tuchinsky, UMAP 211. A model is developed showing how our bodies contract the windpipe during a cough to maximize the velocity of the airflow (making the cough maximally effective). Complete the module and prepare a short report for classroom discussion.

2. "An Application of Calculus in Economics: Oligopolistic Competition," by Donald R. Sherbert, UMAP 518. The author analyzes mathematical models that investigate the competitive structure of a market in which a small number of firms compete. Thus a change in price or production level by one firm will cause a reaction in the others. Complete the module and prepare a short report for classroom discussion.

3. "Five Applications of Max-Min Theory from Calculus," UMAP 341 by Thurmon Whitley. In this module several unconstrained optimization problems are solved using the calculus. Scenarios include maximizing profit, minimizing cost, minimizing travel time of light as it passes through several mediums (Snell's Law), minimizing the surface area of a bee's cell, and the surgeon's problem of attaching an artery in a way to minimize the resistance of blood flow and strain on the heart.

8.4 CONSTRAINED CONTINUOUS OPTIMIZATION

Example 1 An Oil Transfer Company

Sometimes in modeling an optimization situation we need to find a maximum or minimum value of a function when the independent variables are constrained to lie within some particular subset of the plane (such as a disk, or along a straight line, or a closed triangular region). In this section we present two examples of such constrained problems and explore a powerful method for finding the extreme values known as the method of *Lagrange Multipliers*.

Consider a situation in which you are hired as a consultant for a small oil transfer company. The management desires a policy of minimum cost due to restricted tank storage space.

Problem identification *Minimize the costs associated with dispensing and holding the oil to maintain sufficient oil to satisfy demand while meeting the restricted tank storage space constraint.*

Assumptions Many factors determine the total cost of transferring oil. For our model we include the following variables: holding costs of the oil in the storage tank, withdrawal rate of the oil from the tank per unit time, the cost of the oil, and the size of the tank.

Model formulation We define the following variables for two types of oil (so $i = 1$ or 2):

x_i = amount of oil type i available

a_i = cost of oil type i

b_i = withdrawal rate per unit time of oil type i

h_i = holding (storage) costs per unit time for oil type i

t_i = space in cubic feet to store one unit of oil type i

T = total amount of storage space available

Historical records have been studied and a formula has been derived that describes the system costs in terms of our variables. Our objective is to minimize the sum of the cost variables:

$$\text{Minimize } f(x_1, x_2) = \left. \left(\frac{a_1 b_1}{x_1} + \frac{h_1 x_1}{2} \right) + \left(\frac{a_2 b_2}{x_2} + \frac{h_2 x_2}{2} \right) \right\} \qquad \textbf{(8.8)}$$
$$\text{such that } g(x_1, x_2) = t_1 x_1 + t_2 x_2 = T \text{ (space constraint)}$$

We are provided the following data:

Oil type	a_i (\$)	b_i	h_i (\$)	t_i (ft^3)
1	9	3	0.50	2
2	4	5	0.20	4

You measure the storage tank and find only 24 ft^3 of available space. After substituting the data, the formulation for our problem is

$$\text{Minimize } f(X) = 27/x_1 + 0.25x_1 + 20/x_2 + 0.10x_2$$
$$\text{such that } 2x_1 + 4x_2 = 24$$

Lagrange Multipliers

Model solution The method commonly used for solving nonlinear optimization problems with equality constraints is known as the method of **Lagrange Multipliers**. The method involves introducing a new variable λ (called a Lagrange Multiplier) and setting up the function

$$L(x_1, x_2, \lambda) = f(x_1, x_2) + \lambda \left[g(x_1, x_2) - T \right]$$

For our problem, this function is

$$L(x_1, x_2, \lambda) = 27/x_1 + 0.25x_1 + 20/x_2 + 0.10x_2 + \lambda(2x_1 + 4x_2 - 24) \qquad \textbf{(8.9)}$$

The solution methodology is to take the partial derivatives of Equation (8.9) with respect to the variables x_1, x_2, and λ and set them equal to zero; that is,

$$\frac{\partial L}{\partial x_1} = \frac{-27}{x_1^2} + 0.25 + 2\lambda = 0$$
$$\frac{\partial L}{\partial x_2} = \frac{-20}{x_2^2} + 0.10 + 4\lambda = 0$$
$$\frac{\partial L}{\partial \lambda} = 2x_1 + 4x_2 - 24 = 0$$

Using a computer algebra system, we find the solution to be $x_1 = 5.0968$, $x_2 = 3.4516$, $\lambda = 0.3947$, and $f(x_1, x_2) = \$12.71$. By perturbing the values of x_1 and x_2 slightly in either direction (but satisfying the constraint), we find the value of $f(x_1, x_2)$ increases. Thus we conclude the solution is a minimum.

shadow price

Model sensitivity   The variable λ has special significance and is called a **shadow price**. The value of λ represents the amount the objective function would change for a *1 unit increase* in the constraint represented by λ. Therefore, in this problem, the value of $\lambda = 0.3947$ means that if the storage space constraint is changed from 24 to 25 ft³, the value of the objective function would change from \$12.71 to approximately $12.71 + (1)(0.3947)$ or \$13.10. The economic interpretation is that another cubic foot of storage tank capacity increases the dispensing and holding costs by about \$0.40.

Example 2   ***A Space Shuttle Water Container***

Consider a space shuttle and an astronaut's water container that is to be stored within the shuttle's wall. The water container is made in the form of a sphere surmounted by a cone (like an ice cream cone), the base of which is equal to the radius of the sphere (see Figure 8.12). If the radius of the sphere is restricted to exactly 6 ft and a surface area of 450 ft² is all that is allowed in the design, find the dimensions x_1 and x_2 such that the volume of the container is a maximum.

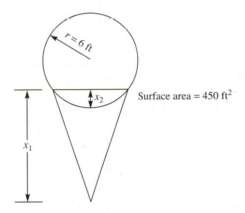

FIGURE 8.12   Space shuttle water container

Problem identification   *Maximize the volume of the water container for the astronauts while meeting the design restrictions.*

Assumptions   There are many factors that can influence the design of a water container. For our model we include the shape and dimensions, volume, surface area, and radius of the sphere corresponding with Figure 8.12.

Model formulation   We define the following variables:

V_c: volume of the conical top, which equals $\dfrac{\pi r^2}{12} x_1$

V_s: volume of the cut sphere, which equals $\left[\dfrac{4}{3}\pi r^3 - \dfrac{1}{6}\pi x_2 \left(\dfrac{3r^2}{4} + x_2^2 \right) \right]$

$V_w = V_c + V_s$: volume of the water container

S_c: surface area of the cone, which equals $\dfrac{\pi r}{2}\sqrt{\dfrac{r^2}{4}+x_1^2}$

S_s: surface area of the sphere $4\pi r^2 - \pi\left(\dfrac{r^2}{4}+x_2^2\right)$

$S_T = S_c + S_s$: total surface area

We want to maximize the volume V_w of the container. The container is constrained by the total available surface area S_T. Thus the problem is to

$$\text{Maximize } f(x_1, x_2) = \frac{\pi r^2}{12}x_1 + \frac{4}{3}\pi r^3 - \frac{1}{6}\pi x_2\left(\frac{3r^2}{4}+x_2^2\right) \qquad \textbf{(8.10)}$$

$$\text{Subject to } \frac{\pi r}{2}\sqrt{\frac{r^2}{4}+x_1^2}+4\pi r^2 - \pi\left(\frac{r^2}{4}+x_2^2\right)=450$$

Model solution   We use the method of Lagrange Multipliers to solve this equality constrained optimization problem. We set up the function

$$L(x_1, x_2, \lambda) = \frac{\pi r^2}{12}x_1 + \frac{4}{3}\pi r^3 - \frac{1}{6}\pi x_2\left(\frac{3r^2}{4}+x_2^2\right)$$
$$- \lambda\left[\frac{\pi r}{2}\sqrt{\frac{r^2}{4}+x_1^2}+4\pi r^2 - \pi\left(\frac{r^2}{4}+x_2^2\right)-450\right]$$

We substitute $r = 6$ and $\pi = 3.14$ into the equation to simplify the expression:

$$L(x_1, x_2, \lambda) = 9.42x_1 - 14.13x_2 - 0.52x_2^3 + 904.32$$
$$- \lambda\left(9.42\sqrt{x_1^2+9} - 3.14x_2^2 - 26.10\right)$$

Taking the partial derivatives of L with respect to x_1, x_2, and λ and setting them equal to zero gives

$$\frac{\partial L}{\partial x_1} = 9.42 - 9.42\lambda \cdot x_1(x_1^2 + 9)^{-1/2} = 0$$

$$\frac{\partial L}{\partial x_2} = -14.13 - 1.57x_2^2 + 2\lambda \cdot 3.14x_2 = 0$$

$$\frac{\partial L}{\partial \lambda} = -9.42\sqrt{x_1^2 + 9} + 3.14x_2^2 + 26.10 = 0$$

Using a computer algebra system, we find the solution to be

$$x_1 = 1.95 \text{ ft}, \ x_2 = 1.56 \text{ ft}, \ \lambda = 1.83, \text{ which implies } f(x_1, x_2) = 898.72 \text{ ft}^3$$

Model sensitivity   The Lagrange Multiplier value $\lambda = 1.8$ means that if the total surface area is increased by 1 unit, the volume of the water container will increase by approximately 1.8 ft³.

8.4 PROBLEMS

1. Resolve the oil transfer problem when the storage capacity is 25 ft^3. How does this result compare with our estimated value?

2. Resolve the water container problem when the surface area available is 500 ft^2 and the radius is 9 ft.

Use the method of Lagrange Multipliers to solve Problems 3–6.

3. Find the minimum distance from the surface $x^2 + y^2 - z^2 = 1$ to the origin.

4. Find three numbers whose sum is 9 and whose sum of squares is as small as possible.

5. Find the hottest point (x, y, z) along the elliptical orbit

$$4x^2 + y^2 + 4z^2 = 16$$

where the temperature function is

$$T(x, y, z) = 8x^2 + 4yz - 16z + 600$$

6. *A Least-Squares Plane*   Given the four points (x_k, y_k, z_k)

$$(0, 0, 0), (0, 1, 1), (1, 1, 1), (1, 0, -1)$$

find the values of $A, B,$ and C to minimize the sum of squared errors

$$\sum_{k=1}^{4} (Ax_k + By_k + C - z_k)^2$$

if the points must lie in the plane

$$z = Ax + By + C$$

7. Resolve the oil transfer problem if we introduce a second storage tank that can be filled to its capacity of 30 ft^3. (*Hint:* You might reformulate with four variables x_{ij}, the amount of oil type i stored in storage tank j.)

8.4 PROJECTS

1. "Lagrange Multipliers and the Design of Multistage Rockets," by Anthony L. Peressini, UMAP 517. The method of Lagrange Multipliers is applied to compute the minimum total mass of an n-stage rocket capable of placing a given payload in an orbit at a given altitude above the Earth's surface. Familiarity with elementary minimization techniques for functions of several variables, the method of Lagrange Multipliers, and the concepts of linear momentum and conservation of momentum are required.

2. "Lagrange Multipliers: Applications to Economics," by Christopher H. Nevison, UMAP 270. The Lagrange Multiplier Method is interpreted and studied as the marginal rate of change of a utility function. Differential calculus through Lagrange Multipliers is required.

3. Research the requirements for the necessary and sufficient conditions for the method of Lagrange Multipliers and prepare a 10-minute talk.

8.5   MANAGING RENEWABLE RESOURCES: THE FISHING INDUSTRY[2]

Consider the plight of the Antarctic baleen whale, which yielded a peak catch of 2.8 million tons in 1937 but only 50,000 tons in 1978. Or the case of the Peruvian anchoveta, which yielded 12.3 million tons in 1970 but only 500,000 tons just 8 years later. The anchoveta fishery was Peru's largest industry and the world's most productive fishery. The seriousness of the economic impact is realized when we consider biologists' estimates that, even without fishing, it will take several years, or even decades in the case of the baleen whale, for these fisheries to reach their former levels of maximum biological productivity.[3]

Resources such as the anchoveta and baleen whale are called *renewable* (as opposed to *exhaustible*) resources. Exhaustible resources can yield only a finite total amount, whereas renewable resources can (theoretically) yield an unlimited total amount and be maintained at some positive level. The management of a renewable resource, such as a fishery, involves several critical considerations. What should the harvest rate be? How sensitive is the survival of the species to population fluctuation caused by harvesting, or to natural disasters such as a temporary alteration in the ocean currents (which contributed to the demise of the anchoveta)? The economist Adam Smith (1723–1790) proclaimed that each individual in pursuing only his or her own selfish good was led by an invisible hand to achieve the best for all. Will that invisible hand really ensure that market forces work in the best interest of humanity and the renewable resources? Or will intervention be required to improve the situation, either for humanity or for the resource? In this section we use graphical submodels to gain a qualitative understanding of these management issues.

Scenario   Consider the harvesting of a common fish, such as haddock, in a large competitive fishing industry. *Given the population level of the fish at some time, what happens to future population levels?* Future population levels depend, among other factors, on the harvesting rate of the fish and their natural reproductive rate (births minus deaths per unit time). Let's develop submodels for harvesting and reproduction separately.

---

[2]Optional
[3]Data from Colin W. Clarke's article, "The Economics of Over-exploration," *Science* 181 (1973a): 630–634.

A harvesting submodel   The classical theory of the firm was presented in Chapter 1, and the graphical models for total profit and marginal profit are reproduced in Figure 8.13. From Figure 8.13a we can see that a firm breaks even only when it can produce at least q_1 items, where total revenue equals total cost. To maximize profits the firm should continue to produce items as long as the marginal revenue exceeds the marginal cost; that is, produce up to the quantity q_2 as depicted in Figure 8.13b.

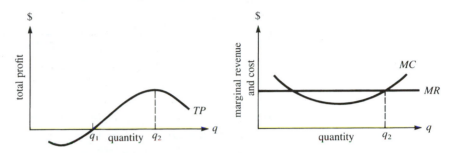

FIGURE 8.13   A graphical model for the theory of the firm

How does the theory of the firm relate to the fishing industry? For a common fish such as the haddock and a competitive fishing industry, it is probably reasonable to assume that price is a constant. If a common fish cannot be marketed at that price, consumers will simply switch to another type of fish. Likewise, in a large industry the quantity of a particular fish marketed by an individual firm should not affect the price of the fish. Hence, over a wide range of values, constant price appears to be a reasonable initial assumption.

Next, let's consider the costs a fishing company encounters. These include salaries, fuel capital tied up in equipment, processing, and refrigeration. As usual, once a time horizon is selected, these costs can be divided into two categories: fixed costs (such as capital costs), which are independent of the yield harvested, and variable costs (such as processing), which depend on the size of the yield. The basic principles underlying the theory of the firm apply reasonably well. For example, we would expect a typical fishing company must harvest some minimum yield at a given level of effort (number of boats, person-hours of labor, and so forth) to break even.

An interesting condition exists with the fishing industry: The cost of harvesting a given number of fish depends on *the size of the fish population*. Certainly less effort is required to catch a given number of fish when they are plentiful, so at a given level of effort we would expect to catch more fish when they are plentiful. Thus we assume the average unit harvesting cost $c(N)$ decreases as the size of the fish population N increases. This assumption is depicted in Figure 8.14, where the harvesting cost per fish is shown as a decreasing function of the size of the fish population. It is important to note that the independent variable N is the size of the fish population, not the size of the yield. The market price p of a fish is also shown in the figure.

The submodel represented by Figure 8.14 suggests that it would be unprofitable to harvest a particular species of fish unless the population level is at least N_L, where the cost of harvesting each fish equals the price paid for it by the con-

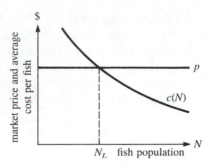

FIGURE 8.14   The (average) harvesting cost per fish decreases as the fish population level increases

sumer. In those instances in which it is economically feasible to harvest the species, harvesting is a force driving the population level to N_L. If the population level lies significantly above N_L, then the excess profit potential causes the fishing of the species to be intensified. On the other hand, if the fish population falls below N_L, the fishing would cease (theoretically), causing the population level to increase.

We have reached an impasse with the harvesting submodel. Although we intended to develop the two submodels independently, we now realize that the average cost of harvesting a fish depends on the size of the fish population and hence its reproductive capacity. Even if we know the current size of the fish population, we still need to know how the magnitudes of the harvesting and reproduction rates compare to determine whether the population will increase or decrease. We turn now to the development of a simple reproduction submodel.

A reproduction submodel   Consider how a species of fish grows in the absence of fishing. Among the many factors affecting its net growth rate are the birthrate, death rate, and environmental conditions. Each of these factors will be developed in detail in Chapter 10, where we construct several population models. For our purposes here, a simple qualitative graphical model will suffice. Let $N(t)$ denote the size of the fish population at any time t, and let $g(N)$ represent the rate of growth of the function $N(t)$. Assume that $g(N)$ is approximated by the continuous model $g(N) = dN/dt$. When $N = 0$, there are neither births nor deaths. Now assume there is a maximum population level N_u that can be supported by the environment. This level might be imposed by the availability of food, the effect of predators, or some similar inhibitor. At N_u we assume births equal deaths, or $g(N_u) = 0$. Thus under our assumptions, $g(N)$ has zeros at $N = 0$ and $N = N_u$ and positive values in between. A graphical representation of g is depicted in Figure 8.15.

Before proceeding, we offer one analytical submodel for reproduction as an illustration. A simple quadratic function having zeros at $N = 0$ and $N = N_u$ is $g(N) = aN(N_u - N)$, where a is a positive constant. Note that the quadratic is positive for $0 < N < N_u$ and negative for $N > N_u$ as required. Because $g'(N) = a(N_u - 2N), N = N_u/2$ is a critical point of g. Also, $g''(N) = -2a$ implies g is everywhere concave downward, as depicted in Figure 8.15.

Let's see what the submodel for reproduction suggests about the population level. Suppose the population level is currently at level N_1 (see Figure 8.15). Because

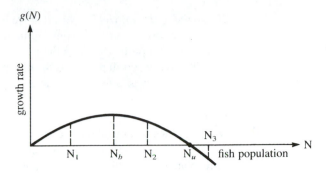

FIGURE 8.15 A submodel for reproduction

$g(N_1) > 0$ and $g(N) = dN/dt$, the function N is increasing. As N increases, $g(N) = dN/dt$ increases as well until it reaches a maximum at $N = N_b$. For $N > N_b$ the derivative remains positive but becomes smaller and approaches zero as N gets closer to N_u. Thus the population approaches $N = N_u$ as suggested in Figure 8.16. Convince yourself that the curves suggested for starting populations of $N = N_2$ and $N = N_3$ are qualitatively correct in Figure 8.16 as well, where $N_1 < N_b < N_2 < N_u < N_3$. Note that the ordinate in Figure 8.15 is the slope of the curves in Figure 8.16. In Chapter 10 we develop an analytical model using differential equations to represent the population growth in Figure 8.16, which is called *logistic* growth.

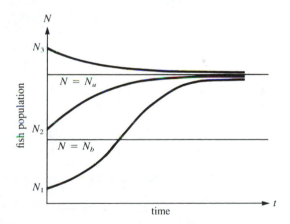

FIGURE 8.16 Regardless of the initial population, the population approaches $N = N_u$ without fishing

The biological optimum population level We now interpret our submodel for reproduction. In the absence of fishing, the population always approaches N_u, which is called the *carrying capacity* of the environment. Moreover, a population level exists where the net biological growth is at a maximum. This population level is called *biological optimum population* and is denoted by N_b. In our submodel, $N_b = N_u/2$, as established earlier and depicted in Figure 8.15. (We establish this result more precisely in Chapter 10.)

The social optimum yield Now that we have developed a submodel for reproduction, let's return to our harvesting submodel. Assume the fishing industry seeks to maximize total profit, which is the product of the yield and the average profit per fish. The average profit per fish is the difference between the price p and the average cost $c(N)$ to harvest a fish. Thus the expression for total profit TP is

$$TP = (\text{yield}) \times [p - c(N)] \qquad (8.11)$$

What shall we assume about the yield? From the fishing company's point of view, the ideal situation is to have a constant yield from year to year. This situation would allow planning for efficient staffing and capitalization of the resources allocated to fishing that particular species. Otherwise, in lean years there would be resources, such as fishing vessels and staff, that would be underused. It is also reasonable to assume that consumer demand for a certain kind of fish is pretty much constant from year to year, say for a 3- or 4-yr period. Let's see where the assumption of constant yield leads us.

First, how is constant yield obtained from year to year? One way is to harvest the difference between the births and deaths over time. Because the function $g(N)$ approximates this difference for a given population N of fish, this idea suggests that $g(N)$ equals the yield. To illustrate the idea, consider Figure 8.17. Suppose the fish population is presently at N_2. If precisely the amount of fish $g(N_2)$ is harvested, the population will remain at N_2 (zero net growth) and the fishing operation could be repeated annually indefinitely. Convince yourself that any population level N with $0 < N < N_u$ can be maintained by harvesting the amount $g(N)$ per year.

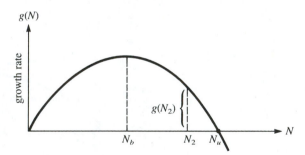

FIGURE 8.17 Harvesting $g(N)$ annually permits a sustainable yield

Note that N_b, the population level for which the biological growth is a maximum, is of particular importance under the interpretation of a sustainable yield. This is true because N_b is the population level that allows the largest sustainable yield, $g(N_b)$. Because this population level has important social implications, we refer to N_b as the *social optimum population level*. Note that the social and biological optimum levels are coincident in this example.

The economic optimum population level Assuming the yield to be $g(N)$, Equation (8.11) for total profit becomes

$$TP = g(N)[p - c(N)] \qquad (8.12)$$

The fishing industry wants to maximize *TP*. What level of fish population permits the greatest profit? Because $g(N)$ reaches a maximum at $N = N_b$, we might be tempted to conclude that profit is maximized there also. After all, the greatest yield occurs there. However, the average profit continues to increase as N increases, based on our submodel depicted in Figure 8.14. Thus, by choosing $N > N_b$ we may increase the profit while simultaneously catching fewer fish because $g(N) < g(N_b)$. Let's see if this possibility is indeed the case.

What does the graph of the total profit function *TP* look like? The factor $g(N)$ has zeros at $N = 0$ and $N = N_u$, and the factor $[p - c(N)]$ has a zero at $N = N_L$. One representation of a continuous function meeting these requirements is shown in Figure 8.18 and suggests that a population level does exist at which profits can be maximized. We call that population level the *economic optimum population level* and denote it by N_p.

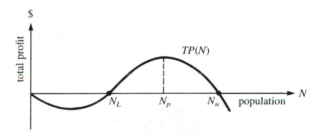

FIGURE 8.18   A continuous total profit function

Relating all optimum population levels   Up to this point several population levels of interest to us have been introduced. They are summarized as follows:

N_L = minimum population level for economic feasibility

N_b = social and biological optimum population levels

N_p = economic optimum population level

N_u = maximum population sustainable by the environment (the environmental carrying capacity)

Next we relate these population levels. To be worthwhile to fish a species on a prolonged basis, it must be the case that $N_L < N_u$, with N_b and N_p somewhere in between. But how are N_b and N_p related? If we believe that the fishing industry will eventually find the population level that maximizes profit, we would hope for social and biological considerations that $N_p = N_b$. Let's see if Adam Smith's invisible hand is at work.

Consider Figure 8.19. For the case $N_L < N_p < N_u$, three possible locations for N_b are shown on the graph for the total profit function *TP*. Study the graph and convince yourself of the following situations:

1. $N_b < N_p$ if and only if $TP'(N_b) > 0$
2. $N_b = N_p$ if and only if $TP'(N_b) = 0$
3. $N_b > N_p$ if and only if $TP'(N_b) < 0$

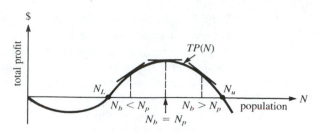

FIGURE 8.19 Three possible locations for N_b are $N_b < N_p$, $N_b = N_p$, $N_b > N_p$

Now, taking the derivative of Equation (8.12), we have

$$TP' = g'(N)[p - c(N)] - g(N)c'(N) \tag{8.13}$$

From the definition of N_b, $g'(N_b) = 0$ and $g(N_b) > 0$, and from the submodel for average cost depicted in Figure 8.14, $c'(N) < 0$ for all N. Substitution of N_b into Equation (8.13) yields $TP'(N_b) > 0$, which implies that $N_b < N_p$ from Figure 8.19. Thus under our assumptions, by allowing the fish population to exceed N_b, the fishing company can catch fewer fish yet reap greater profits than is possible at N_b! Let's interpret our model to assist management in determining the alternatives for controlling the location of the various population levels N_L, N_b, and N_p.

Model interpretation　As we will see in Chapter 10, the assumptions underlying the reproduction submodel are simplistic, so we cannot expect to draw precise conclusions from the graphical models developed here. However, our purpose in constructing the model was to identify and analyze qualitatively some of the key issues in managing a renewable resource. Consider again the baleen whale and Peruvian anchoveta in terms of the models we have developed.

Several considerations must be taken into account with whales. First, the whale population displays a low natural growth rate (typically only 5–10% a year), and even a small harvest can result in a negative net growth rate when the population is low. Second, many conservationists argue that there is a minimum population N_s for the whale below which the species will not survive. If harvesting or natural disasters cause the whale population to fall below N_s, the species will be driven to extinction. These ideas are incorporated in the graphical representations depicted in Figure 8.20, showing both the reproduction and population submodels.

Note from Figure 8.20 how sensitive the future whale population is to levels in the vicinity of N_s. The relatively high market price for the whale causes the minimum population N_L for economic feasibility to be quite low (see Figure 8.14) so that N_L can be quite near the level N_s. Thus the location of N_L becomes critical. (Argue that in the case of the whale, N_L could be smaller than N_s. Management of the fishery in such a case requires separating N_s and N_L.) Ideally, we would like N_L and N_b to coincide. Taxation and the establishment of quotas are two alternatives discussed in the following problem set.

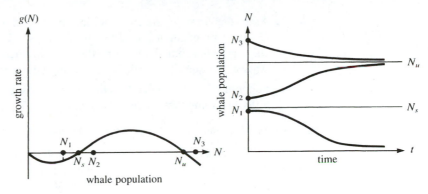

FIGURE 8.20   Reproduction and population models incorporating a minimum survival level

In the case of the Peruvian anchoveta, it is estimated that the maximum sustainable yield is 10 million tons annually. Thus any additional catch (such as the 12.3 million tons caught in 1970) would result in a population decrease, even if the population were at the biological optimum level N_b. From Figure 8.17 we suspect that as the population decreases from the biological optimum population, the sustainable yield decreases, suggesting that the harvest should be reduced. However, from a practical point of view, once the fishing fleet and staff are established to catch 12.3 million tons, there is strong economic motivation to continue harvesting at that level by fishing more intensively. This critical situation existed during the 1970s in Peru, where there were few economic alternatives and government intervention to enforce strict regulation of the fishery was politically volatile. We can well imagine how these practical considerations become substantially more complex when more than one country is involved in managing the resource.

8.5 PROBLEMS

1. Assume that the environmental carrying capacity N_u is determined mainly by the availability of food. Argue that under such an assumption, as N approaches N_u the physical condition of the average fish deteriorates as competition for the food supply becomes more severe. What does this suggest about the survival of the species when natural disasters such as storms, severe winters, and similar circumstances further restrict the food supply? Where should conservationists attempt to maintain the population level?

2. In 1981 and 1982 the deer population in the Florida Everglades was very high. Although the deer were plentiful, they were very thin and on the brink of starvation. Hunting permits were issued to thin out the herd. This action caused much furor on the part of environmentalists and conservationists. Explain the poor health of the deer and the purpose of the special hunting permits in terms of population growth and population submodels.

3. Argue that for many species a minimum population level is required for survival. Give several examples. Call this minimum survival level N_s. Suggest a simple cubic growth submodel meeting these requirements, as depicted in Figure 8.20. Answer the questions in Problem 1 using your graphical submodel.

4. Suppose $N_u < N_L$. What does this inequality suggest about the economic feasibility of fishing that species? Give several examples.

Problems 5 and 6 relate to fishing regulation.

5. One of the key assumptions underlying the models developed in this section is that the harvest rate equals the growth rate for a sustainable yield. The reproduction submodels in Figures 8.17 and 8.20 suggest that if the current population levels are known, it is possible to estimate the growth rate. The implication of this knowledge is that if a quota for the season is established based on the estimated growth rate, then the fish population can be maintained, increased, or decreased as desired. This quota system might be implemented by requiring all commercial fishermen to register their catch daily, and then closing the season when the quota is reached. Discuss the difficulties in determining reproduction models precise enough to be used in this manner. How would you estimate the population level? What are the disadvantages of having a quota that varies from year to year? Discuss the practical political difficulties of implementing such a procedure.

6. One of the difficulties in managing a fishery in a free enterprise system is that excess capacity can be created through overcapitalization. This happened in 1970 when the capacity of the Peruvian anchoveta fishermen was sufficient to catch and process the maximum annual growth rate in less than 3 months. A disadvantage of restricting access to the fishery by closing the season after a quota is reached is that this excess capacity is idle during much of the season, which creates a politically and economically unsatisfying situation. An alternative is to control the capacity in some manner. Suggest several procedures for controlling the capacity that is developed. What difficulties would be involved in implementing a procedure such as restricting the number of commercial fishing permits issued?

Problems 7–9 relate to taxation.

7. Figure 8.14 suggests that market forces tend to drive the population to N_L. Use that figure to show how taxation or subsidization can be used to control the location of N_L. What forms might the taxation and subsidization take? (*Hint:* One cost to the fisherman is the various taxes he pays.) Apply your ideas to the whale fishery.

8. Taxation is appealing from a theoretical perspective because with a properly designed tax desired goals can be achieved through normal market forces rather than by some artificial method (such as restricting the number of commercial permits). Assume a fish population is currently at N_b and you desire to maintain it at that level by harvesting $g(N_b)$. How can a tax be determined for each fish caught to cause N_L, N_b, and N_p to coincide? (*Hint:* Consider Equation (8.11) and the condition required for $N_b = N_p$.)

9. A constant price has been assumed in all the models developed in this section. Suggest some fisheries for which that assumption is not realistic. How might you alter the assumption? How would you determine the appropriate tax to be applied?

8.5 Further Reading

CLARK, Colin W. *Mathematical Bioeconomics: The Optimal Management of Renewable Resources*. New York: Wiley, 1976.

MAY, Robert M., John R. Beddington, Colin W. Clark, Sidney J. Holt, & Richard M. Laws. "Management of Multispecies Fisheries." *Science* 205 (July 1979): 267–277.

Chapter Nine

LINEAR PROGRAMMING AND NUMERICAL SEARCH METHODS

INTRODUCTION

In Chapter 5 we considered three criteria for fitting a selected model to a collection of data:

1. Minimize the sum of the absolute deviations
2. Minimize the largest of the absolute deviations (Chebyshev criterion)
3. Minimize the sum of the squared deviations (least-squares criterion)

Also in Chapter 5 we used calculus to solve the optimization problem resulting from the application of the least-squares criterion. Although we formulated several optimization problems resulting from the first criterion to minimize the sum of the absolute deviations, we were unable to solve the resulting mathematical problem. In Section 9.5 we study several search techniques that allow us to find good solutions to that curve-fitting criterion and examine many other optimization problems as well.

In Chapter 5 we interpreted the Chebyshev criterion for several models. For example, given a collection of m data points (x_i, y_i), $i = 1, 2, \ldots, m$, fit the collection to the line $y = ax + b$ (determined by the parameters a and b) that minimizes the greatest distance r_{max} between any data point (x_i, y_i) and its corresponding point $(x_i, ax_i + b)$ on the line. That is, the largest absolute deviation, $r = \text{Maximum}\{|y_i - y(x_i)|\}$, is minimized over the entire collection of data points. This criterion defines the optimization problem to

$$\text{Minimize } r$$

Subject to

$$\left. \begin{array}{l} r - r_i \geq 0 \\ r + r_i \geq 0 \end{array} \right\} \text{ for } i = 1, 2, \ldots, m$$

which is a *linear program* for many applications. In Chapter 8 we saw several other applications of linear programs as well (such as maximizing profit subject to constraints on labor and capital resources). In Sections 9.1 through 9.3 we will learn how to solve linear programs geometrically and algebraically. We will learn how to determine the sensitivity of the optimal solution to the coefficients appearing in the linear program in Section 9.4. Let's begin by finding geometrical solutions to curve-fitting problems and other linear programs.

9.1   LINEAR PROGRAMMING I: GEOMETRIC SOLUTIONS

Consider using the Chebyshev criterion to fit the model $y = cx$ to the following data set:

x	1	2	3
y	2	5	8

The optimization problem that determines the parameter c to minimize the largest absolute deviation $r_i = |y_i - y(x_i)|$ (residual or error) is the linear program

$$\text{Minimize } r$$

Subject to

$$
\begin{array}{ll}
r - (2 - c) \geq 0 & \text{(constraint 1)} \\
r + (2 - c) \geq 0 & \text{(constraint 2)} \\
r - (5 - 2c) \geq 0 & \text{(constraint 3)} \\
r + (5 - 2c) \geq 0 & \text{(constraint 4)} \\
r - (8 - 3c) \geq 0 & \text{(constraint 5)} \\
r + (8 - 3c) \geq 0 & \text{(constraint 6)}
\end{array}
\right\} \qquad \textbf{(9.1)}
$$

In this section we solve this problem geometrically. But first, let's use a simpler example to define a linear program for which we describe the geometrical solution procedure.

Defining a Linear Program: The Carpenter's Problem

Recall the production schedule example from Section 8.1, which we restate here. A carpenter makes tables and bookcases for a net unit profit he estimates as $25 and $30, respectively. He is trying to determine how many of each piece of furniture to make each week. He has up to 690 board feet of lumber to devote weekly to the project and up to 120 hours of labor. He can use the lumber and labor productively elsewhere if they are not used in the production of tables and bookcases. He estimates that it requires 20 board feet of lumber and 5 hr of labor to complete a table and 30 board feet of lumber and 4 hr of labor for a bookcase. He also estimates that he can sell all the tables and bookcases produced. The carpenter wishes to determine a weekly production schedule for tables and bookcases that maximizes his profits.

Let x_1 denote the number of tables to be produced weekly, and let x_2 denote the number of bookcases. Then the model becomes

$$\text{Maximize } 25x_1 + 30x_2$$

Subject to

$$
\begin{aligned}
20x_1 + 30x_2 &\leq 690 \quad \text{(lumber)} \\
5x_1 + 4x_2 &\leq 120 \quad \text{(labor)} \\
x_1, x_2 &\geq 0 \quad\ \ \text{(nonnegativity)}
\end{aligned}
$$
(9.2)

This model is an example of a linear program as defined in Section 8.1. Remember that $25x_1 + 30x_2$ is called the objective function, and x_1 and x_2 are the decision variables. We restrict our attention to linear programs involving two decision variables.

Interpreting a Linear Program Geometrically

Linear programs can include a set of constraints that are linear equations or linear inequalities. In the case of two decision variables, an equality requires that solutions to the linear program lie precisely on the line representing the equality. What about inequalities?

To gain some insight, consider the constraints

$$
\begin{aligned}
x_1 + 2x_2 &\leq 4 \\
x_1, x_2 &\geq 0
\end{aligned}
$$
(9.3)

The **nonnegativity** constraints $x_1, x_2 \geq 0$ mean that possible solutions lie in the first quadrant. The inequality $x_1 + 2x_2 \leq 4$ divides the first quadrant into two regions. The **feasible region** is the half-plane in which the constraint is satisfied. The feasible region can be found by graphing the equation $x_1 + 2x_2 = 4$ and determining which half-plane is feasible, as shown in Figure 9.1.

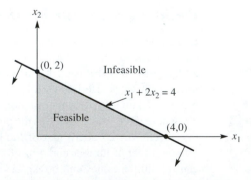

FIGURE 9.1   The feasible region for the constraints $x_1 + 2x_2 \leq 4, \quad x_1, x_2 \geq 0$

If the feasible half-plane fails to be obvious, choose a convenient point (such as the origin) and substitute it into the constraint to determine if it's satisfied. If it is, then all points on the same side of the line as this point will also satisfy the constraint. Now let's find the feasible region and the optimal solution for the carpenter's problem formulated as given.

Example 1   The Carpenter's Problem

A linear program has the important property that the points satisfying the constraints form a **convex set**, which is a set in which any two of its points are joined by a straight-line segment all of whose points lie within the set. The set depicted in Figure 9.2a fails to be convex, whereas the set in Figure 9.2b is convex.

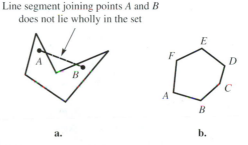

FIGURE 9.2   The set shown in **a.** is not convex, whereas the set shown in **b.** is convex

An **extreme point** (corner point) of a convex set is any boundary point in the convex set that is the unique intersection point of two of the (straight-line) boundary segments. In Figure 9.2b, the points A, B, C, D, E, F are extreme points.

Let's illustrate these ideas with the carpenter's problem. The convex set for the constraints in the carpenter's problem is graphed and given by the polygon region $ABCD$ in Figure 9.3. Note that there are six intersection points of the constraints, but only four of these points (namely, $A, B, C,$ and D) satisfy all the constraints and hence belong to the convex set. The points A, B, C, D are the extreme points of the polygon. The variables y_1 and y_2 will be explained later in the sequel.

If an optimal solution to a linear program exists, it turns out that it must occur among the extreme points of the convex set formed by the set of constraints. The values of the objective function (profit for the carpenter's problem) at the extreme points are

Extreme point	Objective function value
A $(0, 0)$	$0
B $(24, 0)$	600
C $(12, 15)$	750
D $(0, 23)$	690

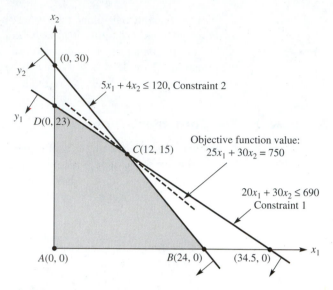

FIGURE 9.3 The set of points satisfying the constraints of a linear program form a convex set

Thus the carpenter should make 12 tables and 15 bookcases each week to earn a maximum weekly profit of $750. We provide further geometrical evidence later in this section that extreme point C is optimal.

Before considering a second example, let's summarize the ideas presented thus far. The constraint set to a linear program is a convex set, which generally contains an infinite number of feasible points to the linear program. If an optimal solution to the linear program exists, it must be taken on at one or more of the extreme points. Thus, to find an optimal solution we choose from among all the extreme points the one with the best value for the objective function.

Example 2 A Data-Fitting Problem

Let's now solve the linear program represented by Equation (9.1). Given the model $y = cx$ and the data set

x	1	2	3
y	2	5	8

we wish to find a value for c such that the resulting largest absolute deviation is as small as possible. In Figure 9.4 we graph the set of six constraints

$$r - (2 - c) \geq 0 \quad \text{(constraint 1)}$$
$$r + (2 - c) \geq 0 \quad \text{(constraint 2)}$$
$$r - (5 - 2c) \geq 0 \quad \text{(constraint 3)}$$
$$r + (5 - 2c) \geq 0 \quad \text{(constraint 4)}$$

$$r - (8 - 3c) \geq 0 \quad \text{(constraint 5)}$$
$$r + (8 - 3c) \geq 0 \quad \text{(constraint 6)}$$

by first graphing the equations

$$r - (2 - c) = 0 \quad \text{(constraint 1 boundary)}$$
$$r + (2 - c) = 0 \quad \text{(constraint 2 boundary)}$$
$$r - (5 - 2c) = 0 \quad \text{(constraint 3 boundary)}$$
$$r + (5 - 2c) = 0 \quad \text{(constraint 4 boundary)}$$
$$r - (8 - 3c) = 0 \quad \text{(constraint 5 boundary)}$$
$$r + (8 - 3c) = 0 \quad \text{(constraint 6 boundary)}$$

We note that constraints 1, 3, and 5 are satisfied above and to the right of the graph of their boundary equations. Similarly, constraints 2, 4, and 6 are satisfied above and to the left of their boundary equations. To convince yourself, pick a point (such as the origin) and determine if the point satisfies the constraint. If it does, it must be in the feasible region determined by the constraint.

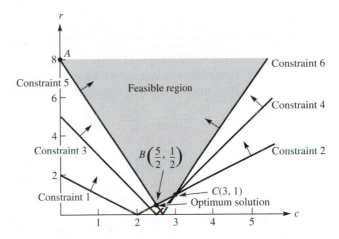

FIGURE 9.4   The feasible region for fitting $y = cx$ to a collection of data

The intersection of all the feasible regions for constraints 1–6 form a convex set in the c, r-plane with extreme points labeled A, B, and C in Figure 9.4. The point A is the intersection of constraint 5 and the r-axis: $r - (8 - 3c) = 0$ and $c = 0$, or $A = (0, 8)$. Similarly, B is the intersection of constraints 5 and 2:

$$r - (8 - 3c) = 0 \quad \text{or} \quad r + 3c = 8$$
$$r + (2 - c) = 0 \quad \text{or} \quad r - c = -2$$

yielding $c = \frac{5}{2}$ and $r = \frac{1}{2}$, or $B = (\frac{5}{2}, \frac{1}{2})$. Finally, C is the intersection of constraints 2 and 4 yielding $C = (3, 1)$. Note that the set is unbounded. (We discuss unbounded convex sets later.) If an optimal solution to the problem exists, at least one extreme point must take on the optimal solution. We now evaluate the objective function $f(r) = r$ at each of the three extreme points.

Extreme point	Objective function value
(c, r)	$f(r) = r$
A	8
B	$\frac{1}{2}$
C	1

The extreme point with the smallest value of r is the extreme point B with coordinates $(\frac{5}{2}, \frac{1}{2})$. Thus $c = \frac{5}{2}$ is the optimal value of c. No other value of c will result in a largest absolute deviation as small as $|r_{max}| = \frac{1}{2}$.

Model interpretation Let's interpret the optimal solution for the preceding data-fitting problem. Resolving the linear program, we obtain a value of $c = \frac{5}{2}$ corresponding to the model $y = \frac{5}{2}x$. Further, the objective function value $r = \frac{1}{2}$ should correspond to the largest deviation resulting from the fit. Let's check to see if that's true.

The data points and the model $y = \frac{5}{2}x$ are plotted in Figure 9.5. Note that a largest deviation of $r_i = \frac{1}{2}$ occurs for both the first and third data points. Fix one end of a ruler at the origin. Rotate the ruler to convince yourself geometrically that no other line passing through the origin can yield a smaller largest absolute deviation. Thus the model $y = \frac{5}{2}x$ is optimal by the Chebyshev criterion.

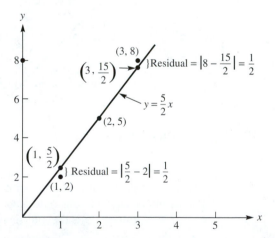

FIGURE 9.5 The line $y = (5/2)x$ results in a largest absolute deviation $r_{max} = \frac{1}{2}$, the smallest possible r_{max}

Empty and unbounded feasible regions We have been careful to say that if an optimal solution to the linear program exists, at least one of the extreme points must take on the optimal value for the objective function. When does an optimal solution fail to exist? Moreover, when does more than one optimal solution exist?

If the feasible region is empty, no feasible solution can exist. For example, given the constraints

$$x_1 \leq 3$$

and

$$x_1 \geq 5$$

there is no value of x_1 that satisfies both of them. We say that such constraint sets are *inconsistent*.

There is another reason an optimal solution may fail to exist. Consider Figure 9.4 and the constraint set for the data-fitting problem in which we noted that the feasible region is *unbounded* (in the sense that either x_1 or x_2 can become arbitrarily large). Then it would be impossible to

$$\text{Maximize } x_1 + x_2$$

over the feasible region because x_1 and x_2 can take on arbitrarily large values. Note, however, that even though the feasible region is unbounded, an optimal solution *does* exist for the objective function we considered in Example 2. So it is not *necessary* for the feasible region to be bounded for an optimal solution to exist.

Level curves of the objective function   Consider again the carpenter's problem. The objective function is $25x_1 + 30x_2$, and in Figure 9.6 we plot the lines

$$25x_1 + 30x_2 = 650$$
$$25x_1 + 30x_2 = 750$$
$$25x_1 + 30x_2 = 850$$

in the first quadrant.

Note that the objective function has constant values along these line segments. The line segments are called **level curves** of the objective function. As we move in a direction perpendicular to these line segments, the objective function either increases or decreases. Now superimpose the constraint set from the carpenter's problem

$$20x_1 + 30x_2 \leq 690 \quad \text{(lumber)}$$
$$5x_1 + 4x_2 \leq 120 \quad \text{(labor)}$$
$$x_1, x_2 \geq 0 \quad \text{(nonnegativity)}$$

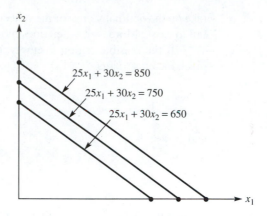

FIGURE 9.6 The level curves of the objective function f are parallel line segments in the first quadrant; the objective function either increases or decreases as we move in a direction perpendicular to the level curves

onto these level curves (see Figure 9.7). Notice that the level curve with value 750 is the one that intersects the feasible region exactly once at the extreme point $C(12, 15)$.

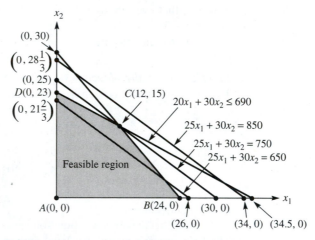

FIGURE 9.7 The level curve $25x_1 + 30x_2 = 750$ is tangent to the feasible region at extreme point C

Can there be more than one optimal solution? Consider the following slight variation of the carpenter's problem in which the labor constraint has been changed:

$$\text{Maximize } 25x_1 + 30x_2$$

Subject to

$$20x_1 + 30x_2 \leq 690 \quad \text{(lumber)}$$
$$5x_1 + 6x_2 \leq 150 \quad \text{(labor)}$$
$$x_1, x_2 \geq 0 \quad \text{(nonnegativity)}$$

The constraint set and the level curve $25x_1 + 30x_2 = 750$ are graphed in Figure 9.8. Notice that the level curve and boundary line for the labor constraint coincide. Thus both extreme points B and C have the same objective function value of 750, which is optimal. In fact, the entire line segment BC coincides with the level curve $25x_1 + 30x_2 = 750$. Thus there are infinitely many optimal solutions to the linear program, all along line segment BC.

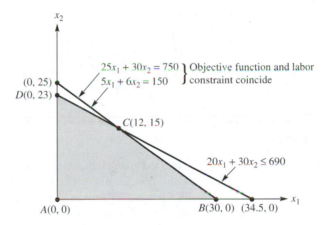

FIGURE 9.8 The line segment BC coincides with the level curve $25x_1 + 30x_2 = 750$; every point between extreme points C and B, as well as extreme points C and B, is an optimal solution

In Figure 9.9 we summarize the general two-dimensional case for optimizing a linear function on a convex set. The figure shows a typical convex set together with the level curves of a linear objective function. Figure 9.9 provides geometrical intuition for the following fundamental theorem of linear programming.

THEOREM 9.1 Suppose the feasible region of a linear program is a nonempty and bounded convex set. Then the objective function must attain both a maximum and minimum value occurring at extreme points of the region. If the feasible region is unbounded, the objective function need not assume its optimal values. If either a maximum or minimum does exist, it must occur at one of the extreme points.

The power of this theorem is that it guarantees an optimal solution to a linear program from among the extreme points of a bounded nonempty convex set.

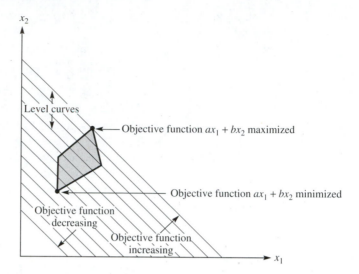

FIGURE 9.9 A linear function assumes its maximum and minimum value on a nonempty and bounded convex set at an extreme point

9.1 PROBLEMS

1. Consider a company that carves wooden soldiers. The company specializes in two main types: Confederate soldiers and Union soldiers. The profit for each is $28 and $30, respectively. A Confederate soldier requires 2 units of lumber, 4 hr of carpentry, and 2 hr of finishing to complete. A Union soldier requires 3 units of lumber, 3.5 hr of carpentry, and 3 hr of finishing to complete. Each week the company has 100 units of lumber delivered, and there are 120 hr of carpentry time and 90 hr of finishing time available. Determine the number of each type of wooden soldier to produce to maximize weekly profits.

2. A local company restores cars and trucks for resale. Each vehicle must be processed in the refinishing-paint shop and the machine-body shop. Each car contributes on average $3,000 to profit, and each truck contributes on average $2,000 to profit. The refinishing-paint shop has 2400 work-hours available and the machine-body shop has 2500 work-hours available. A car requires 50 work-hours in the machine-body shop and 40 work-hours in the refinishing-paint shop. A truck requires 50 work-hours in the machine-body shop and 60 work-hours in the refinishing-paint shop. Use graphical linear programming to determine a daily production schedule that will maximize the company's profits.

3. A Montana farmer owns 45 acres of land. She is planning to plant each acre with wheat or corn. Each acre of wheat yields $200 in profits; each acre of corn yields $300 in profits. The labor and fertilizer requirements for each follows. The farmer has 100 workers and 120 tons of fertilizer available. Determine how many acres of wheat and corn need to be planted to maximize profits.

	Wheat	Corn
Labor (workers)	3	2
Fertilizer (tons)	2	4

4. Solve the following problems using graphical analysis:

   a. Maximize $x + y$

   Subject to

   $$x + y \leq 6$$
   $$3x - y \leq 9$$
   $$x, y \geq 0$$

   b. Minimize $x + y$

   Subject to

   $$x + y \geq 6$$
   $$3x - y \geq 9$$
   $$x, y \geq 0$$

   c. Maximize $10x + 35y$

   Subject to

   $$8x + 6y \leq 48 \quad \text{(board-feet of lumber)}$$
   $$4x + y \leq 20 \quad \text{(hr of carpentry)}$$
   $$y \geq 5 \quad \text{(demand)}$$
   $$x, y \geq 0 \quad \text{(nonnegativity)}$$

5. Fit the given model to the data using Chebyshev's criterion to minimize the largest deviation.

   a. $y = cx$

y	11	25	54	90
x	5	10	20	30

   b. $y = cx^2$

y	10	90	250	495
x	1	3	5	7

9.1 Further Reading

BAZARRA, M., Hanif D. Sherali, & C. M. Shetty. *Nonlinear Programming: Theory and Algorithms*. 2nd ed. New York: Wiley, 1993.

RAO, S. S. *Optimization: Theory and Applications*. New Delhi: Wiley Eastern Limited, 1979.

WINSTON, Wayne. *Mathematical Programming: Applications and Algorithms*. 2nd ed. Boston: PWS, 1995.

WINSTON, Wayne. *Operations Research: Applications and Algorithms*. 3rd ed. Belmont, CA: Duxbury Press, 1994.

9.2   LINEAR PROGRAMMING II: ALGEBRAIC SOLUTIONS

The graphical solution to the carpenter's problem suggests a rudimentary procedure for finding an optimal solution to a linear program with a nonempty and bounded feasible region:

1. Find all intersection points of the constraints.
2. Determine which intersection points, if any, are feasible to obtain the extreme points.
3. Evaluate the objective function at each extreme point.
4. Choose the extreme point(s) with the largest (or smallest) value for the objective function.

To implement this procedure algebraically, we must characterize the intersection points and the extreme points.

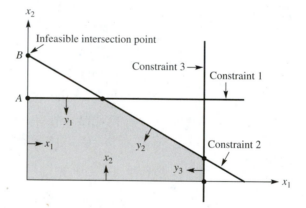

FIGURE 9.10   The variables x_1, x_2, y_1, y_2, and y_3 measure the satisfaction of each of the constraints; intersection point A is characterized by $y_1 = x_1 = 0$; intersection point B is not feasible because y_1 is negative; the circled intersection points surrounding the shaded region are all feasible because none of the five variables is negative there

The convex set depicted in Figure 9.10 consists of three linear constraints (plus the two nonnegativity constraints). The nonnegative variables y_1, y_2, and y_3 in the figure measure the degree by which a point satisfies constraints 1, 2, and 3, respectively. The variable y_i is added to the left side of inequality constraint i to convert it to an equality. Thus $y_2 = 0$ characterizes those points that lie precisely on constraint 2, and a negative value for y_2 indicates the violation of constraint 2.

Likewise, the decision variables x_1 and x_2 are constrained to nonnegative values. Thus the values of the decision variables x_1 and x_2 measure the degree of satisfaction of the nonnegativity constraints $x_1 \geq 0$ and $x_2 \geq 0$. Note that along the x_1-axis, the decision variable x_2 is 0. Now consider the values for the entire set of variables $\{x_1, x_2, y_1, y_2, y_3\}$. If two of the variables simultaneously have the value 0, then we

intersection point   have characterized an **intersection point** in the x_1x_2-plane. All (possible) intersection points can be determined systematically by setting all possible distinguishable pairs of the five variables to zero and solving for the remaining three dependent variables. If a solution to the resulting system of equations exists, then it must be

feasible solution   an intersection point, which may or may not be a **feasible solution**. A negative value for any of the five variables indicates that a constraint is not satisfied. Such an intersection point would be **infeasible**. For example, the intersection point B where $y_2 = 0$ and $x_1 = 0$ gives a negative value for y_1 and hence is not feasible. Let's illustrate the procedure by solving the carpenter's problem algebraically.

Example 1   ***Solving the Carpenter's Problem Algebraically***

Consider again the carpenter's model:

$$\text{Maximize } 25x_1 + 30x_2$$

Subject to

$$20x_1 + 30x_2 \leq 690 \quad \text{(lumber)}$$
$$5x_1 + 4x_2 \leq 120 \quad \text{(labor)}$$
$$x_1, x_2 \geq 0 \quad \text{(nonnegativity)}$$

We convert each of the first two inequalities to equations by adding new nonnegative slack variables y_1 and y_2. If either y_1 or y_2 is negative, the constraint is not satisfied. Thus the problem becomes

$$\text{Maximize } 25x_1 + 30x_2$$

Subject to

$$20x_1 + 30x_2 + y_1 = 690$$
$$5x_1 + 4x_2 + y_2 = 120$$
$$x_1, x_2, y_1, y_2 \geq 0$$

We now consider the entire set of four variables $\{x_1, x_2, y_1, y_2\}$, which are interpreted geometrically in Figure 9.11. To determine a possible intersection point in the x_1x_2-plane, set two of the four variables equal to zero. There are $\frac{4!}{2!\,2!} = 6$ possible intersection points to consider in this way (four variables taken two at a time). Let's begin by setting the variables x_1 and x_2 equal to zero, resulting in the

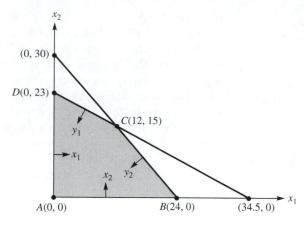

FIGURE 9.11 The variables $\{x_1, x_2, y_1, y_2\}$ measure the satisfaction of each constraint; an intersection point is characterized by setting two of the variables to zero

following set of equations:

$$y_1 = 690$$

$$y_2 = 120$$

which is a feasible intersection point $A(0, 0)$ because all four variables are nonnegative.

For the second intersection point we choose the variables x_1 and y_1 and set them to zero, resulting in the system

$$30x_2 = 690$$

$$4x_2 + y_2 = 120$$

that has solution $x_2 = 23$ and $y_2 = 28$, which is also a feasible intersection point $D(0, 23)$.

For the third intersection point we choose x_1 and y_2 and set them to zero yielding the system

$$30x_2 + y_1 = 690$$

$$4x_2 = 120$$

with solution $x_2 = 30$ and $y_1 = -210$. Thus the first constraint is violated by 210 units, indicating that the intersection point $(0, 30)$ is infeasible.

In a similar manner, choosing y_1 and y_2 and setting them to zero gives $x_1 = 12$ and $x_2 = 15$, corresponding to the intersection point $C(12, 15)$, which is feasible. Our fifth choice is to select the variables x_2 and y_1 and set them to zero, giving values of $x_1 = 34.5$ and $y_2 = -52.5$; therefore, the second constraint is not satisfied. Thus the intersection point $(34.5, 0)$ is infeasible.

Finally we determine the sixth intersection point by setting the variables x_2 and y_2 to zero to determine $x_1 = 24$ and $y_1 = 210$; therefore, the intersection point $B(24, 0)$ is feasible.

In summary, of the six possible intersection points in the x_1x_2-plane, four were found to be feasible. For the four, we find the value of the objective function by substitution:

Extreme point	Value of objective function
A $(0,0)$	$0
D $(0,23)$	690
C $(12,15)$	750
B $(24,0)$	600

Our procedure determines that the optimum solution to maximize the profit is $x_1 = 12$ and $x_2 = 15$. That is, the carpenter should make 12 tables and 15 bookcases for a maximum profit of $750.

Computational Complexity: Intersection Point Enumeration

We now generalize the procedure presented in the carpenter example. Suppose we have a linear program with m nonnegative decision variables and n constraints, where each constraint is an inequality of the form $\leq$. First, we convert each inequality to an equation by adding a nonnegative slack variable y_i to the ith constraint. We now have a total of $m + n$ nonnegative variables. To determine an intersection point, choose m of the variables (because we have m decision variables) and set them to zero. There are $\frac{(m+n)!}{m!\,n!}$ possible choices to consider. Obviously, as the size of the linear program increases (in terms of the numbers of decision variables and constraints), this technique of enumerating all possible intersection points becomes unwieldy, even for powerful computers. How can we improve the procedure?

Note that we enumerated some intersection points in the carpenter example that turned out to be infeasible. Is there a way to quickly identify that a possible intersection point is infeasible? Moreover, if we find an extreme point (i.e., a feasible intersection point) and know the corresponding value of the objective function, can we quickly determine if another proposed extreme point will improve the value of the objective function? In conclusion, we desire a procedure that does not enumerate infeasible intersection points and enumerates only those extreme points that improve the value of the objective function for the best solution found so far in the search. We study one such procedure in the next section.

9.2 PROBLEMS

1.–5. Using the method discussed in this section, re-solve Problems 1–5 in Section 9.1.

6. How many possible intersection points are there in the following cases?

  a. 2 decision variables and $5 \leq$ inequalities
  b. 2 decision variables and $10 \leq$ inequalities

    c. 5 decision variables and 12 ≤ inequalities

    d. 25 decision variables and 50 ≤ inequalities

    e. 2000 decision variables and 5000 ≤ inequalities

9.3 LINEAR PROGRAMMING III: THE SIMPLEX METHOD

So far we have learned to find an optimal extreme point by searching among all possible intersection points associated with the decision and slack variables. Can we reduce the number of intersection points we actually consider in our search? Certainly, once finding an initial feasible intersection point, we need not consider a potential intersection point that fails to improve the value of the objective function. Can we test the optimality of our current solution against other possible intersection points? But even if an intersection point promises to be more optimal than the current extreme point, it is of no interest if it violates one or more of the constraints. Is there a test to determine whether a proposed intersection point is feasible? The **Simplex Method**, developed by George Dantzig, incorporates both *optimality* and *feasibility* tests to find the optimal solution(s) to a linear program (if one exists):

> An **optimality test** shows whether an intersection point corresponds to a value of the objective function better than the best value found so far.

> A **feasibility test** determines whether the proposed intersection point is feasible.

    To implement the Simplex Method we first separate the decision and slack variables into two nonoverlapping sets, which we call the **independent** and **dependent** sets. For the particular linear programs we consider, the original independent set will consist of the decision variables, and the slack variables will belong to the dependent set. The Simplex Method consists of the following steps, which we amplify in the discussion to follow.

Steps of the Simplex Method

    1. Tableau Format: Place the linear program in Tableau Format, as explained later.

    2. Initial Extreme Point: The Simplex Method begins with a known extreme point, usually the origin $(0, 0)$.

    3. Optimality Test: Determine whether an adjacent intersection point improves the value of the objective function. If not, the current extreme point is optimal. If an improvement is possible, the optimality test determines which variable currently in the independent set (having value zero) should *enter* the dependent set and become nonzero.

    4. Feasibility Test: To find a new intersection point, one of the variables in the dependent set must *exit* to allow the entering variable from Step 3 to become dependent. The feasibility test determines which current dependent variable to choose for exiting, ensuring feasibility.

    5. Pivot: Form a new equivalent system of equations by eliminating the new dependent variable from the equations that do not contain the variable that exited in Step 4. Then set the new independent variables to zero in the new system to find the values of the new dependent variables, thereby determining an intersection point.

    6. Repeat Steps 3–5 until an optimal extreme point is found.

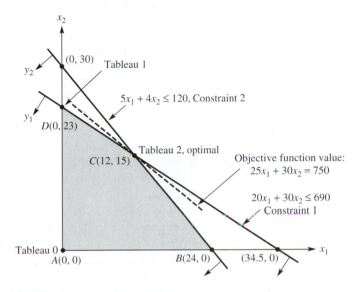

FIGURE 9.12   The set of points satisfying the constraints of a linear program (the shaded region) form a convex set

    Before detailing each of the preceding steps, let's consider again the carpenter's problem (see Figure 9.12). The origin is an extreme point, so we choose it as our starting point. Thus x_1 and x_2 are the current arbitrary independent variables and are assigned the value zero, and y_1 and y_2 are the current dependent variables with values 690 and 120, respectively. The optimality test determines whether a current independent variable assigned the value zero could improve the value of the objective function if it is made dependent and positive. For example, either x_1 or x_2, if made positive, would improve the objective function value. (They have positive coefficients in the objective function we are trying to maximize.) Thus the optimality test determines a promising variable to enter the dependent set. Later we give a general rule for choosing which independent variable to enter when more than one candidate exists. In the carpenter's problem at hand, we select x_2 as the new dependent variable.

    The variable chosen for entry into the dependent set by the optimality condition replaces one of the current dependent variables. The feasibility condition determines which exiting variable this entering variable replaces. Basically, the entering variable replaces whichever current dependent variable can assume a zero value while maintaining nonnegative values for all the remaining dependent variables. That is, the feasibility condition ensures that the new intersection point will be feasible and hence an extreme point. In Figure 9.12, the feasibility test would lead us to the intersection point $(0, 23)$, which is feasible, and not to $(0, 30)$, which is infeasible. Thus x_2 replaces y_1 as a dependent or nonzero variable; x_2 enters and y_1 exits the set of dependent variables.

Computational Efficiency

The feasibility test does not require actual computation of the values of the dependent variables when selecting an exiting variable for replacement. Instead, we will see that an appropriate exiting variable is selected by quickly determining whether any variable becomes negative if the dependent variable being considered for replacement is assigned the value zero (a ratio test that will be explained later). If any variable would become negative, then the dependent variable under consideration cannot be replaced by the entering variable if feasibility is to be maintained. When a set of dependent variables corresponding to a more optimal extreme point is found from the optimality and feasibility tests, the values of the new dependent variables are determined by pivoting. The pivoting process essentially solves an equivalent system of equations for the new dependent variables after the exchange of the entering and exiting variables in the dependent set. The values of the new dependent variables are obtained by setting the independent variables equal to zero. Note that only one dependent variable is replaced at each stage. *Geometrically, the Simplex Method proceeds from an initial extreme point to an adjacent extreme point until no adjacent extreme point is more optimal.* At that time, the current extreme point is an optimal solution. We now detail the steps of the Simplex Method.

Step 1. Tableau Format

Many formats exist for implementing the Simplex Method. The format we use assumes the objective function is to be maximized and the constraints are less than or equal to inequalities. (If the problem is not expressed initially in this format, it can easily be changed to this format.) For the carpenter's example, the problem is to

$$\text{Maximize } 25x_1 + 30x_2$$

Subject to

$$20x_1 + 30x_2 \leq 690$$
$$5x_1 + 4x_2 \leq 120$$
$$x_1, x_2 \geq 0$$

Next we adjoin a new constraint to ensure that any solution improves the best value of the objective function found so far. Take the initial extreme point as the origin, where the value of the objective function is 0. We want to constrain the objective function to be better than its current value, so we require

$$25x_1 + 30x_2 \geq 0$$

Because all the constraints must be less than or equal to inequalities, multiply the new constraint by -1 and adjoin it to the original constraint set:

$$20x_1 + 30x_2 \le 690 \quad \text{(constraint 1, lumber)}$$
$$5x_1 + 4x_2 \le 120 \quad \text{(constraint 2, labor)}$$
$$-25x_1 - 30x_2 \le 0 \quad \text{(objective function constraint)}$$

The Simplex Method implicitly assumes that all variables are nonnegative, so we do not repeat the nonnegativity constraints in the remainder of the presentation.

Next, we convert each inequality to an equality by adding a *nonnegative* new variable y_i (or z), called a *slack variable* because it measures the slack or degree of satisfaction of the constraint. A negative value for y_i indicates the constraint is not satisfied. (We use the variable z for the objective function constraint to avoid confusion with the other constraints.) This process gives the *augmented constraint set*

$$20x_1 + 30x_2 + y_1 = 690$$
$$5x_1 + 4x_2 + y_2 = 120$$
$$-25x_1 - 30x_2 + z = 0$$

where the variables x_1, x_2, y_1, y_2 are nonnegative. The value of the variable z represents the value of the objective function as we shall see. (Note from the last equation, $z = 25x_1 + 30x_2$, is the value of the objective function.)

Step 2. Initial Extreme Point

Because there are two decision variables, all possible intersection points lie in the $x_1 x_2$-plane and can be determined by setting two of the variables $\{x_1, x_2, y_1, y_2\}$ to 0. (The variable z is *always* a dependent variable and represents the value of the objective function at the extreme point in question.) The origin is feasible and corresponds to the extreme point characterized by $x_1 = x_2 = 0$, $y_1 = 690$, and $y_2 = 120$. Thus x_1 and x_2 are independent variables assigned the value 0; y_1, y_2, and z are dependent variables whose values are then determined. As we will see, z conveniently records the current value of the objective function at the extreme points of the convex set in the $x_1 x_2$-plane as we compute them by elimination.

Step 3. Optimality Test for Choosing an Entering Variable

In the preceding format, a negative coefficient in the last (or objective function) equation indicates that the corresponding variable could improve the current objective function value. Thus the coefficients -25 and -30 indicate that either x_1 or x_2 could enter and improve the current objective function value of $z = 0$. (The current constraint corresponds to $z = 25x_1 + 30x_2 \ge 0$, with x_1 and x_2 currently independent and 0.) When more than one candidate exists for the entering variable, a general rule for selecting the variable to enter the dependent set is that variable with the largest (in absolute value) negative coefficient in the objective function row. If no negative coefficients exist, the current solution is optimal. In the case at hand, we choose x_2 as the new entering variable. (The procedure is inexact because at this stage we do not know what values the entering variable can assume.)

Step 4. Feasibility Test for Choosing an Exiting Variable

The entering variable x_2 (in our example) must replace either y_1 or y_2 as a dependent variable (because z *always* remains the third dependent variable). To determine which of these variables is to exit the dependent set, first divide the right-hand side values 690 and 120 (associated with the original constraint inequalities) by the components for the entering variable in each inequality (30 and 4, respectively, in our example) to obtain the ratios $\frac{690}{30} = 23$ and $\frac{120}{4} = 30$. From the subset of ratios that are positive (both in this case), the variable corresponding to the minimum ratio is chosen for replacement (y_1, which corresponds to 23 in this case). *The ratios represent the value the entering variable would obtain if the corresponding exiting variable were assigned the value* 0. Thus only positive values are considered, and the smallest positive value is chosen to not drive any variable negative. For instance, if y_2 were chosen as the exiting variable and assigned the value 0, then x_2 would assume a value 30 as the new dependent variable. But then y_1 would be negative, indicating that the intersection point $(0, 30)$ does not satisfy the first constraint. Note that the intersection point $(0, 30)$ is not feasible in Figure 9.12. The **minimum positive ratio rule** illustrated previously obviates enumeration of any infeasible intersection points. In the case at hand the dependent variable corresponding to the smallest ratio 23 is y_1, so it becomes the exiting variable. Thus x_2, y_2, and z form the new set of dependent variables, and x_1 and y_1 form the new set of independent variables.

Step 5. Pivoting to Solve for the New Dependent Variable Values

Next we derive a new (equivalent) system of equations by eliminating the entering variable x_2 in all the equations of the previous system that do not contain the exiting variable y_1. There are numerous ways to execute this step, such as the method of elimination used in Section 9.2. Then we find the values of the dependent variables x_2, y_2, and z when the independent variables x_1 and y_1 are assigned the value 0 in the new system of equations. This is called the **pivoting procedure**. The values of x_1 and x_2 give the new extreme point (x_1, x_2), and z is the (improved) value of the objective function at that point.

After performing the pivot, apply the optimality test again to determine whether another entering variable candidate exists. If so, choose an appropriate one and apply the feasibility test to choose an exiting variable. Then perform the pivoting procedure again. Repeat the process over and over until no variable has a negative coefficient in the objective function row. We now summarize the procedure and use it to solve the carpenter's problem.

Summary of the Simplex Method

Step 1. *Place the problem in Tableau Format.* Adjoin slack variables as needed to convert inequality constraints to equalities. Remember that all variables are nonnegative. Include the objective function constraint as the last constraint, including its slack variable z.

Step 2. *Find one initial extreme point.* (For the problems we consider, the origin will be an extreme point.)

Step 3. *Apply the optimality test.* Examine the last equation (which corresponds to the objective function). If all its coefficients are nonnegative, then stop; the current extreme point is optimal. Otherwise, some variables have negative coefficients, so choose the variable with the largest (in absolute value) negative coefficient as the new entering variable.

Step 4. *Apply the feasibility test.* Divide the current right-hand side values by the corresponding coefficient values of the entering variable in each equation. Choose the exiting variable to be the one corresponding to the smallest positive ratio after this division.

Step 5. *Pivot.* Eliminate the entering variable from all the equations that do not contain the exiting variable. (For example, we can use the elimination procedure presented in Section 9.2.) Then assign the value 0 to the variables in the new independent set (consisting of the exited variable and the variables remaining after the entering variable has left to become dependent). The resulting values give the new extreme point (x_1, x_2) and objective function value z for that point.

Step 6. *Repeat Steps 3–5* until an optimal extreme point is found.

Example 1   The Carpenter's Problem

Step 1.   The Tableau Format gives

$$20x_1 + 30x_2 + y_1 = 690$$
$$5x_1 + 4x_2 + y_2 = 120$$
$$-25x_1 - 30x_2 + z = 0$$

Step 2.   The origin $(0, 0)$ is an initial extreme point for which the independent variables are $x_1 = x_2 = 0$; the dependent variables are $y_1 = 690$, $y_2 = 120$, and $z = 0$.

Step 3.   We apply the optimality test to choose x_2 as the variable entering the dependent set because it corresponds to the negative coefficient with the largest absolute value.

Step 4.   Applying the feasibility test, we divide the right-hand side values 690 and 120 by the components for the entering variable x_2 in each equation (30 and 4, respectively), yielding the ratios $\frac{690}{30} = 23$ and $\frac{120}{4} = 30$. The smallest positive ratio is 23, corresponding to the first equation that has the slack variable y_1. Thus we choose y_1 as the exiting dependent variable.

Step 5.   We pivot to find the values of the new dependent variables x_2, y_2, and z when the independent variables x_1 and y_1 are set to the value 0. After eliminating the new dependent variable x_2 from each previous equation that does not contain the exiting variable y_1, we obtain the equivalent system

$$\frac{2}{3}x_1 + x_2 + \frac{1}{30}y_1 \qquad\qquad = 23$$
$$\frac{7}{3}x_1 \qquad - \frac{2}{15}y_1 + y_2 \qquad = 28$$
$$-5x_1 \qquad + \quad y_1 \qquad + z = 690$$

Setting $x_1 = y_1 = 0$, we determine $x_2 = 23$, $y_2 = 28$, and $z = 690$. These results give the extreme point $(0, 23)$ where the value of the objective function is $z = 690$.

Applying the optimality test again, we see that the current extreme point $(0, 23)$ is not optimal (because there is a negative coefficient -5 in the last equation corresponding to the variable x_1). Before continuing, observe that we really do not need to write out the entire symbolism of the equations in each step. We merely need to know the coefficient values associated with the variables in each of the equations together with the right-hand side. A table format, or *tableau*, is commonly used to record these numbers. We illustrate the completion of the carpenter's problem using this format, where the headers of each column designate the variables; the abbreviation RHS is the value of the right-hand side. We begin with Tableau 0, corresponding to the initial extreme point at the origin.

Tableau 0 (Original Tableau)

x_1	x_2	y_1	y_2	z	RHS
20	30	1	0	0	$690 (= y_1)$
5	4	0	1	0	$120 (= y_2)$
-25	-30	0	0	1	$0 (= z)$

Dependent variables: $\{y_1, y_2, z\}$
Independent variables: $x_1 = x_2 = 0$
Extreme point: $(x_1, x_2) = (0, 0)$
Value of objective function: $z = 0$

Optimality test   The entering variable is x_2 (corresponding to -30 in the last row).

Feasibility test   Compute the ratios for the RHS divided by the coefficients in the column labeled x_2 to determine the minimum positive ratio.

Entering variable

x_1	x_2	y_1	y_2	z	RHS	RATIO
20	30	1	0	0	690	$23 \;(= 690/30)$ ←Exiting variable
5	4	0	1	0	120	$30 \;(= 120/4)$
-25	-30	0	0	1	0	*

Choose y_1 corresponding to the minimum positive ratio 23 as the exiting variable.

Pivot   Divide the row containing the exiting variable (the first row in this case) by the coefficient of the entering variable in that row (the coefficient of x_2 in this case), giving a coefficient of 1 for the entering variable in this row. Then eliminate the entering variable x_2 from the remaining rows (which do not contain the exiting variable y_1 and have a zero coefficient for it). The results are summarized in the next tableau, which uses five-place decimal approximations to the numerical values.

Tableau 1

x_1	x_2	y_1	y_2	z	RHS
0.66667	1	0.03333	0	0	23 ($= x_2$)
2.33333	0	−0.13333	1	0	28 ($= y_2$)
−5.00000	0	1.00000	0	1	690 ($= z$)

Dependent variables: $\{x_2, y_2, z\}$
Independent variables: $x_1 = y_1 = 0$
Extreme point: $(x_1, x_2) = (0, 23)$
Value of objective function: $z = 690$

The pivot determines that the new dependent variables have the values $x_2 = 23$, $y_2 = 28$, and $z = 690$.

Optimality test   The entering variable is x_1 (corresponding to the coefficient -5 in the last row).

Feasibility test   Compute the ratios for the RHS.

Entering variable

x_1	x_2	y_1	y_2	z	RHS	RATIO	
0.66667	1	0.03333	0	0	23	34.5 ($= 23/0.66667$)	
2.33333	0	−0.13333	1	0	28	12.0 ($= 28/2.33333$)	←Exiting variable
−5.00000	0	1.00000	0	1	690	*	

Choose y_2 as the exiting variable because it corresponds to the minimum positive ratio 12.

Pivot   Divide the row containing the exiting variable (the second row in this case) by the coefficient of the entering variable in that row (the coefficient of x_1 in this case), giving a coefficient of 1 for the entering variable in this row. Then eliminate the entering variable x_1 from the remaining rows (which do not contain the exiting variable y_2 and have a zero coefficient for it). The results are summarized in the next tableau.

Tableau 2

x_1	x_2	y_1	y_2	z	RHS
0	1	0.071429	−0.28571	0	15 (= x_2)
1	0	−0.057143	0.42857	0	12 (= x_1)
0	0	0.714286	2.14286	1	750 (= z)

Dependent variables: $\{x_2, x_1, z\}$
Independent variables: $y_1 = y_2 = 0$
Extreme point: $(x_1, x_2) = (12, 15)$
Value of objective function: $z = 750$

Optimality test   Because there are no negative coefficients in the bottom row, $x_1 = 12$ and $x_2 = 15$ gives the optimal solution $z = \$750$ for the objective function. Note that starting with an initial extreme point, only two of the possible six intersection points had to be enumerated. The power of the Simplex Method is its reduction of the computations required to find an optimal extreme point.

Example 2

$$\text{Maximize } 3x_1 + x_2$$

Subject to

$$2x_1 + x_2 \le 6$$
$$x_1 + 3x_2 \le 9$$
$$x_1, x_2 \ge 0$$

The problem in Tableau Format is

$$2x_1 + x_2 + y_1 = 6$$
$$x_1 + 3x_2 + y_2 = 9$$
$$-3x_1 - x_2 + z = 0$$

where $x_1, x_2, y_1, y_2,$ and $z \ge 0$.

Tableau 0 (Original Tableau)

x_1	x_2	y_1	y_2	z	RHS
2	1	1	0	0	6 (= y_1)
1	3	0	1	0	9 (= y_2)
−3	−1	0	0	1	0 (= z)

Dependent variables: $\{y_1, y_2, z\}$
Independent variables: $x_1 = x_2 = 0$
Extreme point: $(x_1, x_2) = (0, 0)$
Value of objective function: $z = 0$

Optimality test   The entering variable is x_1 (corresponding to -3 in the bottom row).

Feasibility test   Compute the ratios of the RHS divided by the column labeled x_1 to determine the minimum positive ratio.

x_1	x_2	y_1	y_2	z	RHS	RATIO
2	1	1	0	0	6	③ $(= 6/2)$ ← —Exiting variable
1	3	0	1	0	9	9 $(= 9/1)$
⟨-3⟩	-1	0	0	1	0	*

—Entering variable

Choose y_1 corresponding to the minimum positive ratio 3 as the exiting variable.

Pivot   Divide the row containing the exiting variable (the first row in this case) by the coefficient of the entering variable in that row (the coefficient of x_1 in this case), giving a coefficient of 1 for the entering variable in this row. Then eliminate the entering variable x_1 from the remaining rows (which do not contain the exiting variable y_1 and have a zero coefficient for it). The results are summarized in the next tableau.

Tableau 1

x_1	x_2	y_1	y_2	z	RHS
1	$\frac{1}{2}$	$\frac{1}{2}$	0	0	$3 (= x_1)$
0	$\frac{5}{2}$	$-\frac{1}{2}$	1	0	$6 (= y_2)$
0	$\frac{1}{2}$	$\frac{3}{2}$	0	1	$9 (= z)$

Dependent variables: $\{x_1, y_2, z\}$
Independent variables: $x_2 = y_1 = 0$
Extreme point: $(x_1, x_2) = (3, 0)$
Value of objective function: $z = 9$

The pivot determines that the dependent variables have the values $x_1 = 3$, $y_2 = 6$, and $z = 9$.

Optimality test   There are no negative coefficients in the bottom row. Thus $x_1 = 3$ and $x_2 = 0$ is an extreme point giving the optimal objective function value $z = 9$.

Remarks   We have assumed that the origin is a feasible extreme point. If it is not, then an extreme point must be found before the Simplex Method as presented can be used. We have also assumed that the linear program is not degenerate in the sense that no more than two constraints intersect at the same point. These and other issues are studied in more advanced optimization courses.

9.3 PROBLEMS

Use the Simplex Method to re-solve Problems 1–5 in Section 9.1.

9.3 PROJECTS

1. Write a computer code to perform the basic Simplex Algorithm. Solve Problem 3 in Section 9.1 using your code.

9.4 LINEAR PROGRAMMING IV: SENSITIVITY ANALYSIS

A mathematical model typically approximates a problem under study. For example, the coefficients in the objective function of a linear program may only be estimates. Or the amount of the resources constraining production made available by management may vary, depending on the profit returned per unit of resource invested. (Management may be willing to procure additional resources if the additional profit is high enough.) Thus management would like to know if the additional profit to be realized justifies the cost of another unit of resource. If so, over what range of values for the resources is the analysis valid? Hence, in addition to solving a linear program we would like to know how sensitive the optimal solution is to changes in the constants used to formulate the program. In this section we analyze graphically the effect on the optimal solution of changes in the coefficients of the objective function and the amount of resource available. Using the carpenter's problem as an example, we answer the following questions:

1. Over what range of values for the profit per table does the current solution remain optimal?

2. What is the value of another unit of the second resource (labor)? That is, how much will the profit increase if another unit of labor is obtained? Over what range of labor values is the analysis valid? What is required to increase profit beyond this limit?

Sensitivity of the Optimal Solution to Changes in the Coefficients of the Objective Function

Consider again the carpenter's problem. The objective function is to maximize profits where each table nets $25 profit and each bookcase $30. If z represents the amount of profit, then we wish to

$$\text{Maximize } z = 25x_1 + 30x_2$$

Note that z is a function of two variables and we can draw the level curves of z in the x_1x_2-plane. In Figure 9.13, we graph the level curves $z = 650$, $z = 750$, and $z = 850$ for illustrative purposes.

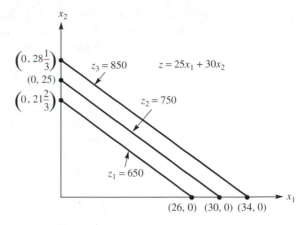

FIGURE 9.13   The level curves of $z = 25x_1 + 30x_2$ in the x_1x_2-plane have a slope of $-5/6$

Note that every level curve is a line with slope $-\frac{5}{6}$. In Figure 9.14, we superimpose on the previous graph the constraint set for the carpenter's problem and see that the optimal solution $(12, 15)$ gives an optimal objective function value of $z = 750$.

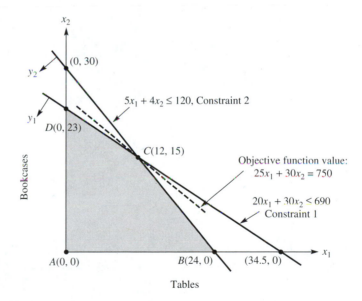

FIGURE 9.14   The level curve $z = 750$ is tangent to the convex set of feasible solutions at extreme point C $(12, 15)$

Now we ask the question, what is the effect of changing the value of the profit for each table. Intuitively, if we increase the profit sufficiently, we eventually make only tables (giving the extreme point of 24 tables and 0 bookcases), instead of the current mix of 12 tables and 15 bookcases. Similarly, if we decrease the profit per table sufficiently, we should make only bookcases at the extreme point $(0, 23)$. Note again that the slope of the level curves of the objective function is $-\frac{5}{6}$. If we let c_1 represent the profit per bookcase, then the objective function becomes

$$\text{Maximize } z = c_1 x_1 + 30 x_2$$

with slope $-c_1/30$ in the $x_1 x_2$-plane. As we vary c_1, the slope of the level curves of the objective function changes. Examine Figure 9.15 to convince yourself that the current extreme point $(12, 15)$ remains optimal as long as the slope of the objective function is between the slopes of the two binding constraints. In this case, the extreme point $(12, 15)$ remains optimal as long as the slope of the objective function is less than $-\frac{2}{3}$ but greater than $-\frac{5}{4}$, the slopes of the lumber and labor constraints, respectively. If we start with the slope for the objective function as $-\frac{2}{3}$, as we increase c_1 we rotate the level curve of the objective function clockwise. If we rotate clockwise, the optimal extreme point changes to $(24, 0)$ if the slope of the objective function is less than $-\frac{5}{4}$. Thus the *range* of values for which the current extreme point remains optimal is given by the inequality

$$-\frac{5}{4} \le -\frac{c_1}{30} \le -\frac{2}{3}$$

or

$$20 \le c_1 \le 37.5$$

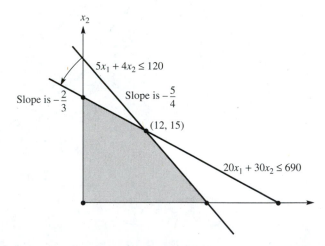

FIGURE 9.15   The extreme point $(12, 15)$ remains optimal for objective functions with a slope between $-5/4$ and $-2/3$.

Interpreting this result, if the profit per table exceeds 37.5, the carpenter should produce only tables (i.e., 24 tables). If the profit per table is reduced below 20, the carpenter should produce only bookcases (i.e., 23 bookcases). If c_1 is between 20 and 37.5, he should produce the mix of 12 bookcases and 15 tables. As we change c_1 over the range $[20, 37.5]$ the value of the objective function changes even though the location of the extreme point does not. Because he is making 12 tables, the objective function changes by a factor of 12 times the change in c_1. Note that at the limit $c_1 = 20$ there are *two* extreme points C and B, which produce the same value for the objective function. Likewise if $c_1 = 37.5$, the extreme points D and C produce the same value for the objective function. In such cases we say that there are *alternative optimal solutions.*

Changes in the Amount of Resource Available

Currently there are 120 units of labor available, all of which are used to produce the 12 tables and 15 bookcases represented by the optimal solution. What is the effect of increasing the amount of labor? If b_2 represents the units of available labor (the second resource constraint), the constraint can be rewritten as

$$5x_1 + 4x_2 \leq b_2$$

What happens geometrically as we vary b_2? To answer this question, graph the constraint set for the carpenter's problem with the original value of $b_2 = 120$ and a second value, say $b_2 = 150$ (see Figure 9.16). Note that the effect of increasing b_2 is to translate the constraint upward and to the right. As this happens, the optimal value of the objective function moves along the line segment AA', which lies on the lumber constraint. As the optimal solution moves along the line segment from A to A', the value of x_1 increases and the value of x_2 decreases. The net effect of increasing b_2 is to increase the value of the objective function. But by how much? One goal is to determine how much the objective function value changes as b_2 *increases by 1 unit.*

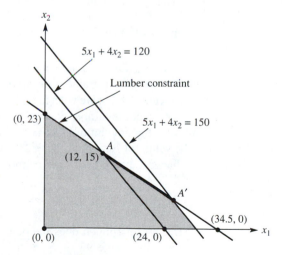

FIGURE 9.16   As the amount of labor resource b_2 increases from 120 to 150 units, the optimal solution moves from A to A' along the lumber constraint, increasing x_1 and decreasing x_2

Note that if b_2 increases beyond $5 \cdot 34.5 = 172.5$, the optimal solution remains at the extreme point $(34.5, 0)$. That is, at $(34.5, 0)$ the lumber constraint must also be increased if the objective function is to be increased further. Thus, increasing the labor constraint to 200 units results in some excess labor that cannot be used unless the amount of lumber is increased beyond its present value of 690 (see Figure 9.17). Following a similar analysis, if b_2 is decreased, the value of the objective function moves along the lumber constraint until the extreme point $(0, 23)$ is reached. Further reductions in b_2 would cause the optimal solution to move from $(0, 23)$ down the y-axis to the origin.

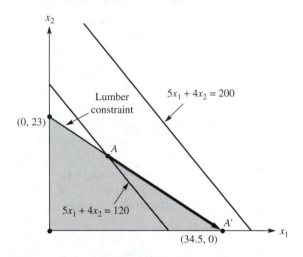

FIGURE 9.17   As resource b_2 increases from 120 to 172.5, the optimal solution moves from A to A' along the line segment AA'; increasing b_2 beyond $b_2 = 172.5$ does not increase the value of the objective function unless the lumber constraint is also increased (moving it upward to the right)

Now let's find the range of values of b_2 for which the optimal solution moves along the lumber constraint as the amount of labor varies. Refer to Figure 9.18 and convince yourself that we wish to find the value for b_2 for which the labor constraint would intersect the lumber constraint on the x_1-axis, or point E $(34.5, 0)$. At point E $(34.5, 0)$, the amount of labor is $5 \cdot 34.5 + 4 \cdot 0 = 172.5$. Similarly, we wish to find the value for b_2 at point D $(0, 23)$, which is $5 \cdot 0 + 4 \cdot 23 = 92$. Summarizing, as b_2 changes, the optimal solution moves along the lumber constraint as long as

$$92 \le b_2 \le 172.5$$

But by how much does the objective function change as b_2 increases by 1 unit within the range $92 \le b_2 \le 172.5$? We will analyze this in several ways. First, suppose $b_2 = 172.5$. The optimal solution is then the new extreme point E $(34.5, 0)$ and the value of the objective function is $34.5 \cdot 25 = 862.5$ at E. Thus the objective function increased by $862.5 - 750 = 112.5$ units when b_2 increased by

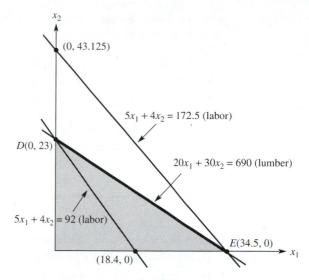

FIGURE 9.18 As b_2 increases from 92 to 172.5, the optimal solution moves from point D $(0, 23)$ to point E $(34.5, 0)$ along the line segment DE, the lumber constraint

$172.5 - 120 = 52.5$ units. Hence the change in the objective function for 1 unit of change in labor is

$$\frac{862.5 - 750}{172.5 - 120} \simeq 2.14$$

Now let's analyze the value of a unit change of labor in another way. If b_2 increases by 1 unit from 120 to 121, then the new extreme point A' represented by the intersection of the constraints

$$20x_1 + 30x_2 = 690$$
$$5x_1 + 4x_2 = 121$$

is the point $A'(12.429, 14.714)$, which has an objective function value of 752.14. Thus the net effect as b_2 increases by 1 hr of labor is to increase the objective function by approximately \$2.14.

Economic Interpretation of a Unit Change in a Resource

In the foregoing analysis, we saw that as 1 more hour of labor is added, the objective function increases by \$2.14 as long as the total amount of labor does not exceed 172.5 hours. Thus in terms of the objective function, an *additional* unit of labor is worth 2.14 units. If management can procure a unit of labor for less than 2.14, it would be profitable to do so. Conversely, if management can sell labor for more than 2.14 (which is valid until labor is reduced to 92 units), it should also consider doing that. Note that our analysis gives the value of a unit of resource in terms of the value of the objective function at the optimum extreme point, which is a *marginal value*.

Sensitivity analysis is a powerful methodology for interpreting linear programs. The information embodied in a carefully accomplished sensitivity analysis is often at least as valuable to the decision maker as the optimal solution to the linear program itself. In advanced courses in optimization, you can learn how to perform a sensitivity analysis algebraically. Moreover, the coefficients in the constraint set, as well as the right-hand side of the constraints, can be analyzed for their sensitivity.

9.4 PROBLEMS

1. Consider the example in this section. Determine the sensitivity of the optimal solution to a change in c_2 using the objective function $25x_1 + c_2x_2$.

2. Perform a complete sensitivity analysis (objective function coefficients and right-hand side values) of the wooden toy soldier problem in Section 9.1 (Problem 1).

3. Why is sensitivity analysis important in linear programming?

9.5   NUMERICAL SEARCH METHODS

In Chapter 8, we discussed optimization methods for continuous models using calculus. If we want to maximize a function $f(x)$ over some interval $[a, b]$, we know that setting the first derivative equal to zero yields its critical points. The second derivative test may then be used to characterize the nature of these critical points. We also know we may have to check the endpoints and points where the first derivative fails to exist.

It may be impossible algebraically to solve the equation resulting from setting the first derivative equal to zero. For example, consider a machine that earns revenue at the rate of e^{-t} dollars a year for t years, and then can be sold for $\frac{1}{1+t}$ dollars. If we model maximizing revenues as a function of years, the problem is to maximize

$$f(T) = \int_0^T e^{-t}\, dt + \frac{1}{1 + T}$$

Setting the first derivative equal to zero yields

$$f'(T) = e^{-T} - (1 + T)^{-2} = 0$$

which cannot be algebraically solved for T in closed form. In such cases, we can use a search procedure to approximate the optimal solution.

Various search methods permit us to approximate solutions to nonlinear optimization problems with a single independent variable. Two search methods commonly used are the Dichotomous and Golden Section Search methods. Both share several features common to most search methods. We discuss these features and the methods.

A *unimodal function* on an interval has exactly one point where a maximum or minimum occurs in the interval. If the function is known (or assumed to be) multimodal, then it must be subdivided into separate unimodal functions. (In most practical problems, the optimal solution is known to lie in some restricted range of the independent variable.) More precisely, $f(x)$ is a **unimodal function** with an interior local maximum on an interval $[a, b]$ if for some point x^* on $[a, b]$, the function is strictly increasing on $[a, x^*]$ and strictly decreasing on $[x^*, b]$. A similar statement holds for $f(x)$ being unimodal with an interior local minimum. These concepts are illustrated in Figure 9.19.

unimodal function

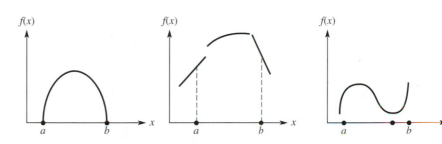

a. Unimodal functions on the interval $[a, b]$

b. A function that is not unimodal on the interval $[a, b]$

FIGURE 9.19   Examples of unimodal functions

The unimodal assumption is important to find the subset of the interval $[a, b]$ that contains the optimal point x to maximize (or minimize) $f(x)$.

Search Method Paradigm

With most search methods, we divide the region $[a, b]$ into two overlapping intervals $[a, x_2]$ and $[x_1, b]$ after placing two test points x_1 and x_2 in the original interval $[a, b]$ according to some criterion of our chosen search method, as illustrated in Figure 9.20. We then determine the subinterval where the optimal solution lies and use that subinterval to continue the search based on the function evaluations $f(x_1)$ and $f(x_2)$. There are three cases (illustrated in Figure 9.21) in the maximization problem (the minimization problem is analogous) with experiments x_1 and x_2 placed between $[a, b]$ according to the chosen search method (fully discussed later):

Case 1: $f(x_1) < f(x_2)$. Because $f(x)$ is unimodal, the solution cannot occur in the interval $[a, x_1]$. The solution must lie in the interval $(x_1, b]$.

Case 2: $f(x_1) > f(x_2)$. Because $f(x)$ is unimodal, the solution cannot occur in the interval $(x_2, b]$. The solution must lie in the interval $[a, x_2)$.

Case 3: $f(x_1) = f(x_2)$. The solution must lie somewhere in the interval (x_1, x_2).

We next present two commonly used search techniques.

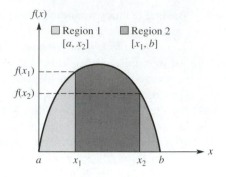

FIGURE 9.20   Location of test points for search methods (overlapping intervals)

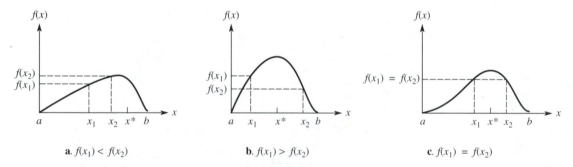

a. $f(x_1) < f(x_2)$ **b.** $f(x_1) > f(x_2)$ **c.** $f(x_1) = f(x_2)$

FIGURE 9.21   Cases where $f(x_1) < f(x_2)$, $f(x_1) > f(x_2)$, and $f(x_1) = f(x_2)$

Dichotomous Search Method

Assume we have a function $f(x)$ to maximize over a specified interval $[a, b]$. The Dichotomous Search Method computes the midpoint $\frac{a+b}{2}$, and then moves slightly to either side of the midpoint to compute two test points: $\frac{a+b}{2} \pm \varepsilon$, where ε is some very small real number. In practice, the number ε is chosen as small as the accuracy of the computational device will permit, the objective being to place the two test points as close together as possible. Figure 9.22 illustrates this procedure for a maximization problem. The procedure continues until it gets within some small interval containing the optimal solution. Here are the steps in this algorithm.

Dichotomous Search Algorithm to Maximize $f(x)$ over the Interval $a \leq x \leq b$

STEP 1   Initialize: Choose a small number $\varepsilon > 0$, such as 0.01. Select a small $t > 0$ between $[a, b]$ called the *length of uncertainty* for the search. Calculate the number of iterations n using the formula

$$(0.5)^n = t/[b - a]$$

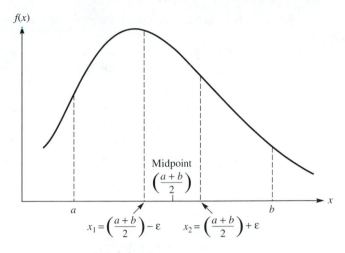

FIGURE 9.22 Dichotomous Search

STEP 2 For $k = 1$ to n, do Steps 3 and 4.

STEP 3 $x_1 = \left(\dfrac{a + b}{2}\right) - \varepsilon$

and

$$x_2 = \left(\dfrac{a + b}{2}\right) + \varepsilon$$

STEP 4 (For a maximization problem)

   a. If $f(x_1) \geq f(x_2)$, then let
      $a = a$
      $b = x_2$
      $k = k + 1$
     Return to Step 3.

   b. If $f(x_1) < f(x_2)$, then let
      $b = b$
      $a = x_1$
      $k = k + 1$
     Return to Step 3.

STEP 5 Let $x^* = \dfrac{a + b}{2}$ and $MAX = f(x^*)$.
STOP

In this presentation, the number of iterations to perform is determined by the reduction in the length of uncertainty desired. Alternatively, we may wish to continue to iterate until the change in the dependent variable is less than some predetermined amount, say Δ. That is, continue to iterate until $f(a) - f(b) \leq \Delta$.

For example, in an application where $f(x)$ represents the profit realized by producing x items, it might make more sense to stop when the change in profit is less than some acceptable amount. To minimize a function $y = f(x)$, either maximize $-y$ or switch the directions of the signs in Steps 4a and 4b.

Example 1 Suppose we want to maximize $f(x) = -x^2 - 2x$ over the interval $-3 \leq x \leq 6$. Assume we want the optimal tolerance to be less than 0.2. We arbitrarily choose ε (the distinguishability constant) to be 0.01. Next we determine the number n of iterations using the relationship $(0.5)^n = 0.2/[6 - (-3)]$, or $n \ln(0.5) = \ln(0.2/9)$, which implies that $n = 5.49$ (and we round to the next higher integer, $n = 6$). Table 9.1 gives the results for implementing the search in the algorithm.

TABLE 9.1 Results of a Dichotomous Search for Example 1*

a	b	x_1	x_2	$f(x_1)$	$f(x_2)$
-3	6	1.49	1.51	-5.2001	-5.3001
-3	1.51	-0.755	-0.735	0.9400	0.9298
-3	-0.735	-1.8775	-1.8575	0.2230	0.2647
-1.8775	-0.735	-1.3163	-1.2963	0.9000	0.9122
-1.3163	-0.735	-1.0356	-1.0156	0.9987	0.9998
-1.0356	-0.735	-0.8953	-0.8753	0.9890	0.9845
-1.0356	-0.8753				

*The numerical results in this section were computed carrying the 13-place accuracy of the computational device being used. The results were then rounded to four decimal places for presentation.

The length of the final interval is less than the 0.2 tolerance initially specified. From Step 5 we estimate the location of a maximum at

$$x^* = \frac{-1.0356 - 0.8753}{2} = -0.9555$$

with $f(-0.9555) = 0.9980$. (Examining our table we see that $f(-1.0156) = 0.9998$, a better estimate.) We note that the number $n = 6$ refers to the number of intervals searched. For this example, we can use calculus to find the optimal solution $f(-1) = 1$ at $x = -1$.

Golden Section Search Method

The Golden Section Search Method is a procedure that uses the **golden ratio**. To better understand the golden ratio, divide the interval $[0, 1]$ into two separate subintervals of lengths r and $1 - r$, as shown in Figure 9.23. These subintervals are said to be divided into the golden ratio if the length of the whole interval to the length of the longer segment equals the length of the longer segment to the length of the smaller segment. Symbolically, we can write it as $1/r = r/(1 - r)$ or $r^2 + r - 1 = 0$, because $r > 1 - r$ in the figure.

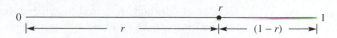

FIGURE 9.23 Golden ratio using a line segment

Solving this last equation gives the two roots

$$r_1 = \frac{\sqrt{5} - 1}{2} \quad \text{and} \quad r_2 = \frac{-\sqrt{5} - 1}{2}$$

Only the positive root r_1 lies in the given interval $[0, 1]$. The numerical value of r_1 is approximately 0.618 and is known as the golden ratio.

The Golden Section Search Method incorporates the following assumptions:

1. The function $f(x)$ must be unimodal over the specified interval $[a, b]$.
2. The function must have a maximum (or minimum) value over a known interval of uncertainty.
3. The method gives an approximation to the minimum rather than the exact maximum.

The method will determine a final interval containing the optimal solution. The length of the final interval can be controlled and made arbitrarily small by the selection of a tolerance value. The length of the final interval will be less than our specified tolerance level.

The search procedure to find an approximation to the maximum value is iterative. It requires evaluations of $f(x)$ at the test points $x_1 = a + (1 - r)(b - a)$ and $x_2 = a + r(b - a)$ and then determines the new interval of search (see Figure 9.24). If $f(x_1) < f(x_2)$, then the new interval is $[x_1, b]$; if $f(x_1) > f(x_2)$, then the new interval is $[a, x_2]$, as in the Dichotomous Search Method. The iterations continue until the final interval length is less than the tolerance imposed, and the final interval contains the optimal solution point. The length of this final interval determines the accuracy in finding the approximate optimal solution point. The

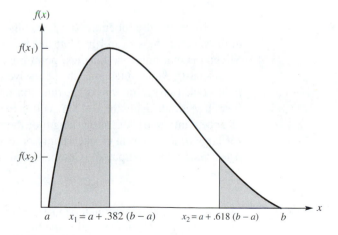

FIGURE 9.24 Location of x_1 and x_2 for the Golden Section Search

number of iterations required to achieve the tolerance length can be found as the integer greater than k, where $k = \ln[(\text{tolerance})/(b - a)]/\ln[0.618]$. Alternatively, the method can be stopped when an interval $[a, b]$ is less than the required tolerance. We now summarize the steps of the Golden Section Search Method.

Golden Section Search Method to Maximize $f(x)$ over the Interval $a \le x \le b$

STEP 1   Initialize: Choose a tolerance $t > 0$.

STEP 2   Set $r = 0.618$ and define the test points:

$$x_1 = a + (1 - r)(b - a)$$
$$x_2 = a + r(b - a)$$

STEP 3   Calculate $f(x_1)$ and $f(x_2)$.

STEP 4   (For a maximization problem) Compare $f(x_1)$ with $f(x_2)$:

   a. If $f(x_1) \le f(x_2)$, then the new interval is $[x_1, b]$:
      a becomes the previous x_1
      b does not change
      x_1 becomes the previous x_2
      Find the new x_2 using the formula in Step 2.

   b. If $f(x_1) > f(x_2)$, then the new interval is $[a, x_2]$:
      a remains unchanged
      b becomes the previous x_2
      x_2 becomes the previous x_1
      Find the new x_1 using the formula in Step 2.

STEP 5   If the length of the new interval from Step 4 is less than the tolerance t specified, then stop. Otherwise go back to Step 3.

STEP 6   Estimate x^* as the midpoint of the final interval $x^* = \frac{a+b}{2}$ and compute $MAX = f(x^*)$.

STOP

To minimize a function $y = f(x)$, either maximize $-y$ or switch the directions of the signs in Steps 4a and 4b. Note that the advantage of the Golden Section Search Method is that only one new test point (and evaluation of the function at the test point) must be computed at each successive iteration, compared with two new test points (and two evaluations of the function at those test points) for the Dichotomous Search Method. Using the Golden Section Search Method, the length of the interval of uncertainty is 61.8% of the length of the previous interval of uncertainty. Thus for large n, the interval of uncertainty is reduced by approximately $(0.618)^n$ after n test points are computed. (Compare with $(0.5)^{n/2}$ for the Dichotomous Search Method.)

Example 2   Suppose we want to maximize $f(x) = -3x^2 + 21.6x + 1$ over $0 \le x \le 25$, with a tolerance of $t = 0.25$. Determine the first two test points and evaluate $f(x)$ at each test point:

$$x_1 = a + 0.382(b - a) \rightarrow x_1 = 0 + 0.382(25 - 0) = 9.55$$

and

$$x_2 = a + 0.618(b - a) \rightarrow x_2 = 0 + 0.618(25 - 0) = 15.45$$

Then

$$f(x_1) = -66.3275 \quad \text{and} \quad f(x_2) = -381.3875$$

Because $f(x_1) > f(x_2)$, we discard all values in $(x_2, b]$ and select the new interval $[a, b] = [0, 15.45]$. Then $x_2 = 9.55$, which is the previous x_1, and $f(x_2) = -66.2972$. We must now find the position of the new test point x_1, and evaluate $f(x_1)$:

$$x_1 = 0 + (1 - r)(15.45 - 0) = 5.9017$$
$$f(x_1) = 23.9865$$

Again, $f(x_1) > f(x_2)$, so the new interval $[a, b]$ is $[0, 9.55]$. Then the new $x_2 = 5.9017$ and $f(x_2) = 23.9865$. We find a new x_1 and $f(x_1)$:

$$x_1 = 0 + (1 - r)(9.55 - 0) = 3.6475$$
$$f(x_1) = 39.8732$$

Because $f(x_1) > f(x_2)$, we discard $(x_2, b]$ and our new search interval is $[a, b] = [0, 5.9017]$. Then $x_2 = 3.6475$ with $f(x_2) = 39.8732$. We find a new x_1 and $f(x_1)$:

$$x_1 = 0 + (1 - r)(5.9017 - 0) = 2.2542$$
$$f(x_1) = 34.4469$$

Because $f(x_2) > f(x_1)$, we discard $[a, x_1)$ and the new interval is $[a, b] = [2.2542, 5.9017]$. The new $x_1 = 3.6475$ with $f(x_1) = 39.8732$. We find a new x_2 and $f(x_2)$:

$$x_2 = 2.2545 + r(5.9017 - 2.2542) = 4.5085$$
$$f(x_2) = 37.4039$$

This process continues until the length of the interval of uncertainty, $b - a$, is less than the tolerance, $t = 0.25$. This requires 10 iterations. The results of the Golden Section Search Method for Example 2 are summarized in Table 9.2.

The final interval $[a, b] = [3.4442, 3.6475]$ is the first interval of our $[a, b]$ intervals with length less than our 0.25 tolerance. The value of x that maximizes

TABLE 9.2 Golden Section Search Method results for Example 2

k	a	b	x_1	x_2	$f(x_1)$	$f(x_2)$
0	0	25	9.5491	15.4509	−66.2972	−381.4479
1	0	15.4506	5.9017	9.5491	23.9865	−66.2972
2	0	9.5592	3.6475	5.9017	39.8732	23.9865
3	0	5.9017	2.2542	3.6475	34.4469	39.8732
4	2.2542	5.9017	3.6475	4.5085	39.8732	37.4039
5	2.2542	4.5085	3.1153	3.6475	39.1752	39.8732
6	3.1153	4.5085	3.6475	3.9763	39.8732	39.4551
7	3.1153	3.9763	3.4442	3.6475	39.8072	39.8732
8	3.4442	3.9763	3.6475	3.7731	39.8732	39.7901
9	3.4442	3.7731	3.5698	3.6475	39.8773	39.8732
10	3.4442	3.6475				

the given function over the interval must lie within this final interval of uncertainty [3.4442, 3.6475]. We estimate $x^* = (3.4442 + 3.6475)/2 = 3.5459$ and $f(x^*) = 39.8712$. The actual maximum, which in this case can be found by calculus, occurs at $x^* = 3.60$, where $f(3.60) = 39.88$.

As illustrated, we stopped when the interval of uncertainty was less than 0.25. Alternatively, we can compute the number of iterations required to attain the accuracy specified by the tolerance. Because the interval of uncertainty is 61.8% of the length of the interval of uncertainty at each stage, we have

$$\frac{\text{length of final interval (tolerance } t)}{\text{length of initial interval}} = 0.618^k$$

$$\frac{0.25}{25} = 0.618^k$$

$$k = \frac{\ln 0.01}{\ln 0.618} = 9.57, \text{ or 10 iterations}$$

In general, the number of iterations k required is given by

$$k = \frac{\ln (\text{tolerance}/(b - a))}{\ln 0.618}$$

Example 3 Revisit Model-Fitting Criterion

Recall the curve-fitting procedure from Chapter 5 using the criterion

$$\text{Minimize } \sum |y_i - y(x_i)|$$

Let's use the Golden Section Search Method to fit the model $y = cx^2$ to the following data for this criterion:

x	1	2	3
y	2	5	8

The function to be minimized is

$$f(c) = |2 - c| + |5 - 4c| + |8 - 9c|$$

and we will search for an optimal value of c in the closed interval $[0, 3]$. We choose a tolerance $t = 0.2$. We apply the Golden Section Search Method until an interval of uncertainty is less than 0.2. The results are summarized in Table 9.3.

TABLE 9.3 Golden Section Search Method used in model fitting for Example 3

k	a	b	c_1	c_2	$f(c_1)$	$f(c_2)$
0	0	3	1.1459	1.8541	3.5836	11.2492
1	0	1.8541	0.7082	1.1459	5.0851	3.5836
2	0.7082	1.8541	1.1459	1.4164	3.5836	5.9969
3	0.7082	1.4164	0.9787	1.1459	2.9149	3.5836
4	0.7082	1.1459	0.8754	0.9787	2.7446	2.9149
5	0.7082	0.9787	0.8115	0.8754	3.6386	2.7446
6	0.8115	0.9787				

The length of the final interval is less than 0.2. We can estimate $c^* = (0.8115 + 0.9787)/2 = 0.8951$, with $f(0.8951) = 2.5804$. In the problem set, we ask you to show analytically that the optimal value for c is $c = \frac{8}{9}$.

Example 4   *Optimizing Industrial Flow*

Figure 9.25 represents a physical system engineers might need to consider for an industrial flow process. As shown, let x represent the flow rate of dye into the coloring process of cotton fabric. Based on this rate, the reaction differs with the other substances in the process as evidenced by the function shown in Figure 9.25. The function is defined as

$$f(x) = \begin{cases} 2 + 2x - x^2 & \text{for } 0 \leq x \leq \frac{3}{2} \\ -x + \frac{17}{4} & \text{for } \frac{3}{2} < x \leq 4 \end{cases}$$

The function defining the process is unimodal. The company wants to find the flow rate x that maximizes the reaction of the other substances $f(x)$. Through experimentation the engineers found that the process is sensitive to within about 0.20 of the actual value of x. They also found that the flow is either *off* $(x = 0)$ or *on* $(x > 0)$. The process will not allow for turbulent flow that occurs above $x = 4$ for this process. Thus $x \leq 4$ and we use a tolerance of 0.20 to maximize $f(x)$ over $[0, 4]$. Using the Golden Section Search Method, we locate the first two test points

$$x_1 = 0 + 4(1 - r) = 1.5279$$
$$x_2 = 0 + 4(r) = 2.4721$$

and evaluate the function at the test points

$$f(x_1) = 2.7221$$
$$f(x_2) = 1.7779$$

The results of the search are given in Table 9.4.

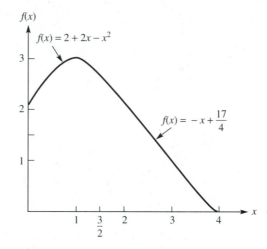

$f(x)$

$f(x) = 2 + 2x - x^2$

$f(x) = -x + \dfrac{17}{4}$

FIGURE 9.25   Industrial flow process function

TABLE 9.4 Results of Golden Section Search Method for Example 4

a	b	x_1	x_2	$f(x_1)$	$f(x_2)$
0	4	1.5279	2.4721	2.7221	1.7779
0	2.4721	0.9443	1.5279	2.9969	2.7221
0	1.5279	0.5836	0.9443	2.8266	2.9969
0.5836	1.5279	0.9443	1.1672	2.9969	2.9720
0.5836	1.1672	0.8065	0.9443	2.9626	2.9969
0.8065	1.1672	0.9443	1.0294	2.9969	2.9991
0.9443	1.1672	1.0294	1.0820	2.9991	2.9933
0.9443	1.0820				

We stop because $(1.0820 - 0.9443) < 0.20$. The midpoint of the interval is 1.0132, where the value of the function is 2.9998. In the problem set, you are asked to show analytically that a maximum value of $f(x^*) = 3$ occurs at $x = 1$.

9.5 PROBLEMS

1. Use the Dichotomous Search Method with a tolerance of $t = 0.2$ and $\varepsilon = 0.01$.

   a. Minimize $f(x) = x^2 + 2x$, $-3 \le x \le 6$

   b. Maximize $f(x) = -4x^2 + 3.2x + 3$, $[-2 \le x \le 2]$

2. Use the Golden Section Search Method with a tolerance of $t = 0.2$.

   a. Minimize $f(x) = x^2 + 2x$, $-3 \le x \le 6$

   b. Maximize $f(x) = -4x^2 + 3.2x + 3$, $[-2 \le x \le 2]$

3. Use the curve-fitting criterion to minimize the sum of the absolute deviations for the following models and data set:

   a. $y = ax$ **b.** $y = ax^2$ **c.** $y = ax^3$

x	7	14	21	28	35	42
y	8	41	133	250	280	297

4. For Example 2, show that the optimal value of c is $c^* = \frac{8}{9}$. *Hint:* Apply the definition of the absolute value to obtain a piecewise continuous function. Then find the minimum value of the function over the interval $[0, 3]$.

5. For Example 3, show that the optimal value of x is $x^* = 1$.

9.5 PROJECTS

1. *Fibonacci Search*—One of the more interesting search techniques uses the Fibonacci sequence. This search method can be used even if the function is not continuous. The method uses the Fibonacci numbers to place test points for the search. These Fibonacci numbers are defined as follows: $F_0 = F_1 = 1$ and $F_n = F_{n-1} + F_{n-2}$, for $n = 2, 3, 4, \ldots$, yielding the sequence 1, 1, 2, 3, 5, 8, 13, 21, 34, 55, 89, 144, 233, 377, 510, 887, 1397, and so forth.

   a. Find the ratio between successive Fibonacci numbers using the preceding sequences. Then find the numerical limit as n gets large. How is this limiting ratio related to the Golden Section Search Method?

   b. Research and make a short presentation on the Fibonacci Search Method. Present your results to the class.

2. *Methods Using Derivatives: Newton's Method*—One of the best-known interpolation methods is Newton's Method, which exploits a quadratic approximation to the function $f(x)$ at a given point x_1. The quadratic approximation q is given by

$$q(x) = f(x_1) + f'(x_1)(x - x_1) + \frac{1}{2}f''(x_1)(x - x_1)^2$$

The point x_2 is then taken to be the point where q' equals 0. Continuing this procedure yields the sequence

$$x_{k+1} = x_k - \frac{f'(x_k)}{f''(x_k)}$$

where $k = 1, 2, 3, \ldots$. This procedure is terminated when either $|x_{k+1} - x_k| < \varepsilon$ or $|f'(x_k)| < \varepsilon$, where ε is some small number. This procedure can only be applied to twice-differentiable functions if $f''(x)$ never equals 0.

a. Starting with $x = 4$ and a tolerance of $\varepsilon = 0.01$, use Newton's Method to minimize $f(x) = x^2 + 2x$, over $-3 \le x \le 6$.

b. Use Newton's Method to minimize

$$f(x) = \begin{cases} 4x^3 - 3x^4 & \text{for } x > 0 \\ 4x^3 + 3x^4 & \text{for } x < 0 \end{cases}$$

Let the tolerance be $\varepsilon = 0.01$ and start with $x = 0.4$.

c. Repeat part **b.**, starting at $x = 0.6$. Discuss what happens when you apply the method.

9.5 Further Reading

BAZARRA, M., Hanif D. Sherali, & C. M. Shetty. *Nonlinear Programming: Theory and Algorithms*. 2nd ed. New York: Wiley, 1993.

RAO, S. S. *Optimization: Theory and Applications*. New Delhi: Wiley Eastern Limited, 1979.

WINSTON, Wayne. *Mathematical Programming: Applications and Algorithms*. 2nd ed. Boston: PWS, 1995.

WINSTON, Wayne. *Operations Research: Applications and Algorithms*. 3rd ed. Belmont, CA: Duxbury Press, 1994.

Chapter Ten

MODELING WITH A DIFFERENTIAL EQUATION

INTRODUCTION

Quite often we have information relating a rate of change of a dependent variable with respect to one or more independent variables and are interested in discovering the function relating the variables. For example, if P represents the number of people in a large population at some time t, then it is reasonable to assume that the rate of change of the population with respect to time depends on the current size of P as well as other factors that are discussed in Section 10.1. For ecological, economical, and other important reasons, it is desirable to determine a relationship between P and t to make predictions about P. If the present population size is denoted by $P(t)$ and the population size at time $t + \Delta t$ is $P(t + \Delta t)$, then the change in population ΔP during that time period Δt is given by

$$\Delta P = P(t + \Delta t) - P(t) \tag{10.1}$$

The factors affecting the population growth are developed in detail in Section 10.1. For now, let's assume a simple proportionality: $\Delta P \propto P$. For example, if immigration, emigration, age, and gender are all neglected, we can assume that during a unit time period a certain percentage of the population reproduces while a certain percentage dies. Suppose the constant of proportionality k is expressed as a percentage per unit time. Then our proportionality assumption gives

$$\Delta P = P(t + \Delta t) - P(t) = kP\Delta t \tag{10.2}$$

difference equation

discrete mathematics
finite mathematics

Equation (10.2) is called a **difference equation**. Note that we are treating a discrete set of time periods rather than allowing t to vary *continuously* over some interval. Difference equations belong to an important area of mathematics, called **discrete** or **finite mathematics**. In this situation the discrete set of times may give the

345

population in future years at those distinct times (perhaps after the spring spawn in a fish population). Referring to Figure 10.1, observe that the horizontal distance between the points $(t_0, P(t_0))$ and $(t_0 + \Delta t, P(t_0 + \Delta t))$ is Δt, which may represent the time between spawning periods in a fish population growth problem or the length of a fiscal period in a budget growth problem. The time t_0 refers to a particular time. The vertical distance, ΔP in this case, represents the change in the dependent variable.

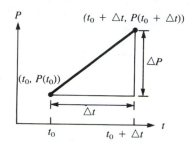

FIGURE 10.1 The size of the population at future time intervals of length Δt gives a discrete set of points

Assume that t does vary continuously so we can take advantage of the calculus. Division of Equation (10.2) by Δt gives

$$\frac{\Delta P}{\Delta t} = \frac{P(t + \Delta t) - P(t)}{\Delta t} = kP \qquad \textbf{(10.3)}$$

We can interpret $\Delta P/\Delta t$ physically as the *average rate of change* in P during the time period Δt. For example, $\Delta P/\Delta t$ may represent the average daily growth of the budget. In other scenarios, however, it may have no physical interpretation: If fish spawn only in the spring, it is somewhat meaningless to talk about the average daily growth in the fish population. Again in Figure 10.1, $\Delta P/\Delta t$ can be interpreted geometrically as the slope of the line segment connecting the points $(t_0, P(t_0))$ and $(t_0 + \Delta t, P(t_0 + \Delta t))$. Next allow Δt to approach zero. The definition of the derivative gives the *differential equation*

$$\lim_{\Delta t \to 0} \frac{\Delta P}{\Delta t} = \frac{dP}{dt} = kP$$

where dP/dt represents the *instantaneous rate of change*. In many situations the instantaneous rate of change has an identifiable physical interpretation, such as in the flow of heat from a space capsule after entering the ocean or the reading of a car speedometer as the car accelerates. However, in the case of a fish population that has a discrete spawning period or the budget process that has a discrete fiscal period, the instantaneous change may be somewhat meaningless. These latter scenarios are more appropriately modeled using difference equations, but it is occasionally advantageous to approximate a difference equation with a differential equation.

The derivative is used in two distinct roles:

1. To represent the instantaneous rate of change in *continuous* problems
2. To approximate an average rate of change in *discrete* problems

The advantage of approximating an average rate of change by a derivative is that the calculus often helps in uncovering a functional relationship between the variables under investigation. For instance, the solution to the Model (10.3) is $P = P_0 e^{kt}$, where P_0 is the population at time $t = 0$. However, many differential equations cannot be solved so easily using analytical techniques. In such cases the solutions are approximated using discrete methods. An introduction to numerical techniques is presented in Section 10.4. In case the solution being approximated is to a differential equation that is itself an approximation to a difference equation, the modeler should consider using a discrete method with the finite difference equation directly (see Chapter 3).

The interpretation of the derivative as an instantaneous rate of change is useful in many modeling applications. The geometrical interpretation of the derivative as the slope of the line tangent to the curve is useful for constructing numerical solutions. Let's review more carefully these important concepts from the calculus.

The Derivative as a Rate of Change

The origins of the derivative lie in humankind's curiosity about motion and our need to develop a deeper understanding of motion. The search for the laws governing planetary motion, the study of the pendulum and its application to clock building, and the laws governing the flight of a cannonball were the kinds of problems stimulating the minds of mathematicians and scientists in the 16th and 17th centuries. Such problems motivated the development of the calculus.

To remind ourselves of one interpretation of the derivative, consider a particle whose distance s from a fixed position depends on time t. Let the graph in Figure 10.2 represent the distance s as a function of time t, and let (t_1, s_1) and (t_2, s_2) denote two points on the graph.

Define $\Delta t = t_2 - t_1$ and $\Delta s = s_2 - s_1$, and form the ratio $\Delta s / \Delta t$. Note that this ratio represents a rate: an increment of distance traveled Δs over some increment of time Δt. That is, the ratio $\Delta s / \Delta t$ represents the average velocity during the time period in question. Now remember how the derivative ds/dt evaluated at $t = t_1$ is defined:

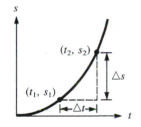

FIGURE 10.2   Graph of distance s as a function of time t

$$\left. \frac{ds}{dt} \right|_{t=t_1} = \lim_{\Delta t \to 0} \frac{\Delta s}{\Delta t} \tag{10.4}$$

Physically, what occurs as $\Delta t \to 0$? Using the interpretation of average velocity, we can see that at each stage of using a smaller Δt we are computing the average velocity over smaller and smaller intervals with left endpoint at t_1 until, in the limit, we have the instantaneous velocity at $t = t_1$. If we think of the motion of a moving vehicle, this instantaneous velocity would correspond to the exact reading of its (perfect) speedometer at the instant t_1.

More generally, if $y = f(x)$ is a differentiable function, then the derivative dy/dx at any given point can be interpreted as the *instantaneous rate of change* of *y* with respect to *x* at that point. Interpreting the derivative as an instantaneous rate of change is useful in many modeling applications.

The Derivative as the Slope of the Tangent Line

Let's consider another interpretation of the derivative. As scholars sought knowledge about the laws of planetary motion, their chief need was to observe and measure the heavenly bodies. However, the construction of lenses for use in telescopes was a difficult task. To grind a lens to the correct curvature to achieve the desired light refraction requires knowing the tangent to the curve describing the lens surface.

Let's examine the geometrical implications of the limit in Equation (10.4). We consider $s(t)$ simply as a curve. Let's examine a set of secant lines, each emanating from the point $A = (t_1, s(t_1))$ on the curve. To each secant there corresponds a pair of increments $(\Delta t_i, \Delta s_i)$ as shown in Figure 10.3. The lines AB, AC, and AD are secant lines. As $\Delta t \to 0$, these secant lines approach the line tangent to the curve at the point A. Because the slope of each secant is $\Delta s / \Delta t$, we can interpret the derivative as *the slope of the line tangent to the curve $s(t)$ at the point A*. The interpretation of the derivative evaluated at a point as the slope of the line tangent to the curve at that point is useful in constructing numerical approximations to solutions of differential equations. Numerical approximations are discussed in Section 10.4.

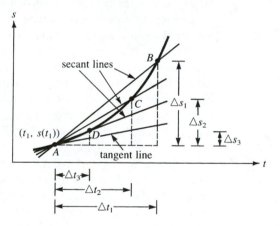

FIGURE 10.3 The slope of each secant line approximates the slope of the tangent line to the curve at the point A

10.1   POPULATION GROWTH

Interest in how populations tend to grow was stimulated in the late 18th century when Thomas Malthus (1766–1834) published *An Essay on the Principle of Population as it Affects the Future Improvement of Society*. In his book Malthus put forth an exponential growth model for human population and concluded that eventually the population would exceed the capacity to grow an adequate food supply. Although

the assumptions of the Malthusian model leave out factors important to population growth (so the model has proved to be inaccurate for technologically developed countries), it is instructive to examine this model as a basis for later refinement.

Problem identification   Suppose we know the population at some given time, say it is P_0 at time $t = t_0$, and we are interested in predicting the population P at some future time $t = t_1$. In other words, we want to find a population function $P(t)$ for $t_0 \leq t \leq t_1$ satisfying $P(t_0) = P_0$.

Assumptions   Consider some factors that pertain to population growth. Two obvious ones are the *birthrate* and the *death rate*. The birthrate and death rate are determined by different factors. The birthrate is influenced by infant mortality rate, attitudes toward and availability of contraceptives, attitudes toward abortion, health care during pregnancy, and so forth. The death rate is affected by sanitation and public health, wars, pollution, medicines, diet, psychological stress and anxiety, and so forth. Other factors that influence population growth in a given region are immigration and emigration, living space restrictions, availability of food and water, and epidemics. For our model, let's neglect all these latter factors. (If we are dissatisfied with our results, we can include these factors later in a more refined model, possibly in a simulation model.) Now we'll consider only the birthrate and death rate. Since knowledge and technology have helped humankind diminish the death rate below the birthrate, human populations have tended to grow.

Let's begin by assuming that during a small unit time period a percentage b (given as a decimal equivalent) of the population is newly born. Similarly, a percentage c of the population dies. In other words, the new population $P(t + \Delta t)$ is the old population $P(t)$ plus the number of births minus the number of deaths during the time period Δt. Symbolically,

$$P(t + \Delta t) = P(t) + bP(t)\Delta t - cP(t)\Delta t$$

or

$$\frac{\Delta P}{\Delta t} = bP - cP = kP$$

Thus we are really assuming that the average rate of change of the population over an interval of time is proportional to the size of the population. Using the instantaneous rate of change to approximate the average rate of change, we have the following model:

$$\frac{dP}{dt} = kP, \quad P(t_0) = P_0, \quad t_0 \leq t \leq t_1 \tag{10.5}$$

where (for growth) k is a positive constant.

Solving the model   We can separate the variables and rewrite Equation (10.5) by moving all terms involving P and dP to one side of the equation and all terms in t

and *dt* to the other. This gives

$$\frac{dP}{P} = k\,dt$$

Integration of both sides of this last equation yields

$$\ln P = kt + C \tag{10.6}$$

for some constant C. Applying the condition $P(t_0) = P_0$ to Equation (10.6) to find C results in

$$\ln P_0 = kt_0 + C$$

or

$$C = \ln P_0 - kt_0$$

Then substitution for C into (10.6) gives

$$\ln P = kt + \ln P_0 - kt_0$$

or, simplifying algebraically,

$$\ln \frac{P}{P_0} = k(t - t_0)$$

Finally, by taking the exponential of both sides of the preceding equation and multiplying the result by P_0, we obtain the solution

$$P(t) = P_0 e^{k(t-t_0)} \tag{10.7}$$

Malthusian model of population growth   Equation (10.7), which is known as the **Malthusian model of population growth**, predicts that population grows exponentially with time.

Verifying the model   Because $\ln(P/P_0) = k(t - t_0)$, our model predicts that if we plot $\ln P/P_0$ versus $t - t_0$, a straight line passing through the origin with slope k should result. However, if we plot the population data for the United States for several years, the model does not fit very well, especially in the later years. In fact, the 1970 census for the population of the United States was 203,211,926, and in 1950 it was 150,697,000. Substituting these values into Equation (10.7) and dividing the first result by the second gives

$$\frac{203{,}211{,}926}{150{,}697{,}000} = e^{k(1970-1950)}$$

Thus

$$k = \left(\frac{1}{20}\right) \ln \frac{203{,}211{,}926}{150{,}697{,}000} \approx 0.015$$

That is, during the 20-year period from 1950 to 1970, population in the United States was increasing at the average rate of 1.5% per year. We can use this information together with Equation (10.7) to predict the population for 1980. In this case, $t_0 = 1970$, $P_0 = 203{,}211{,}926$, and $k = 0.015$ yields

$$P(1980) = 203{,}211{,}926 e^{0.015(1980-1970)} = 236{,}098{,}574$$

The 1980 census for the population of the United States was 226,505,000 (rounded to the nearest thousand). Thus our prediction is off the mark by roughly 4%. We can probably live with that magnitude error, but let's look into the distant future. Our model predicts that the population of the United States will be 28,688 billion in the year 2300, a population that exceeds current estimates of the maximum sustainable population of the entire planet! We are forced to conclude that our model is unreasonable over the long run.

Some populations do grow exponentially provided that the population is not too large. In most populations, however, individual members eventually compete with one another for food, living space, and other natural resources. Let's refine our Malthusian model of population growth to reflect this competition.

Refining the model to reflect limited growth   Let's consider that the proportionality factor k, measuring the rate of population growth in Equation (10.5), is now no longer constant but a function of the population. As the population increases and gets closer to the maximum population M, the rate k decreases. One simple submodel for k is the linear one

$$k = r(M - P), \ r > 0$$

where r is a constant. Substitution into Equation (10.5) leads to

$$\frac{dP}{dt} = r(M - P)P \tag{10.8}$$

or

$$\frac{dP}{P(M - P)} = r \, dt \tag{10.9}$$

Again we assume the *initial condition* $P(t_0) = P_0$. (Model (10.8) was first introduced by the Dutch mathematical biologist Pierre-Francois Verhulst (1804–1849) and is

logistic growth   referred to as **logistic growth**.) It follows from elementary algebra that

$$\frac{1}{P(M - P)} = \frac{1}{M}\left(\frac{1}{P} + \frac{1}{M - P}\right)$$

Thus Equation (10.9) can be rewritten as

$$\frac{dP}{P} + \frac{dP}{M - P} = rM\,dt$$

which integrates to

$$\ln P - \ln|M - P| = rMt + C \tag{10.10}$$

for some arbitrary constant C. Using the initial condition, we evaluate C in the case $P < M$:

$$C = \ln\frac{P_0}{M - P_0} - rMt_0$$

Substituting into (10.10) and simplifying gives

$$\ln\frac{P}{M - P} - \ln\frac{P_0}{M - P_0} = rM(t - t_0)$$

or

$$\ln\frac{P(M - P_0)}{P_0(M - P)} = rM(t - t_0)$$

Exponentiating both sides of this equation gives

$$\frac{P(M - P_0)}{P_0(M - P)} = e^{rM(t - t_0)}$$

or

$$P_0(M - P)e^{rM(t - t_0)} = P(M - P_0)$$

Then,

$$P_0 M e^{rM(t - t_0)} = P(M - P_0) + P_0 P e^{rM(t - t_0)}$$

so that solving for the population P gives

$$P(t) = \frac{P_0 M e^{rM(t - t_0)}}{M - P_0 + P_0 e^{rM(t - t_0)}}$$

To estimate P as $t \to \infty$, we rewrite this last equation as

$$P(t) = \frac{MP_0}{[P_0 + (M - P_0)e^{-rM(t-t_0)}]} \qquad (10.11)$$

Notice from Equation (10.11) that $P(t)$ approaches M as t tends to infinity. Moreover, from Equation (10.8) we calculate the second derivative

$$P'' = rMP' - 2rPP' = rP'(M - 2P)$$

so that $P'' = 0$ when $P = M/2$. In words, when the population P reaches half the limiting population M, the growth dP/dt is most rapid, and then it starts to diminish toward zero. One advantage of recognizing that the maximum rate of growth occurs at $P = M/2$ is that the information can be used to estimate M. In a situation in which the modeler is satisfied that the growth involved is essentially logistic, if the point of maximum rate of growth has been reached, then $M/2$ can be estimated. The graph of the limited growth Equation (10.11) is depicted in Figure 10.4 for the case $P < M$ (see Problem 2 in the 10.1 problem set for the case $P > M$). Such a curve is called a **logistic curve**.

logistic curve

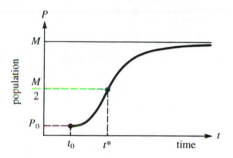

P

M

$\dfrac{M}{2}$

P_0

population

t_0 $t*$ time

t

FIGURE 10.4   Graph of the limited growth model

Verifying the limited growth model   Let's test our Model (10.11) against some real-world data. Equation (10.10) suggests a straight-line relationship of $\ln[P/(M - P)]$ versus t. Let's test this model using the data given in Table 10.1 for the growth of yeast in a culture. To plot $\ln[P/(M - P)]$ versus t, we need an estimate for the limiting population M. From the data in Table 10.1 we see that the population never exceeds 661.8. We estimate $M \approx 665$ and plot $\ln[P/(665 - P)]$ versus t. The graph is shown in Figure 10.5 and does approximate a straight line. Thus we accept the assumptions of logistic growth for bacteria. Now Equation (10.10) gives

$$\ln \frac{P}{M - P} = rMt + C$$

and from the graph in Figure 10.5 we can estimate the slope $rM \approx 0.55$, so that $r \approx 0.0008271$ from our estimate for $M \approx 665$.

TABLE 10.1 Growth of yeast in a culture

Time in hours	Observed yeast biomass	Biomass calculated from logistic Equation (10.13)	Percent error
0	9.6	8.9	−7.3
1	18.3	15.3	−16.4
2	29.0	26.0	−10.3
3	47.2	43.8	−7.2
4	71.1	72.5	2.0
5	119.1	116.3	−2.4
6	174.6	178.7	2.3
7	257.3	258.7	0.5
8	350.7	348.9	−0.5
9	441.0	436.7	−1.0
10	513.3	510.9	−4.7
11	559.7	566.4	1.2
12	594.8	604.3	1.6
13	629.4	628.6	−0.1
14	640.8	643.5	0.4
15	651.1	652.4	0.2
16	655.9	657.7	0.3
17	659.6	660.8	0.2
18	661.8	662.5	0.1

Data from R. Pearl, "The Growth of Population," *Quart. Rev. Biol.* 2 (1927): 532–548.

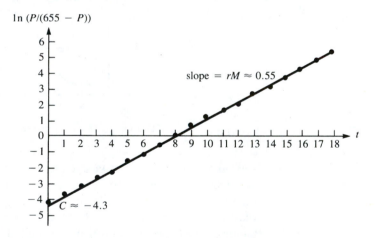

FIGURE 10.5 Plot of $\ln[P/(665 - P)]$ versus t for the data in Table 10.1

It is often convenient to express the logistic Equation (10.11) in another form. To this end, let t^* denote the time when the population P reaches half the limiting value; that is, $P(t^*) = M/2$. It follows from Equation (10.11) that

$$t^* = t_0 - \frac{1}{rM} \ln \frac{P_0}{M - P_0}$$

(see Problem 1a at the end of this section). Solving this last equation for t_0, substituting the result into (10.11), and simplifying algebraically gives

$$P(t) = \frac{M}{1 + e^{-rM(t-t^*)}} \tag{10.12}$$

(see Problem 1b at the end of this section).

We can estimate t^* for the yeast culture data presented in Table 10.1 using Equation (10.10) and our graph in Figure 10.5:

$$t^* = -\frac{C}{rM} \approx \frac{4.3}{0.55} \approx 7.82$$

This calculation gives the logistic equation

$$P(t) = \frac{665}{1 + 73.8e^{-0.55t}} \tag{10.13}$$

by substituting $M = 665$, $r = 0.0008271$, and $t^* = 7.82$ in (10.12).

The logistic model is known to agree quite well for populations of organisms that have very simple life histories; for instance, yeast growing in a culture where space is limited. Table 10.1 shows the calculations for the logistic Equation (10.13), and we can see from the calculated error there is very good agreement with the original data. A plot of the curve is shown in Figure 10.6.

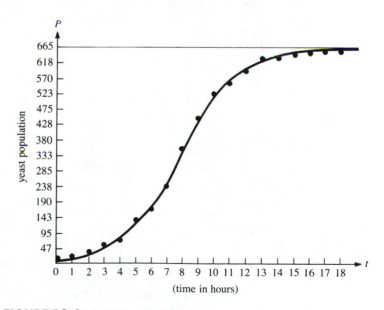

FIGURE 10.6   Logistic curve showing the growth of yeast in a culture based on the data from Table 10.1 and the Model (10.13); the small circles indicate the observed values

Next let's consider some data for human populations. A logistic equation for the population growth of the United States was formulated by Pearl and Reed in 1920. One form of their logistic curve is given by

$$P(t) = \frac{197{,}273{,}522}{1 + e^{-0.03134(t-1914.32)}} \qquad \textbf{(10.14)}$$

where $M = 197{,}273{,}522$, $r = 1.5887 \times 10^{-10}$, and $t^* = 1914.32$ were determined using the census figures for the years 1790, 1850, and 1910 (we ask you to estimate M, r, and t^* in Problem 4 of the 10.1 problem set).

TABLE 10.2 Population of the United States from 1790 to 1980, with predictions from Equation (10.14)

Year	Observed population	Predicted population	Percent error
1790	3,929,000	3,929,000	0.0
1800	5,308,000	5,336,000	0.5
1810	7,240,000	7,227,000	−0.2
1820	9,638,000	9,756,000	1.2
1830	12,866,000	13,108,000	1.9
1840	17,069,000	17,505,000	2.6
1850	23,192,000	23,191,000	−0.0
1860	31,443,000	30,410,000	−3.3
1870	38,558,000	39,370,000	2.1
1880	50,156,000	50,175,000	0.0
1890	62,948,000	62,767,000	−0.3
1900	75,995,000	76,867,000	1.1
1910	91,972,000	91,970,000	−0.0
1920	105,711,000	107,393,000	1.6
1930	122,755,000	122,396,000	−0.3
1940	131,669,000	136,317,000	3.5
1950	150,697,000	148,677,000	−1.3
1960	179,323,000	159,230,000	−11.2
1970	203,212,000	167,943,000	−17.4
1980	226,505,000	174,941,000	−22.8
1990	248,710,000	180,440,000	−27.5

Table 10.2 compares the values predicted in 1920 by the logistic Equation (10.14) with the observed values of the population of the United States. The predicted values agree quite well with the observations up to the year 1950, but as we can see the predicted values are much too small for the years 1970, 1980, and 1990. This should not be too surprising because our model fails to take into account such factors as immigration into the United States, wars, and advances in medical technology. In populations of higher plants and animals, which have complicated life histories and long periods of individual development, there are likely to be numerous responses that greatly modify the population growth.

10.1 PROBLEMS

1. a. Show that the population P in the logistic equation reaches half the maximum population M at time t^* given by

$$t^* = t_0 - (1/rM)\ln[P_0/(M - P_0)]$$

b. Derive the form given by Equation (10.12) for population growth according to the logistic law.

c. Derive the equation $\ln[P/(M - P)] = rMt - rMt^*$ from Equation (10.12).

2. Consider the solution of Equation (10.8). Evaluate the constant C in (10.10) in the case that $P > M$ for all t. Sketch the solutions in this case. Also sketch a solution curve for the case $M/2 < P < M$.

3. The following data were obtained for the growth of a sheep population introduced into a new environment on the island of Tasmania. (Adapted from J. Davidson, "On the Growth of the Sheep Population in Tasmania," *Trans. Roy. Soc. S. Australia* 62(1938): 342–346.)

t (year)	1814	1824	1834	1844	1854	1864
$P(t)$	125	275	830	1200	1750	1650

a. Make an estimate of M by graphing $P(t)$.

b. Plot $\ln[P/(M - P)]$ against t. If a logistic curve seems reasonable, estimate rM and t^*.

4. Using the data for the U.S. population in Table 10.2, estimate M, r, and t^* using the same technique as in the text. Assume you are making the prediction in 1951 using previous census. Use the data from 1960 to 1990 to check your model.

5. The modern philosopher Jean-Jacques Rousseau formulated a simple model of population growth for 18th-century England based on the following assumptions:

The birthrate in London is less than that in rural England.

The death rate in London is greater than that in rural England.

As England industrializes, more and more people migrate from the countryside to London.

Rousseau then reasoned that because London's birthrate was lower and its death rate higher, and rural people tended to migrate there, the population of England would eventually decline to zero. Criticize Rousseau's conclusion.

6. Consider the spreading of a highly communicable disease on an isolated island with population size N. A portion of the population travels abroad and returns to the island infected with the disease. You would like to predict the number of people X who will have been infected by some time t. Consider the following model:

$$\frac{dX}{dt} = kX(N - X)$$

a. List two major assumptions implicit in the preceding model. How reasonable are your assumptions?

b. Graph dX/dt versus X.

c. Graph X versus t if the initial number of infections is $X_1 < N/2$. Graph X versus t if the initial number of infections is $X_2 > N/2$.

d. Solve the model given earlier for X as a function of t.

e. From part **d**, find the limit of X as t approaches infinity.

f. Consider an island with a population of 5000. At various times during the epidemic the number of people infected was recorded as follows:

t (days)	2	6	10
X (people infected)	1887	4087	4853
$\ln[X/(N - X)]$	−0.5	1.5	3.5

Do the collected data support the given model?

g. Use the results in part **f** to estimate the constants in the model, and predict the number of people who will be infected by $t = 12$ days.

7. Assume we are considering the survival of whales and that if the number of whales falls below a minimum survival level m the species will become extinct. Assume also that the population is limited by the carrying capacity M of the environment. That is, if the whale population is above M, then it will experience a decline because the environment cannot sustain that high a population level.

a. Discuss the following model for the whale population:

$$\frac{dP}{dt} = k(M - P)(P - m)$$

where $P(t)$ denotes the whale population at time t and k is a positive constant.

b. Graph dP/dt versus P and P versus t. Consider the cases in which the initial population $P(0) = P_0$ satisfies $P_0 < m$, $m < P_0 < M$, and $M < P_0$.

c. Solve the model in part **a**, assuming that $m < P < M$ for all time. Show that the limit of P as t approaches infinity is M.

d. Discuss how you would test the model in part **a**. How would you determine M and m?

e. Assuming that the model reasonably estimates the whale population, what implications are suggested for fishing? What controls would you suggest?

8. Sociologists recognize a phenomenon called *social diffusion*, which is the spreading of a piece of information, a technological innovation, or a cultural fad among a population. The members of the population can be divided into two classes: those who have the information and those who do not. In a fixed population whose size is known, it is reasonable to assume that the rate of diffusion is proportional to the number who have the information times the number yet to receive it. If X denotes the number of individuals who have the information in a population of N people, then a mathematical model for social diffusion is given by $dX/dt = kX(N - X)$, where t represents time and k is a positive constant.

a. Solve the model and show that it leads to a logistic curve.

b. At what time is the information spreading fastest?

c. How many people will eventually receive the information?

10.1 Projects

1. Complete the requirements of the UMAP module, "The Cobb-Douglas Production Function," by Robert Geitz, UMAP 509. A mathematical model relating the output of an economic system to labor and capital is constructed from the assumptions that (a) marginal productivity of labor is proportional to the amount of production per unit of labor, (b) marginal productivity of capital is proportional to the amount of production per unit of capital, and (c) if either labor or capital tends to zero, then so does production.

2. Complete the UMAP module, "The Diffusion of Innovation in Family Planning," by Kathryn N. Harmon, UMAP 303. This module gives an interesting application of finite difference equations to study the process through which public policies are diffused to understand how national governments might adopt family planning policies.

3. Complete the UMAP module, "Difference Equations With Applications," by Donald R. Sherbert, UMAP 322. This module presents a good introduction to solving first- and second-order linear difference equations, including the method of undetermined coefficients for nonhomogeneous equations. Applications to problems in population and economic modeling are included.

10.1 Further Reading

FRAUENTHAL, James C. *Introduction to Population Modeling.* Lexington, MA: COMAP, 1979.

HUTCHINSON, G. Evelyn. *An Introduction to Population Ecology.* New Haven, CT: Yale University Press, 1978.

LEVINS, R. "The Strategy of Model Building in Population Biology." *American Scientist 54* (1966): 421–431.

LOTKA, A. J. *Elements of Mathematical Biology.* New York: Dover, 1956.

ODUM, E. P. *Fundamentals of Ecology.* Philadelphia: Saunders, 1971.

PEARL, R. & L. J. Reed. "On the Rate of Growth of the Population of the United States since 1790." *Proceedings of the National Academy of Science 6* (1920): 275–288.

10.2 Prescribing Drug Dosage[1]

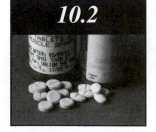

The problem of what dosage to prescribe for a drug and how often the dosage should be administered is an important one in pharmacology. For most drugs there is a concentration below which the drug is ineffective and a concentration above which the drug is dangerous.

[1]This section is adapted from UMAP Unit 72, based on the work of Brindell Horelick and Sinan Koont. The adaptation is presented with the permission of COMAP, 57 Bedford St., Lexington, MA 02173.

Problem identification   *How can the doses and the time between doses be adjusted to maintain a safe but effective concentration of the drug in the blood?*

The concentration in the blood resulting from a single dose of a drug normally decreases with time as the drug is eliminated from the body (see Figure 10.7). We are interested in what happens to the concentration of the drug in the blood as doses are given at regular intervals. If H denotes the highest safe level of the drug and L its lowest effective level, it would be desirable to prescribe a dose C_0 with time T between doses so that the concentration of the drug in the bloodstream remains between L and H over each dose period.

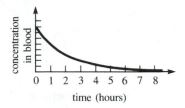

FIGURE 10.7   The concentration of a drug in the bloodstream decreases with time

Let's consider several ways in which the drug might be administered. In Figure 10.8a the time between doses is such that effectively there is no buildup of the drug in the system. In other words, the residual concentration from previous doses is approximately zero. On the other hand, in Figure 10.8b the interval between doses relative to the amount administered and the decay rate of the concentration is such that a residual concentration exists at each time the drug is taken (after the first dose). Furthermore, as depicted in the graph, this residual level seems to be approaching a limit. We will be concerned with determining if this situation is indeed the case and, if so, what that limit must be. Our ultimate goal in prescribing drugs is to determine dose *amounts* and *intervals* between doses such that the lowest effective level L is reached quickly and thereafter the concentration is maintained between the lowest effective level L and the highest safe level H, as depicted in Figure 10.9. We begin by determining the limiting residual level, which depends on our assumptions for the rate of assimilation of the drug in the bloodstream and the rate of decay after assimilation.

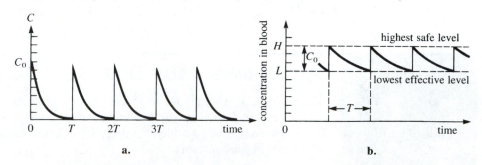

FIGURE 10.8   Residual buildup depends on the time interval between administration of drug doses

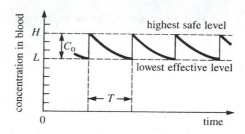

FIGURE 10.9   Safe but effective levels of the drug in the blood; C_0 is the change in concentration produced by one dose and T is the time interval between doses

Assumptions   To solve the problem we have identified, let's consider the factors that determine the concentration $C(t)$ of the drug in the bloodstream at any time t. We begin with

$$C(t) = f(\text{decay rate, assimilation rate, dosage amount, dosage interval,} \ldots)$$

and various other factors including body weight and blood volume. To simplify our assumptions, let's assume that body weight and blood volume are constants (say, an average over some specific age group) and that concentration level is the critical factor in determining the effect of a drug. Next we determine submodels for decay rate and assimilation rate.

Submodel for decay rate   Consider the elimination of the drug from the bloodstream. Probably this is a discrete phenomenon, but let's approximate it by a continuous function. Clinical experiments have revealed that the decrease in the concentration of a drug in the bloodstream will be proportional to the concentration itself. Mathematically, this assumption means that if we assume the concentration of the drug in the blood at time t is a differentiable function $C(t)$, then

$$C'(t) = -kC(t) \tag{10.15}$$

elimination constant   In this formula k is a positive constant, called the **elimination constant** of the drug. Notice $C'(t)$ is negative, as it should be if it is to describe a decreasing concentration. Usually the quantities in Equation (10.15) are measured as follows: the time t is given in hours, $C(t)$ is milligrams per milliliter of blood (mg/ml), $C'(t)$ is mg ml^{-1}hr^{-1}, and k is hr^{-1}.

Assume that the concentrations H and L can be determined experimentally for a given population, such as an age group. (We will be saying more about this assumption in the following discussion.) Then set the drug concentration for a single dose at the level

$$C_0 = H - L \tag{10.16}$$

If we assume C_0 is the concentration at $t = 0$, then we have the model

$$\frac{dC}{dt} = -kC, C(0) = C_0 \tag{10.17}$$

The variables can be separated in Equation (10.17) and the model solved in the same way as the Malthusian model of population growth presented in the preceding section. Solution of the model gives

$$C(t) = C_0 e^{-kt} \tag{10.18}$$

To obtain the concentration at time $t > 0$, multiply the *initial concentration* C_0 by e^{-kt}. The graph of $C(t)$ looks like the one in Figure 10.10.

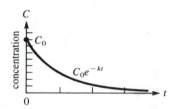

FIGURE 10.10 Exponential model for decay of drug concentration with time

Submodel for assimilation rate Having made an assumption about how drug concentrations decrease with time, let's consider how they increase again when drugs are administered. Our initial assumption is that when a drug is taken, it is diffused so rapidly throughout the blood that the graph of the concentration for the absorption period is, for all practical purposes, vertical. That is, we assume an instantaneous rise in concentration whenever a drug is administered. This assumption may not be as reasonable for a drug taken by mouth as it is for a drug injected directly into the bloodstream. Now let's see how the drug accumulates in the bloodstream with repeated doses.

Drug accumulation with repeated doses Consider what happens to the concentration $C(t)$ when a dose that is capable of raising the concentration by C_0 mg/ml each time it is given is administered regularly at fixed time intervals of length T.

Suppose at time $t = 0$ the first dose is administered. According to our Model (10.18), after T hours have elapsed the residual $R_1 = C_0 e^{-kT}$ remains in the blood, and then the second dose is administered. Because of our assumption concerning the increase in drug concentration as previously discussed, the level of concentration instantaneously jumps to $C_1 = C_0 + C_0 e^{-kT}$. Then after T hours elapse again, the residual $R_2 = C_1 e^{-kT} = C_0 e^{-kT} + C_0 e^{-2kT}$ remains in the blood. This possibility of accumulation of the drug in the blood is depicted in Figure 10.11.

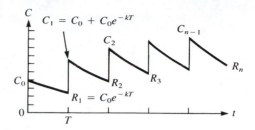

FIGURE 10.11 One possible effect of repeating equal doses

Next, we determine a formula for the nth residual R_n. If we let C_{i-1} be the concentration at the beginning of the ith interval and R_i the *residual concentration* at the end of it, we can easily obtain Table 10.3.

TABLE 10.3 Calculation of residual concentration of drug

i	C_{i-1}	R_i
1	C_0——multiply by e^{-kT}——→C_0e^{-kT} ——add C_0——	
2	$C_0 + C_0e^{-kT}$	$C_0e^{-kT} + C_0e^{-2kT}$
3	$C_0 + C_0e^{-kT} + C_0e^{-2kT}$	$C_0e^{-kT} + C_0e^{-2kT} + C_0e^{-3kT}$
$\vdots$	$\vdots$	$\vdots$
n		$C_0e^{-kT} + \cdots + C_0e^{-nkT}$

From the table,

$$R_n = C_0e^{-kT} + C_0e^{-2kT} + \cdots + C_0e^{-nkT} \qquad \textbf{(10.19)}$$
$$= C_0e^{-kT}(1 + r + r^2 + \cdots + r^{n-1})$$

where $r = e^{-kT}$. Algebraically it is easy to verify that

$$1 + r + r^2 + \cdots + r^{n-1} = \frac{1 - r^n}{1 - r}$$

so substitution for r in Equation (10.19) gives the result

$$R_n = \frac{C_0e^{-kT}(1 - e^{-nkT})}{1 - e^{-kT}} \qquad \textbf{(10.20)}$$

Notice that the number e^{-nkT} is close to 0 when n is large. In fact, the larger n becomes, the closer e^{-nkT} gets to 0. As a result, the sequence of R_n's has a limiting value, which we call R:

$$R = \lim_{n \to \infty} R_n = \frac{C_0 e^{-kT}}{1 - e^{-kT}}$$

or

$$R = \frac{C_0}{e^{kT} - 1} \qquad (10.21)$$

In summary, if a dose capable of raising the concentration by C_0 mg/ml is repeated at intervals of T hours, then the limiting value R of the residual concentrations is given by Formula (10.21). The number k in the formula is the elimination constant of the drug.

Determining the dose schedule From Table 10.3 the concentration C_{n-1} at the beginning of the nth interval is given by

$$C_{n-1} = C_0 + R_{n-1} \qquad (10.22)$$

If the desired dose level is required to approach the highest safe level H as depicted in Figure 10.9, then we want C_{n-1} to approach H as n becomes large. That is,

$$H = \lim_{n \to \infty} C_{n-1} = \lim_{n \to \infty}(C_0 + R_{n-1}) = C_0 + R$$

Combining this last result with $C_0 = H - L$ yields

$$R = L \qquad (10.23)$$

A meaningful way to examine what happens to the residual concentration R for different intervals T between doses is to look at R in comparison with C_0, the change in concentration as a result of each dose. To make this comparison, we form the dimensionless ratio

$$\frac{R}{C_0} = \frac{1}{e^{kT} - 1} \qquad (10.24)$$

Equation (10.24) says that R/C_0 will be close to 0 whenever the time T between doses is long enough to make $e^{kT} - 1$ sufficiently large. As for the intermediate values of R_n, we can see from Table 10.3 that each R_n is obtained from the previous R_{n-1} by adding a positive quantity $C_0 e^{-nkT}$. This means that all R_n's are positive, because R_1 is positive. It also means that R is larger than each of the R_n's. In symbols,

$$0 < R_n < R, \text{ for all } n$$

The implication of this for drug dosage is that whenever R is small, the R_n's are even smaller. In particular, whenever T is long enough to make $e^{kT} - 1$ significantly large, the residual concentration from each dose is almost nil. The various administrations of the drug are then essentially independent, and the graph of $C(t)$ looks like the one depicted in Figure 10.12.

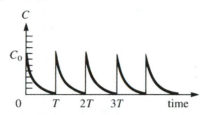

FIGURE 10.12   Drug concentration for long intervals between doses

On the other hand, suppose the length of time T between doses is so short that e^{kT} is not very much larger than 1, so that R/C_0 is significantly greater than 1. As R_n becomes larger, the concentration C_n after each dose becomes larger. The loss during the time period after each dose increases with larger C_n from Equation (10.17). Finally, the drop in concentration after each dose becomes imperceptibly close to the rise in concentration C_0 resulting from each dose. When this condition prevails (the loss in concentration equaling the gain), the concentration will oscillate between R at the end of each period and $R + C_0$ at the start of each period. This situation is depicted in Figure 10.13.

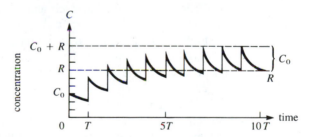

FIGURE 10.13   Buildup of drug concentration when the interval between doses is short

Suppose a drug is ineffective below the concentration L and harmful above some higher concentration H, as discussed previously. Assume now that L and H are safe guidelines so that a person would not suffer a severe overdose if the drug concentration rises somewhat above H and that it is not necessary to begin the buildup process all over again if the concentration falls slightly below L. Then for patient convenience we might opt for the strategy of maximizing the time between drug doses by setting $R = L$ and $C_0 = H - L$, as we have indicated previously. Substitution of $R = L$ and $C_0 = H - L$ in Equation (10.21) yields

$$L = \frac{H - L}{e^{kT} - 1}$$

We then solve the preceding equation for e^{kT} to obtain

$$e^{kT} = H/L$$

Taking the logarithm of both sides of this last equation and dividing the result by k gives the desired dose schedule:

$$T = \frac{1}{k} \ln \frac{H}{L} \tag{10.25}$$

To reach an effective level rapidly, administer a dose, often called a *loading dose*, that will immediately produce a blood concentration of H mg/ml. (For example, this loading dose might equal $2C_0$.) This medication can be followed every $T = (1/k) \ln(H/L)$ hr by a dose that raises the concentration by $C_0 = H - L$ mg/ml.

Verifying the model Our model for prescribing a safe and effective dosage of drug concentration appears to be a good one. It is in accord with the common medical practice of prescribing an initial dose several times larger than the succeeding periodic doses. Also, the model is based on the assumption that the decrease in the concentration of the drug in the bloodstream is proportional to the concentration itself, which has been verified clinically. Moreover, the elimination constant k, which is the positive constant of proportionality in that relationship, is an easily measured parameter (see Problem 1 in the 10.2 problem set). Equation (10.21) permits the prediction of concentration levels under varying conditions for dose rates. Thus the drug may be tested to determine experimentally the lowest effective level L and the highest safe level H with appropriate safety factors to allow for inaccuracies in the modeling process. Then Formulas (10.16) and (10.25) can be used to prescribe a safe and effective dose of the drug (assuming the loading dose is several times larger than C_0). So our model is useful.

One deficiency in the model is the assumption of an instantaneous rise in concentration whenever a drug is administered. A drug, such as aspirin, taken orally requires a finite time to diffuse into the bloodstream; the assumption, therefore, is not realistic for such a drug. For such cases the graph of concentration versus time for a single dose might resemble Figure 10.14.

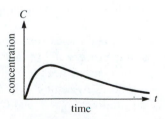

FIGURE 10.14 The concentration of a drug in the bloodstream for a single dose taken orally

10.2 PROBLEMS

1. Discuss how the elimination constant k in Equation (10.15) could be obtained experimentally for a given drug.

2. **a.** If $k = 0.05 \, \text{hr}^{-1}$ and the highest safe concentration is e times the lowest effective concentration, find the length of time between repeated doses that will ensure safe but effective concentrations.

   b. Does part **a** give enough information to determine the size of each dose?

3. Suppose $k = 0.01 \, \text{hr}^{-1}$ and $T = 10$ hr. Find the smallest n such that $R_n > 0.5R$.

4. Given $H = 2 \, \text{mg/ml}$, $L = 0.5 \, \text{mg/ml}$, and $k = 0.02 \, \text{hr}^{-1}$, suppose concentrations below L are not only ineffective but also harmful. Determine a scheme for administering this drug (in terms of concentration and times of dosage).

5. Suppose $k = 0.2 \, \text{hr}^{-1}$ and that the smallest effective concentration is 0.03 mg/ml. A single dose that produces a concentration of 0.1 mg/ml is administered. Approximately how many hours will the drug remain effective?

6. Suggest other phenomena for which the model described in the text might be used.

7. Sketch how a series of doses might accumulate based on the concentration curve given in Figure 10.14.

8. A patient is given a dose Q of a drug at regular intervals of time T. The concentration of the drug in the blood has been shown experimentally to obey the law

$$\frac{dC}{dt} = -ke^{C}$$

   a. If the first dose is administered at $t = 0$ hr, show that after T hr have elapsed, the residual

$$R_1 = -\ln(kT + e^{-Q})$$

   remains in the blood.

   b. Assuming an instantaneous rise in concentration whenever the drug is administered, show that after the second dose and T hr have elapsed again, the residual

$$R_2 = -\ln[kT(1 + e^{-Q}) + e^{-2Q}]$$

   remains in the blood.

   c. Show that the limiting value R of the residual concentrations for doses of Q mg/ml repeated at intervals of T hr is given by the formula

$$R = -\ln\frac{kT}{1 - e^{-Q}}$$

d. Assuming the drug is ineffective below a concentration L and harmful above some higher concentration H, show that the dose schedule T for a safe and effective concentration of the drug in the blood satisfies the formula

$$T = \frac{1}{k}(e^{-L} - e^{-H})$$

where k is a positive constant.

10.2 PROJECTS

1. Write a summary report on the article "Case Studies in Cancer and Its Treatment by Radiotherapy," by J. R. Usher and D. A. Abercrombie, *International Journal of Mathematics Education in Science and Technology 12*, no. 6 (1981), pp. 661–682. Present your report to the class.

In Projects 2–5, complete the requirements of the designated UMAP module.

2. "Selection in Genetics," by Brindell Horelick and Sinan Koont, UMAP 70. This module introduces genetic terminology and basic results about genotype distribution in successive generations. A recurrence relationship is obtained from which the nth generation frequency of a recessive gene can be determined. Calculus is used to derive a technique for approximating the number of generations required for this frequency to fall below any given positive value.

3. "Epidemics," by Brindell Horelick and Sinan Koont, UMAP 73. This unit poses two problems: (1) At what rate must infected persons be removed from a population to keep an epidemic under control? (2) What portion of a community will become infected during an epidemic? Threshold removal rate is discussed, and the extent of an epidemic is discussed when the removal rate is slightly below threshold.

4. "Tracer Methods in Permeability," by Brindell Horelick and Sinan Koont, UMAP 74. This module describes a technique for measuring the permeability of red corpuscle surfaces to K^{42} ions using radioactive tracers. Students learn how radioactive tracers can be used to monitor substances in the body and some limitations and strengths of the model described in this unit.

5. "Modeling the Nervous System. Reaction Time and the Central Nervous System," by Brindell Horelick and Sinan Koont, UMAP 67. The process by which the central nervous system reacts to a stimulus is modeled, and the predictions of the model are compared with experimental data. Students learn what conclusions can be drawn from the model about reaction time and are given an opportunity to discuss the merits of various assumptions about the relationship between intensity of excitation and stimulus intensity.

10.3   SCENARIO REVISITED

Example 1   *Braking Distance*

In our model for vehicular total stopping distance (see Section 4.2) one of the submodels is braking distance:

$$\text{braking distance} = h(\text{weight, speed})$$

Using an argument based on the result that the work done by the braking system must equal the change in kinetic energy, we found that the braking distance d_b is proportional to the square of the velocity. We now use an argument based on the derivative to establish that same result.

Let's assume the braking system is designed in such a way that the maximum braking force increases in proportion to the mass of the car. Basically, this means that if the force per unit area applied by the braking hydraulic system remains constant, the surface area in contact with the brakes would have to increase in proportion to the mass of the car. From an engineering standpoint this assumption seems reasonable.

The implication of the assumption is that the deceleration felt by the passengers is constant, which is probably a plausible design criterion. If it is further assumed that under a panic stop the maximum braking force F is applied continuously, then we obtain

$$F = -km$$

for some positive proportionality constant k. Because F is the only force acting on the car under our assumptions, this gives

$$ma = m\frac{dv}{dt} = -km$$

(the negative sign signals deceleration). Thus

$$\frac{dv}{dt} = -k$$

which integrates to

$$v = -kt + C_1$$

If v_0 denotes the velocity at $t = 0$ when the brakes are initially applied, substitution gives $C_1 = v_0$ so that

$$v = -kt + v_0 \tag{10.26}$$

If t_s denotes the time it takes for the car to stop after the brakes have been applied, then $v = 0$ when $t = t_s$ substituted into Equation (10.26) gives

$$t_s = \frac{v_0}{k} \tag{10.27}$$

If x represents the distance traveled by the car after the brakes are applied, then x is the integral of $v = dx/dt$. Thus from Equation (10.26)

$$x = -0.5kt^2 + v_0 t + C_2$$

When $t = 0$, $x = 0$, which implies $C_2 = 0$ so

$$x = -0.5kt^2 + v_0 t \tag{10.28}$$

Next, let d_b denote the braking distance; that is, $x = d_b$ when $t = t_s$. Substitution of these results into Equation (10.28) yields

$$d_b = -0.5kt_s^2 + v_0 t_s$$

Using Equation (10.27) in this last equation, we have

$$d_b = \frac{-v_0^2}{2k} + \frac{v_0^2}{k} = \frac{v_0^2}{2k} \tag{10.29}$$

Therefore d_b is proportional to the square of the velocity, in accordance with the submodel obtained in Section 4.2.

In Chapter 4 we tested the submodel $d_b \propto v^2$ against some data and found reasonable agreement. The constant of proportionality was estimated to be 0.054 ft · hr^2/mi^2, which corresponds to a value of k in Equation (10.29) of approximately 19.9 ft/sec^2 (see Problem 1 in the 10.3 problem set). If we interpret k as the deceleration felt by a passenger in the vehicle (because $F = -km$ by assumption), we will find it useful to interpret this constant as $0.6g$ (where g is the acceleration of gravity).

10.3 PROBLEMS

1. **a.** Using the estimate that $d_b = 0.054v^2$, where 0.054 has dimension ft · hr^2/mi^2, show that the constant k in Equation (10.29) has the value 19.9 ft/sec^2.

   b. Using the data in Table 4.4, plot d_b in feet versus $v^2/2$ in ft^2/sec^2 to estimate $1/k$ directly.

2. Consider launching a satellite into orbit using a single-stage rocket. The rocket is continuously losing mass, which is being propelled away from it at significant speeds. We are interested in predicting the maximum speed the rocket can attain.[2]

   a. Assume the rocket of mass m is moving with speed v. In a small increment of time Δt it loses a small mass Δm_p, which leaves the rocket with speed u in a direction opposite to v. Here Δm_p is the small propellant mass. The resulting speed of the rocket is $v + \Delta v$. Neglect all external forces (gravity, atmospheric drag, etc.) and assume Newton's Second Law of Motion

   $$\text{force} = \frac{d}{dt}(\text{momentum of system})$$

   where momentum is mass times velocity. Derive the model

   $$\frac{dv}{dt} = \left(\frac{-c}{m}\right)\frac{dm}{dt}$$

   where $c = u + v$ is the relative exhaust speed (the speed of the burnt gases relative to the rocket).

   b. Assume initially at time $t = 0$ the velocity $v = 0$ and the mass of the rocket is $m = M + P$, where P is the mass of the payload satellite and $M = \epsilon M + (1 - \epsilon)M$ $(0 < \epsilon < 1)$ is the initial fuel mass ϵM plus the mass $(1 - \epsilon)M$ of the rocket casings and instruments. Solve the model in part **a** to obtain the speed

   $$v = -c \ln \frac{m}{M + P}$$

   c. Show that when all the fuel is burned, the speed of the rocket is given by

   $$v_f = -c \ln \left(1 - \frac{\epsilon}{1 + \beta}\right)$$

   where $\beta = P/M$ is the ratio of the payload mass to the rocket mass.

   d. Find v_f if $c = 3$ km/sec, $\epsilon = 0.8$, and $\beta = 1/100$. (These are fairly typical values in satellite launchings.)

   e. Suppose scientists plan to launch a satellite in circular orbit h km above the Earth's surface. Assume that the gravitational pull toward the center of the Earth is given by Newton's inverse square law of attraction

   $$\frac{\gamma m M_e}{(h + R_e)^2}$$

---

[2]This problem was suggested by D. N. Burghes and M. S. Borrie, *Modelling with Differential Equations*, West Sussex, England: Ellis Horwood, 1981.

where γ is the universal gravitational constant, m the mass of the satellite, M_e the Earth's mass, and R_e the radius of the Earth. Assume that this force must be balanced by the centrifugal force $mv^2/(h + R_e)$, where v is the speed of the satellite. What speed must be attained by a rocket to launch a satellite into an orbit 100 km above the Earth's surface? From your computation in part **d**, can a single-stage rocket launch a satellite into an orbit of that height?

3. The Gross National Product (GNP) represents the sum of consumption purchases of goods and services, government purchases of goods and services, and gross private investment (which is the increase in inventories plus buildings constructed and equipment acquired). Assume that the GNP is increasing at the rate of 3% per year and that the national debt is increasing at a rate proportional to the GNP.

   a. Construct a system for two ordinary differential equations modeling the GNP and national debt.

   b. Solve the system in part **a**, assuming the GNP is M_0 and the national debt is N_0 at year 0.

   c. Does the national debt eventually outstrip the GNP? Consider the ratio of the national debt to the GNP.

10.3 PROJECTS

Complete the requirements of the indicated UMAP module.

1. "Kinetics of Single Reactant Reactions," by Brindell Horelick and Sinan Koont, UMAP 232. The unit discusses reaction orders of irreversible single reactant reactions. The equation $a'(t) = -k(a(t))^n$ is solved for selected values of n; reaction orders of various reactions are found from experimental data, and the notion of half-life is discussed. Some background knowledge of chemistry is required.

2. "Radioactive Chains: Parents and Daughters," by Brindell Horelick and Sinan Koont, UMAP 234. When a radioactive substance A decays into a substance B, A and B are called parent and daughter. It may happen that B itself is radioactive and is the parent of a new daughter C, and so on. There are three radioactive chains that together account for all naturally occurring radioactive substances beyond thallium on the periodic table. This unit develops models for calculating the amounts of the substances in radioactive chains and discusses transient and secular states of equilibrium between parent and daughter.

3. "The Relationship Between Directional Heading of an Automobile and Steering Wheel Deflection," by John E. Prussing, UMAP 506. This unit develops a model relating the compass heading and the steering wheel deflection using basic geometric and kinematic principles.

10.4 NUMERICAL APPROXIMATION METHODS[3]

In the models developed in the preceding sections of this chapter, we found an equation relating a derivative to some function of the independent and dependent variables; that is,

$$\frac{dy}{dx} = g(x, y)$$

where g is some function in which either x or y may not appear explicitly. Moreover, we were given some starting value; that is, $y(x_0) = y_0$. Finally, we were interested in the values of y for a specific set of x values; that is, $x_0 \le x \le b$. In summary we determined models of the general form

$$\frac{dy}{dx} = g(x, y), \ y(x_0) = y_0, \ x_0 \le x \le b$$

We call first-order ordinary differential equations with the preceding conditions *first-order initial value problems*. As seen from our previous models, they constitute an important class of problems. We now discuss the three parts of the model.

First-Order Initial Value Problems

The differential equation $dy/dx = g(x, y)$ As discussed in our models, we are interested in finding a function $y = f(x)$ whose derivative satisfies an equation $dy/dx = g(x, y)$. Although we do not know f, we can compute its derivative given particular values of x and y. As a result, we can find the slope of the tangent line to the solution curve $y = f(x)$ at specified points (x, y).

The initial value $y(x_0) = y_0$ The initial value equation states that at the initial point x_0, we know the y value is $f(x_0) = y_0$. Geometrically, this means the point (x_0, y_0) lies on the solution curve (Figure 10.15). Thus we know where our solution

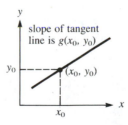

FIGURE 10.15   The solution curve passes through the point (x_0, y_0) and has slope $g(x_0, y_0)$

[3]This section is adapted from UMAP Unit 625, which was written by the authors and David H. Cameron. The adaptation is presented with the permission of COMAP, 57 Bedford St., Lexington, MA, 02173.

curve begins. Moreover, from the differential equation $dy/dx = g(x, y)$ we know that the slope of the solution curve at (x_0, y_0) is the number $g(x_0, y_0)$. This is also depicted in Figure 10.15.

The interval $x_0 \leq x \leq b$ The condition $x_0 \leq x \leq b$ gives the particular interval of the x-axis with which we are concerned. Thus we would like to relate y with x over the interval $x_0 \leq x \leq b$ by finding the solution function $y = f(x)$ passing through the point (x_0, y_0) with slope $g(x_0, y_0)$ (Figure 10.16). Note that the function $y = f(x)$ is continuous over $x_0 \leq x \leq b$ because its derivative exists there.

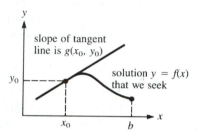

FIGURE 10.16   The solution $y = f(x)$ to the initial value problem is a continuous function over the interval from x_0 to b

Approximating Solutions to Initial Value Problems

We shall now study a method that uses the three parts of the initial value problem together with the geometrical interpretation of the derivative to construct a sequence of discrete points in the plane that numerically approximate the points on the actual solution curve $y = f(x)$. We begin with an example illustrating the method.

Example 1   ***Interest Compounded Continuously***

Suppose at the end of year 1 we have $1,000 invested at 7% annual interest compounded continuously. We would like to know how much money we will have at the end of year 20. Letting $Q(t)$ represent the amount of money at any time t, we have

$$\frac{dQ}{dt} = 0.07Q, \ Q(1) = 1000, \ 1 \leq t \leq 20$$

We would like to find the function $Q(t)$ that solves this initial value problem. Using a dashed curve to represent the unknown function Q, we sketch the preceding information in Figure 10.17.

We know the derivative of $Q(t)$ for known values of Q and t. In particular, we know that at the point $(1, 1000)$ the derivative is $dQ/dt = 0.07(1000) = 70$. Because the derivative can be interpreted as the slope of the line tangent to the

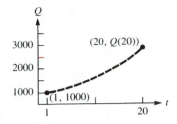

FIGURE 10.17 The curve satisfying $dQ/dt = 0.07Q$, $Q(1) = 1000$, $1 \leq t \leq 20$

curve, we sketch a tangent line to our unknown function Q at $t = 1$ with slope 70. This situation is depicted in Figure 10.18. Because we do not know $Q(t)$, we cannot determine exactly the value of $Q(20)$. We can, however, approximate it by the value on the tangent line when $t = 20$. Now the equation of the tangent line T in point-slope form is given by

$$T - 1000 = 70(t - 1)$$

In other words,

$$T = Q_0 + \frac{dQ}{dt}\Big|_{t=t_0} \Delta t$$

where $Q_0 = 1000$, $dQ/dt_{t=t_0} = 70$, and $\Delta t = t - 1$. When $t = 20$, we see that

$$T(20) = 1000 + 70(20 - 1) = 2330$$

Then we make the approximation

$$Q(20) \approx 2330 = Q_1$$

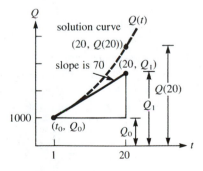

FIGURE 10.18 The point $(20, Q_1)$ on the tangent line approximates the actual solution point $(20, Q(20))$

to the value of the unknown function at $t = 20$. Thus, starting with \$1,000 at the end of year 1, we estimate that we will have \$2,330 at the end of year 20 if interest is compounded continuously at an annual rate of 7%.

Estimating with two steps   We emphasize that we have used the known starting value $Q(1) = 1000$ to calculate the estimate $Q(20) \approx 2330$. How can we improve the approximation and get a more accurate picture of the solution curve? We assumed that the derivative Q' is the constant 70 over the interval $1 \leq t \leq 20$, but we know it actually changes as Q and t change. Perhaps our estimate of $Q(20)$ will be more accurate if we make another estimate at an intermediate point. Setting $\Delta t = (20 - 1)/2 = 9.5$ in that same problem, we obtain

$$Q(10.5) \approx Q_1 = Q_0 + \frac{dQ}{dt}\Big|_{t=1} \Delta t = 1000 + 70(9.5) = 1665$$

This is depicted in Figure 10.19.

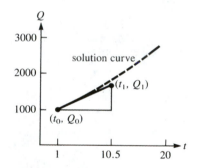

FIGURE 10.19   The point (t_1, Q_1) estimates the solution curve at the halfway value $t = 10.5$

Next we use the estimate $Q(10.5) \approx 1665$ to approximate the derivative at $t = 10.5$ from the formula

$$\frac{dQ}{dt}\Big|_{t=10.5} = 0.07Q(10.5)$$

Note an important difference from our first calculation for $Q'(1)$. We *know* the value of the derivative at $t = 1$ exactly because we know $Q(1) = 1000$, but we must *estimate* the derivative at $t = 10.5$ because $Q(10.5) \approx 1665$ is only an estimate. Now we calculate our estimate $Q(20)$:

$$Q(20) \approx Q_2 = Q_1 + \frac{dQ}{dt}\Big|_{t=10.5} \Delta t$$
$$= Q_1 + 0.07Q(10.5)\Delta t$$
$$\approx 1665 + (0.07)(1665)(9.5) = 2772.23$$

This two-step process is shown in Figure 10.20. You will see shortly that the approximation $Q(20) \approx 2772.23$ is closer to the actual value of the solution at $t = 20$.

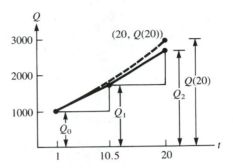

FIGURE 10.20 The value Q_2 approximates the solution $Q(20)$ in a two-step process

In this way a table of approximate values to the solution is built in a step-by-step fashion. The situation is depicted in Figure 10.21. Notice that an error is produced at each step. As these errors accumulate with more and more steps, the approximations $y_1, y_2, \ldots, y_n$ get farther and farther away from the actual solution curve.

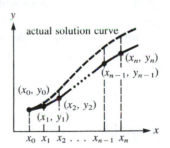

FIGURE 10.21 The points (x_i, y_i) approximate the solution curve

Improving the estimate of the graph   Euler's algorithm can easily be coded for computer implementation to facilitate reductions in step size. Because there was a significant change in the estimate to $Q(20)$ (from 2330 to 2772.23) when we reduced the step size Δt in the compound interest example, we would not be too confident in our results at this point. Using a calculator program, we applied Euler's algorithm with step sizes $\Delta t = 19$, $\Delta t = 9.5$, and $\Delta t = 1.0$ and obtained approximations for Q every integer value of t between 1 and 20. To get an idea of what the unknown solution function looks like, we plotted the points obtained from the various step sizes on a single graph. The graph is displayed in Figure 10.22.

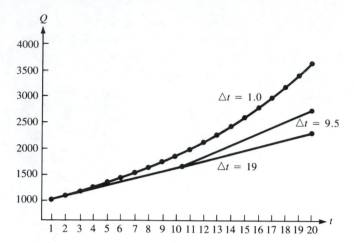

FIGURE 10.22 Plot of the approximate solution for the various step sizes

It might be tempting to reduce the step size even further to obtain greater accuracy. However, each additional calculation not only requires additional computer time but, more important, introduces round-off error. Because these errors accumulate, an ideal method would improve the accuracy of the approximations yet minimize the number of calculations. For this reason Euler's method may prove unsatisfactory. More refined numerical methods for solving the initial value problem are investigated in courses in the numerical methods of differential equations. We will not pursue those methods here.

Finding the solution analytically   Just as we did for the population problem in Section 10.1, we can separate the variables and integrate the initial value problem

$$\frac{dQ}{dt} = 0.07Q, \; Q(1) = 1000, \; 1 \leq t \leq 20$$

to obtain

$$Q = C_1 e^{0.07t}$$

Finally we apply the initial condition that $Q(1) = 1000$ to evaluate the constant C_1; substituting $t = 1$ into the last equation gives

$$1000 = Q(1) = C_1 e^{0.07}$$

so $C_1 = 1000/e^{0.07} \approx 932.39382$. We now have our desired solution function:

$$Q = \left(\frac{1000}{e^{0.07}} \right) e^{0.07t}$$

If we evaluate this solution at $t = 20$, we obtain $Q(20) = 3781.04$. Thus our approximations for $Q(20)$ (2330 with $\Delta t = 19$, 2772.23 with $\Delta t = 9.5$, 3616.53 with $\Delta t = 1$) became more accurate as Δt decreased. Note that even with $\Delta t = 1$, an error of 164.51 results. We can further reduce this error by decreasing Δt. With the aid of a calculator, $\Delta t = 0.01$ in Euler's method gives the approximation $Q(20) \approx 3763.56$.

Separation of variables The method of direct integration used in this chapter works only in those cases in which the dependent and independent variables can be algebraically separated. Any such differential equation can be written in the form

$$p(y)\,dy = q(x)\,dx$$

For example, given the differential equation

$$u(x)v(y)\,dx + q(x)p(y)\,dy = 0$$

we can arrange the equation in the following form:

$$\frac{p(y)}{v(y)}\,dy = \frac{-u(x)}{q(x)}\,dx$$

Because the right-hand side is a function of x and the left-hand side is a function of y only, the solution is obtained by integrating both sides directly. Remember from calculus that even if we are successful in separating the variables, it may not be possible to find the integrals in closed form. The method just described is called *separation of variables* and is one of the most elementary methods available for solving a differential equation.

10.4 PROBLEMS

1. When interest is compounded, the interest earned is added to the principal amount so it may also earn interest. For a one-year period, the principal amount P is given by

$$P = \left(1 + \frac{i}{n}\right)^n P(0)$$

where i is the annual interest rate (given as a decimal) and n is the number of times during the year the interest is compounded.

To lure depositors, banks offer to compound interest at different intervals: semiannually, quarterly, or daily. A certain bank advertises that it compounds interest continuously. If \$100 is deposited initially, formulate a mathematical model describing the growth of the initial deposit during the first year. Assume an annual interest rate of 10%.

2. Use the differential equation model formulated in the preceding problem to answer the following:

   a. From the derivative evaluated at $t = 0$, determine an equation of the tangent line T passing through the point $(0, 100)$.

   b. Estimate $Q(1)$ by finding $T(1)$, where $Q(t)$ denotes the amount of money in the bank at time t (assuming no withdrawals).

   c. Estimate $Q(1)$ using a step size of $\Delta t = 0.5$.

   d. Estimate $Q(1)$ using a step size of $\Delta t = 0.25$.

   e. Plot the estimates you obtained for $\Delta t = 1.0$, 0.5, and 0.25 to approximate the graph of $Q(t)$.

3. **a.** For the differential equation model obtained in Problem 1, find $Q(t)$ by separating the variables and integrating.

   b. Evaluate $Q(1)$.

   c. Compare your previous estimates of $Q(1)$ with its actual value.

   d. Find the effective annual interest rate when an annual rate of 10% is compounded continuously.

   e. Compare the effective annual interest rate computed in part **d** with interest compounded

      i. Semiannually: $(1 + 0.10/2)^2$

     ii. Quarterly: $(1 + 0.10/4)^4$

    iii. Daily $(1 + 0.10/365)^{365}$

   f. Estimate the limit of $(1 + 0.10/n)^n$ as $n \to \infty$ by evaluating the expression for $n = 1000$; 10,000; 100,000.

   g. What is $\lim_{n \to \infty}(1 + 0.1/n)^n$?

10.4 PROJECTS

Complete the requirements of the indicated UMAP module.

1. "Feldman's Model," by Brindell Horelick and Sinan Koont, UMAP 75. This unit develops a version of G. A. Feldman's model of growth in a planned economy in which all of the means of production are owned by the state. Originally the model was developed by Feldman in connection with planning the economy of the former Soviet Union. Students compute numerical values for rates of output, national income, their rates of change, and the propensity to save, and discuss the effects of changes in the parameters of the model and in the units of measurement.

2. "The Digestive Process of Sheep," by Brindell Horelick and Sinan Koont, UMAP 69. This unit introduces a differential equation model for the digestive processes of sheep. The model is tested and fit using collected data and the least-squares criterion.

Chapter Eleven

MODELING WITH SYSTEMS OF DIFFERENTIAL EQUATIONS

INTRODUCTION

Interactive situations occur in the study of economics, ecology, electrical circuits, mechanical systems, celestial mechanics, control systems, and so forth. For example, the study of the dynamics of population growth of various plants and animals is an important ecological application of mathematics. Different species interact in a variety of ways. One animal may serve as the primary food source for another, commonly referred to as a predator–prey relationship. Two species may depend on one another for mutual support, such as a bee's using a plant's nectar as food while simultaneously pollinating that plant; such a relationship is referred to as mutualism. Another possibility occurs when two or more species compete against one another for a common food source or even compete for survival. In this chapter we develop some elementary models to explain these interactive situations and analyze the models using graphical techniques.

In modeling interactive situations involving the dynamics of population growth, we are interested in the answers to questions concerning the species under investigation. For instance, will one species eventually dominate the other and drive it to extinction? Can the species coexist? If so, will their populations reach equilibrium levels, or will they vary in some predictable fashion? Moreover, how sensitive are the answers to the preceding questions relative to the initial population levels or to external perturbations (e.g., natural disasters or development of chemical or biological agents used to control the populations)?

Because we are modeling the rates of change with respect to time, the models invariably involve differential equations (or in a discrete analysis, difference equations). Even with simple assumptions, these equations are often nonlinear and generally cannot be solved analytically, although numerical techniques exist. Nevertheless, qualitative information about the behavior of the variables can often be obtained by simple graphical analysis. We will demonstrate how graphical analysis can be used to answer questions such as those posed in the preceding paragraph. We will also point out limitations for such an analysis and conditions requiring a more sophisticated mathematical analysis.

11.1 GRAPHICAL SOLUTIONS

Consider again the first-order differential equation

$$\frac{dy}{dx} = g(x, y)$$

direction field

which we investigated in Chapter 10. Each time an initial value $y(x_0) = y_0$ is specified, the solution curve must pass through the point (x_0, y_0) and have the slope value $g(x_0, y_0)$ there. Graphically, we can draw a short line segment with the proper slope through each point (x, y) in the plane. The resultant configuration is known as the **direction field** of the first-order differential equation and is illustrated in Figure 11.1. Thus the direction field gives a visual indication of the family of possible solutions to the differential equation. A solution curve is tangent to the direction line at each point through which the curve passes. One solution curve is depicted for the family of curves indicated by the direction field in Figure 11.1.

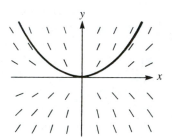

FIGURE 11.1 The direction field of a first-order differential equation assigns to each point in the plane the slope $y' = g(x, y)$

Often the solutions to a first-order differential equation can be expressed in the form $y = f(x)$, where each solution is distinguished by a different constant resulting from integration. For example, if we separate the variables in the differential equation

$$\frac{dy}{dx} = \frac{2y}{x}$$

we obtain

$$\frac{dy}{y} = 2\frac{dx}{x}$$

Integration of both sides then gives

$$\ln|y| = 2\ln|x| + \ln C$$

where the constant of integration is named $\ln C$ for computational convenience. Applying the exponential to both sides of this last equation yields

$$|y| = Cx^2, \quad C > 0 \tag{11.1}$$

Equation (11.1) represents a family of parabolas, each parabola being distinguished by a different value of the positive constant C. If $y \geq 0$, the curves $y = Cx^2$ open upward; if $y < 0$, the curves $y = -Cx^2$ open downward. This family is depicted in Figure 11.2. Excluding the origin, exactly one of the parabolas passes through each point in the plane. Thus by specifying an initial condition $y(x_0) = y_0$, we select a unique solution curve passing through the point (x_0, y_0).

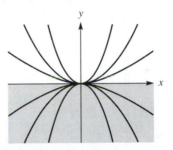

FIGURE 11.2 Family of parabolas $|y| = Cx^2, C > 0$

The question of whether more than one solution curve can pass through a specific point (x_0, y_0) is an important one, but a full discussion is beyond the scope of this book. For uniqueness to occur, certain conditions must be met by the function $g(x, y)$ defining the differential equation. Whenever uniqueness does apply, we know that two solutions cannot cross at the point in question. In Figure 11.2 there is a unique solution parabola through each point in the plane with the exception of the origin; through the origin there are infinitely many solutions. Notice that the constant function $y = 0$ is a solution to the differential equation. This fact is readily seen when the equation is written in the form

$$xy' = 2y$$

which is clearly satisfied by $y = 0$. However, the solution $y = 0$ is not one of the curves represented by the family (11.1). The difficulty lies with the lack of continuity of the function $g(x, y) = 2y/x$ at the origin.[1]

---

[1] A good discussion of the uniqueness problem appears in W. E. Boyce and R. C. DiPrima, *Elementary Differential Equations and Boundary Value Problems*, 3rd ed. (New York: Wiley, 1977), Section 2.11.

Rest Points

Qualitative information concerning the solutions can often be obtained from the differential equation. For example, consider the differential equation describing the Malthusian model of population growth discussed in Chapter 10:

$$\frac{dP}{dt} = kP, \quad k > 0$$

Thus the Malthusian model assumes a simple proportionality between growth rate and population (see Figure 11.3a). Because $P > 0$ and $k > 0$, we can see that $dP/dt > 0$, which means the population curve $P(t)$ is everywhere increasing. Moreover, the smaller the value of k, the less rapid the growth of the population over time. For a fixed value of k, as P increases so does its rate of change dP/dt. Thus the solution curves must appear qualitatively as depicted in Figure 11.3b. Assuming k is constant, at each population level P the derivative dP/dt is constant. Therefore, all the solution curves are horizontal translates of one another. The initial population $P(0) = P_0$ distinguishes these various curves.

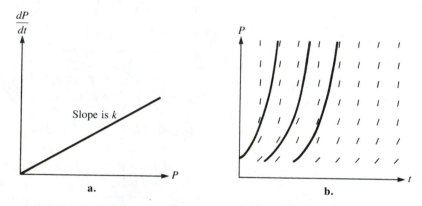

FIGURE 11.3 Graphs of dP/dt versus P and the solution curves for $dP/dt = kP$, where $k > 0$ is fixed

Next, recall the limited growth model studied in Chapter 10:

$$\frac{dP}{dt} = r(M - P)P \qquad (11.2)$$

The graph of dP/dt versus P is the parabola shown in Figure 11.4. Because the second derivative

$$P'' = rP'(M - 2P) \qquad (11.3)$$

is zero when $P = M/2$, positive for $P < M/2$, and negative for $P > M/2$, we conclude that dP/dt is at a maximum at that population level. Note that for $0 < P < M$, the derivative dP/dt is positive and the population $P(t)$ is increasing;

for $M < P$, dP/dt is negative and $P(t)$ is decreasing. Moreover, if $P < M/2$, the second derivative P'' is positive and the population curve is concave upward; for $M > P > M/2$, P'' is negative and the population curve is concave downward.

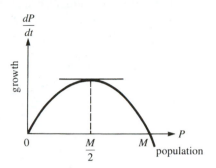

FIGURE 11.4 $dP/dt = r(M - P)P, r > 0, M > 0$

The parabola in Figure 11.4 has zeros at the population levels $P = 0$ and $P = M$. At those levels, $dP/dt = 0$, so no change in the population P can occur. That is, if the population level is at 0, it will remain there for all time; if it is at the level M, it will remain there for all time. Points for which the derivative dP/dt is zero are called **rest points**, **equilibrium points**, **stationary points**, or **critical points** of the differential equation. The behavior of solutions near rest points is of significant interest. Let's examine what happens to the population when P is near the rest points $P = M$ and $P = 0$.

rest points
equilibrium points
stationary points
critical points

Suppose the population P is slightly less than M. Then dP/dt is positive, and the population increases and gets closer to M. On the other hand, if $P > M$, $dP/dt < 0$, and the population will decrease toward M. Thus no matter what positive value is assigned to the starting population, it will tend to the limiting value M as time tends to infinity. We say that M is an **asymptotically stable** rest point because whenever the population level is perturbed away from that level it tends to return there. However, if the starting population is not at the level M, the population $P(t)$ cannot reach M in a finite amount of time. This fact follows from the property that $P = M$ is a solution to the limited growth model (11.2), and two solutions cannot cross.

asymptotically stable

Next, consider the rest point $P = 0$ in Figure 11.4. If P is perturbed slightly away from 0 so that $M > P > 0$, then dP/dt is positive and the population increases toward M. In this situation we say that $P = 0$ is an **unstable** rest point because any population not starting at that level tends to move away from it.

unstable

In general, equilibrium solutions to differential equations are classified as stable or unstable according to whether, graphically, nearby solutions stay close to or converge to the equilibrium, or diverge away from the equilibrium, respectively, as the independent variable t tends to infinity.

At $P = M/2$, P'' is zero and the derivative dP/dt is at a maximum (see Figure 11.4). Thus the population P is increasing most rapidly when $P = M/2$, and a point of inflection occurs in the graph there. These features give each solution curve its characteristic S or sigmoid shape. From the information we have obtained, the family of solutions to the limited growth model must appear (approximately) as shown in Figure 11.5. Notice again that at each population level P, the derivative dP/dt is constant, so that the solution curves are horizontal translates of one another. The curves are distinguished by the initial population level $P(0) = P_0$. In particular, note the solution curve when $M/2 < P_0 < M$ in Figure 11.5.

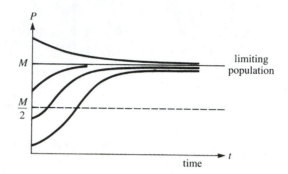

FIGURE 11.5 Population curves for the logistic model $dP/dt = r(M - P)P, r > 0, M > 0$

Systems of First-Order Differential Equations

In Chapter 10, we solved differential equations by separating the variables and integration. Usually, however, it is not so easy to solve a system of differential equations; in fact, it is rare that we can find an analytical solution when the equations are nonlinear, although numerical solution methods exist. It is worthwhile, therefore, to consider a qualitative graphical analysis for solutions to a system of differential equations analogous to our preceding development for single equations. We restrict our discussion to special systems involving only two first-order differential equations.

The system

$$\frac{dx}{dt} = f(x, y) \tag{11.4}$$

$$\frac{dy}{dt} = g(x, y)$$

autonomous   is called an **autonomous** system of differential equations. In such a system the independent variable t is absent (i.e., t does not appear explicitly on the right side of Equations (11.4)). To emphasize the physical significance of autonomous systems, think of the independent variable t as denoting time and the dependent variables as giving position (x, y) in the Cartesian plane. Thus autonomous systems are not time dependent. In order that the system be suitably well behaved, we assume throughout our discussion that the functions f and g together with their first partial derivatives $\partial f/\partial x$, $\partial f/\partial y$, $\partial g/\partial x$, and $\partial g/\partial y$ are continuous over a suitable region of the xy-plane.

It is useful to think of a solution to the autonomous system (11.4) as a curve in the *xy*-plane. That is, a **solution** to (11.4) is a pair of parametric equations $x = x(t)$ and $y = y(t)$ whose derivatives satisfy the system. The solution curve whose coordinates are $(x(t), y(t))$, as t varies over time, is called a **trajectory**, **path**, or **orbit** of the system. The *xy*-plane is referred to as the **phase plane**. It is convenient to think of a trajectory as the path of a moving particle, and we will appeal to this idea throughout this chapter. Note that as the particle moves through the phase plane with increasing t, the direction it moves from a point (x, y) depends only on the coordinates (x, y) and not on the time of arrival.

If (x, y) is a point in the phase plane for which $f(x, y) = 0$ and $g(x, y) = 0$ simultaneously, then both the derivatives dx/dt and dy/dt are zero. Hence there is no motion in either the x or the y direction, and the particle is stationary. Such a point is called a rest point, or equilibrium point, of the system. Notice that whenever (x_0, y_0) is a rest point of the system (11.4), the equations $x = x_0$ and $y = y_0$ give a solution to the system. In fact, this constant solution is the only one passing through the point (x_0, y_0) in the phase plane. The trajectory associated with this solution is the rest point (x_0, y_0) itself. A trajectory $x = x(t)$, $y = y(t)$ is said to approach the rest point (x_0, y_0) if $x(t) \rightarrow x_0$ and $y(t) \rightarrow y_0$ as $t \rightarrow \infty$. In applications it is of interest to see what happens to a trajectory when it comes near a rest point.

The idea of stability is central to any discussion of the behavior of trajectories near a rest point. Roughly, the rest point (x_0, y_0) is **stable** if any trajectory that starts close to the point stays close to it for all future time. It is **asymptotically stable** if it is stable and any trajectory that starts close to (x_0, y_0) approaches that point as t tends to infinity. If it is not stable, the rest point is said to be **unstable**. These notions will be clarified when we examine specific modeling applications later in the chapter. Our goal here is to have a language with which to discuss qualitatively our differential equations models. It is not our intent to study the theoretical aspects of stability, which would require greater mathematical precision than we have presented.

The following results are useful in investigating solutions to the autonomous system (11.4). We offer these results without proof.

1. There is at most one trajectory through any point in the phase plane.
2. A trajectory that starts at a point other than a rest point cannot reach a rest point in a finite amount of time.
3. No trajectory can cross itself unless it is a closed curve. If it is a closed curve, it is a periodic solution.

The implications of these three properties are that from a starting point that is not a rest point, the resulting motion:

a. will move along the same trajectory regardless of the starting time;
b. cannot return to the starting point unless the motion is periodic;
c. can never cross another trajectory; and
d. can only approach (never reach) a rest point.

Therefore, the resulting motion of a particle along a trajectory behaves in one of three possible ways: (1) the particle approaches a rest point, (2) the particle

Margin terms:
solution
trajectory
path
orbit
phase plane

stable
asymptotically stable

unstable

moves along or approaches asymptotically a closed path, or (3) at least one of the trajectory components, $x(t)$ or $y(t)$, becomes arbitrarily large as t tends to infinity. We will apply these ideas to the models we develop in the next several sections.

11.1 PROBLEMS

1. Construct a direction field and sketch a solution curve for the following differential equations:

   a. $dy/dx = y$

   b. $dy/dx = x$

   c. $dy/dx = x + y$

   d. $dy/dx = x - y$

   e. $dy/dx = xy$

   f. $dy/dx = 1/y$

For Problems 2–5, sketch a number of solutions to the equations, showing the correct slope, concavity, and any points of inflection.

2. $dy/dx = (y + 2)(y - 3)$

3. $dy/dx = y^2 - 4$

4. $dy/dx = y^3 - y$

5. $dy/dx = x - 2y$

6. Analyze graphically the equation $dy/dt = ry$ when $r < 0$. What happens to any solution curve as t becomes large?

7. Develop graphically the following models. First graph dP/dt versus P, and then obtain various graphs of P versus t by selecting different initial values $P(0)$ (as in our population example in the text). Identify and discuss the nature of the equilibrium points in each model.

   a. $dP/dt = a - bP, \quad a, b > 0$

   b. $dP/dt = P(a - bP), \quad a, b > 0$

   c. $dP/dt = k(M - P)(P - m), \quad k, M, m > 0$

   d. $dP/dt = kP(M - P)(P - m), \quad k, M, m > 0$

8. Sketch a number of trajectories corresponding to the following autonomous systems and indicate the direction of motion for increasing t. Identify and classify any rest points according to being stable, asymptotically stable, or unstable.

   a. $dx/dt = x, dy/dt = y$

   b. $dx/dt = -x, dy/dt = 2y$

   c. $dx/dt = y, dy/dt = -2x$

   d. $dx/dt = -x + 1, dy/dt = -2y$

9. The Department of Fish and Game in a certain state is planning to issue deer hunting permits. It is known that if the deer population falls below a certain level

m, the deer will become extinct. It is also known that if the deer population goes above the maximum carrying capacity M, the population will decrease to M.

a. Discuss the reasonableness of the following model for the growth rate of the deer population as a function of time

$$\frac{dP}{dt} = kP(M - P)(P - m)$$

where P is the population of the deer and $k > 0$ is a constant of proportionality. Include a graph of dP/dt versus P as part of your discussion.

b. Explain how this growth rate model differs from the logistic model $dP/dt = kP(M - P)$. Is it better or worse than the logistic model? Why?

c. Show that if $P > M$ for all t, then the limit of $P(t)$ as $t \to \infty$ is M.

d. Discuss what happens if $P < m$ for all t.

e. Assuming that $m < P < M$ for all t, explain briefly the steps you would use to solve the differential equation. Do not attempt to solve the differential equation.

f. Graphically discuss the solutions to the differential equation. What are the equilibrium points of the model? Explain the dependence of the equilibrium level of P on the initial conditions. How many deer hunting permits should be issued?

11.1 PROJECTS

1. Complete the requirements of the UMAP module, "Whales and Krill: A Mathematical Model," by Raymond N. Greenwell, UMAP 610. A predator–prey system involving whales and krill is modeled by a system of differential equations. Although the equations are not solvable, information is extracted using dimensional analysis and the study of equilibrium points. The concept of maximum sustainable yield is introduced and used to draw conclusions about fishing strategies. You will learn to construct a differential equations model, remove dimensions from a set of equations, and find equilibrium points of a system of differential equations. You will also learn their significance and practice manipulative skills in algebra and calculus.

11.2   A COMPETITIVE HUNTER MODEL[2]

Up to now we have seen how single-species growth can be modeled as the Malthusian model or the limited growth model. Let's turn our attention to how two different species might compete for common resources.

[2]This section is adapted from UMAP Unit 628, based on the work of Stanley C. Leja and one of the authors. The adaptation is presented with the permission of COMAP, 57 Bedford St., Lexington, MA 02173.

Problem identification Imagine a small pond that is mature enough to support wildlife. We desire to stock the pond with game fish, say trout and bass. Let $x(t)$ denote the population of the trout at any time t, and let $y(t)$ denote the bass population. *Is coexistence of the two species in the pond possible? If so, how sensitive is the final solution of population levels to the initial stockage levels and external perturbations?*

Assumptions The level of the trout population $x(t)$ depends on many variables: the initial level x_0, the amount of competition for limited resources, the existence of predators, and so forth. Initially, we assume the environment can support an unlimited number of trout, so that in isolation

$$\frac{dx}{dt} = ax \quad \text{for } a > 0$$

(Later we may find it desirable to refine the model and use a limited growth assumption.) Next, we modify the preceding differential equation to take into account the competition of the trout with the bass population for living space and a common food supply. The effect of the bass population is to decrease the growth rate of the trout population. This decrease is roughly proportional to the number of possible interactions between the two species, so one submodel is to assume that the decrease is proportional to the product of x and y. These considerations are modeled by the equation

$$\frac{dx}{dt} = ax - bxy = (a - by)x \tag{11.5}$$

The intrinsic growth rate $k = a - by$ decreases as the level of the bass population increases. The constants a and b indicate the degrees of self-regulation of the trout population and its competition with the bass population, respectively. These coefficients must be determined experimentally or by analyzing historical data.

The situation for the bass population is analyzed in the same manner. Thus we obtain the following autonomous system of two first-order differential equations for our model:

$$\frac{dx}{dt} = (a - by)x \tag{11.6}$$

$$\frac{dy}{dt} = (m - nx)y$$

where $x(0) = x_0$, $y(0) = y_0$, and a, b, m, and n are all positive constants. This model is useful in studying the growth patterns of species exhibiting competitive behavior such as the trout and bass.

Graphical analysis of the model One of our concerns is whether the trout and bass populations reach equilibrium levels. If so, then we will know whether coexistence of the two species in the pond is possible. The only way such a state can be achieved is that both populations stop growing; that is, $dx/dt = 0$ and $dy/dt = 0$. Thus we seek the rest points or equilibrium points of the System (11.6).

Setting the right sides of Equations (11.6) equal to zero and solving for x and y simultaneously, we find the rest points $(x, y) = (0, 0)$ and $(x, y) = (m/n, a/b)$ in the phase plane. Along the vertical line $x = m/n$ and the x-axis in the phase plane, the growth dy/dt in the bass population is zero; along the horizontal line $y = a/b$ and the y-axis, the growth dx/dt in the trout population is zero. If the initial stockage is at these rest point levels, there would be no growth in either population. These features are depicted in Figure 11.6.

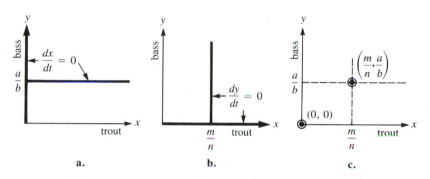

FIGURE 11.6   Rest points of the competitive hunter model given by the System (11.6)

Considering the approximations necessary in any model, it is inconceivable that we would estimate precisely the values for the constants a, b, m, and n in our System (11.6). So the pertinent behavior we need to investigate is what happens to the solution trajectories in the vicinity of the rest points $(0, 0)$ and $(m/n, a/b)$. Specifically, are these points stable or unstable?

To investigate this question graphically, let's analyze the directions of dx/dt and dy/dt in the phase plane. (Although $x(t)$ and $y(t)$ represent the trout and bass populations, respectively, it is helpful to think of the trajectories as paths of a moving particle, in accord with our discussion in the preceding section.) Whenever dx/dt is positive, the horizontal component $x(t)$ of the trajectory is increasing and the particle is moving toward the right; whenever dx/dt is negative, the particle is moving toward the left. Likewise, if dy/dt is positive, the component $y(t)$ is increasing and the particle is moving upward; if dy/dt is negative, the particle is moving downward. In our System (11.6), the vertical line $x = m/n$ divides the phase plane into two half-planes. In the left half-plane dy/dt is positive, and in the right half-plane it is negative. The directions of the associated trajectories are indicated in Figure 11.7. Likewise, the horizontal line $y = a/b$ determines the half-planes where dx/dt is positive or negative. The directions of the associated trajectories are indicated in Figure 11.8. Along the line $y = a/b$ itself, $dx/dt = 0$. Therefore any trajectory crossing this line will do so vertically. Similarly, along the line $x = m/n$, $dy/dt = 0$, so the line will be crossed horizontally. Finally, along the y-axis motion must be vertical, and along the x-axis motion must be horizontal. Combining all

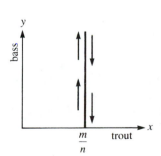

FIGURE 11.7   To the left of $x = m/n$ the trajectories move upward; to the right they move downward

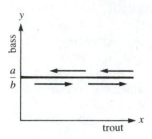

FIGURE 11.8 Above the line $y = a/b$ the trajectories move to the left; below the line they move to the right

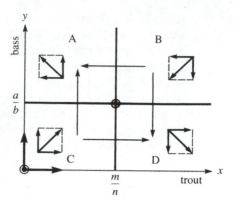

FIGURE 11.9 Composite graphical analysis of the trajectory directions in the four regions determined by $x = m/n$ and $y = a/b$

this information into a single graph gives the four distinct regions A, B, C, D with their respective trajectory directions as depicted in Figure 11.9.

Analyzing the motion in the vicinity of $(0, 0)$, we can see that all motion is away from that rest point: The motion is upward and toward the right. In the vicinity of the rest point $(m/n, a/b)$, the behavior depends on the region in which the trajectory begins. If the trajectory starts in region B, for instance, it will move downward and leftward toward the rest point. As it gets nearer to the rest point, the derivatives dx/dt and dy/dt approach zero. Depending on where the trajectory begins and the relative sizes of the constants a, b, m, and n, either the trajectory will continue moving downward and into region D as it swings past the rest point, or the trajectory will move leftward into region A. Once it enters either of these latter two regions, the trajectory moves away from the rest point. Thus both rest points are unstable. It can be shown that there is exactly one trajectory from region B that gets arbitrarily close to the rest point, slowing down as it gets nearer and nearer to $(m/n, a/b)$. Trajectories that start above this path will cross the line $x = m/n$; those below will cross $y = a/b$. These features are suggested in Figure 11.10.

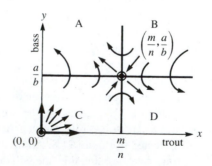

FIGURE 11.10 Motion along the trajectories near the rest points $(0, 0)$ and $(m/n, a/b)$

Model interpretation   The graphical analysis conducted so far leads us to the preliminary conclusion that, under the assumptions of our model, reaching equilibrium levels of both species is highly unlikely. Furthermore, the initial stockage levels turn out to be important in determining which of the two species might survive. Perturbations of the system may also affect the outcome of the competition. Thus mutual coexistence of the species is highly improbable. This phenomenon is known

**Principle of Competitive Exclusion
Gause's Principle**

as the **Principle of Competitive Exclusion**, or **Gause's Principle**. Moreover, the initial conditions completely determine the outcome as depicted in Figure 11.11. We can see from that graph that any perturbation causing a switch from one region (say, below the line joining the two rest points) to the other region (above the line) would change the outcome. Actually the curve separating the starting points in which the bass win from those in which the trout win may not be a straight line as depicted in the figure. One of the limitations of our graphical analysis is that we have not determined that separating curve precisely. If we are satisfied with our model, we may well want to determine that separating boundary.

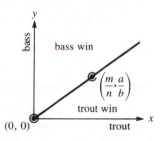

FIGURE 11.11   Qualitative results of analyzing the competitive hunter model

Limitations of a graphical analysis   It is not always possible to determine the nature of the motion near a rest point using only graphical analysis. To understand this limitation, consider the rest point and the direction of motion of the trajectories shown in Figure 11.12. The information in the figure is insufficient to distinguish among the three possible motions shown in Figure 11.13. Moreover, even if we have determined by some other means that Figure 11.13c correctly portrays the motion near the rest point, we might be tempted to deduce that the motion will grow without

FIGURE 11.12

Trajectory direction near a rest point

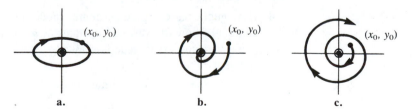

FIGURE 11.13   Three possible trajectory motions: **a.** periodic motion, **b.** motion toward an asymptotically stable rest point, and **c.** motion near an unstable rest point

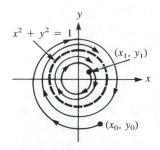

FIGURE 11.14 The solution $x^2 + y^2 = 1$ to (11.7) is a limit cycle

limit cycle

bound in both the x and y directions. However, consider the system given by

$$\frac{dx}{dt} = y + x - x(x^2 + y^2) \tag{11.7}$$

$$\frac{dy}{dt} = -x + y - y(x^2 + y^2)$$

It can be shown that $(0,0)$ is the only rest point for (11.7). Yet any trajectory starting on the unit circle $x^2 + y^2 = 1$ will traverse the unit circle in a periodic solution because in that case $dy/dx = -x/y$ (see Problem 2 in the 11.2 problem set). Moreover, if a trajectory starts inside the circle (provided it does not start at the origin), it will spiral outward asymptotically, getting closer and closer to the circular path as t tends to infinity. Likewise, if the trajectory starts outside the circular region, it will spiral inward and again approach the circular path asymptotically. The solution $x^2 + y^2 = 1$ is called a **limit cycle**. The trajectory behavior is sketched in Figure 11.14. Thus if the System (11.7) models population behavior for two competing species, we would have to conclude that the population levels will eventually be periodic. This example illustrates that the results of a graphical analysis are useful for determining the motion in the immediate vicinity of an equilibrium point only. (Here we have assumed that negative values for x and y have a physical meaning in Figure 11.14 or that the point $(0,0)$ represents the translation of a rest point from the first quadrant to the origin.)

11.2 Problems

1. List three important considerations that are ignored in the development of the competitive hunter model presented in this section.
2. For the System (11.7), show that any trajectory starting on the unit circle $x^2 + y^2 = 1$ will traverse the unit circle in a periodic solution. First introduce polar coordinates and rewrite the system as $dr/dt = r(1 - r^2)$ and $d\theta/dt = 1$.
3. Develop a model for the growth of trout and bass, assuming that in isolation trout demonstrate exponential decay (so that $a < 0$ in Equation (11.6)) and that the bass population actually grows logistically with a population limit M. Analyze graphically the motion in the vicinity of the rest points in your model. Is coexistence possible?
4. How might the competitive hunter model (11.6) be validated? Include a discussion of how the various constants a, b, m, and n might be estimated. How could state conservation authorities use the model to ensure the survival of both species?
5. Consider the competitive hunter model defined by

$$\frac{dx}{dt} = a(1 - x/k_1)x - bxy$$

$$\frac{dy}{dt} = m(1 - y/k_2)y - nxy$$

where x represents the trout population and y the bass population.

a. What assumptions are implicitly being made about the growth of trout and bass in the absence of competition?

b. Interpret the constants a, b, m, n, k_1, k_2 in terms of the physical problem.

c. Perform a graphical analysis and answer the following questions:

   i. What are the possible equilibrium levels?

   ii. Is coexistence possible?

   iii. Pick several typical starting points and sketch typical trajectories in the phase plane.

   iv. Interpret the outcomes predicted by your graphical analysis in terms of the constants a, b, m, n, k_1, and k_2.

Note: When you get to step **i**, you should realize that at least five cases exist. You will need to analyze all five cases. One case is when the lines are coincident.

6. Consider the following economic model: Let P be the price of a single item on the market. Let Q be the quantity of the item available on the market. Both P and Q are functions of time. If we consider price and quantity as two interacting species, the following model might be proposed:

$$\frac{dP}{dt} = aP(b/Q - P)$$

$$\frac{dQ}{dt} = cQ(fP - Q)$$

where $a, b, c,$ and f are positive constants. Justify and discuss the adequacy of the model.

a. If $a = 1, b = 20{,}000, c = 1,$ and $f = 30$, find the equilibrium points of this system. Classify each equilibrium point with respect to its stability, if possible. If a point cannot be readily classified, explain why.

b. Perform a graphical stability analysis to determine what will happen to the levels of P and Q as time increases.

c. Give an economic interpretation of the curves that determine the equilibrium points.

11.2 PROJECTS

Complete the requirements of the referenced UMAP module.

1. "The Budgetary Process: Incrementalism," UMAP 332; "Competition," UMAP 333, by Thomas W. Likens. The politics of budgeting revolve around the allocation of limited resources to agencies and groups competing for them. UMAP 332 develops a model to explain how levels of appropriation change from one period to the next if new budgets are determined by Congress and federal agencies by making marginal adjustments in the status quo. The model assumes that the share received by one agency will not affect or depend on the share received by another agency. In UMAP 333 the model is refined to address the conflictive nature of politics and the necessary interdependence of budgetary decisions.

2. "The Growth of Partisan Support I: Model and Estimation," UMAP 304; "The Growth of Partisan Support II: Model Analytics," UMAP 305, by Carol Weitzel Kohfeld. UMAP 304 presents a simple model of political mobilization, refined to include the interaction between supporters of a particular party and recruitable nonsupporters. UMAP 305 investigates the mathematical properties of the first-order quadratic difference equation model. The model is tested using data from three U.S. counties. An understanding of linear first-order difference equations with constant coefficients is required.

3. "Random Walks: An Introduction to Stochastic Processes," by Ron Barnes, UMAP 520. Random walks are introduced by an example of a gambling game. The associated finite difference equation is developed and solved. The concept of expected gain and the duration of a game are introduced and their usefulness is demonstrated. Generalizations to Markov chains and continuous processes are discussed. Applications in the life sciences and genetics are noted.

11.2 Further Reading

TUCHINSKY, Philip M. *Man in Competition with the Spruce Budworm*, UMAP Expository Monograph. The population of tiny caterpillars periodically explodes in the evergreen forests of Eastern Canada and Maine. They devour the trees' needles and cause great damage to forests central to the economy of the region. The province of New Brunswick is using mathematical models of the budworm–forest interaction in an effort to plan for and control the damage. The monograph surveys the ecological situation and examines the computer simulation and differential equation models that are currently in use.

11.3 A PREDATOR–PREY MODEL[3]

In this section we study a model of population growth of two species in which one species is the primary food source for the other. One example of such a situation occurs in the Southern Ocean, where the baleen whales eat the Antarctic krill, *Euphausia superboa*, as their principal food source. Another example is wolves and rabbits in a closed forest; the wolves eat the rabbits for their principal food source and the rabbits eat vegetation in the forest. Still other examples include sea otters as predators and abalone as prey; and the ladybird beetle, *Novius cardinalis*, as predator and the cottony cushion insect, *Icerya purchasi*, as prey.

---

[3] Optional section

Problem identification   Let's take a closer look at the situation of the baleen whales and Antarctic krill. The whales eat the krill and the krill live on the plankton in the sea. If the whales eat too many krill so that the krill cease to be abundant, the food supply of the whales is greatly reduced. Then the whales will starve or leave the area in search of a new supply of krill. As the population of baleen whales dwindles, the krill population makes a comeback because not so many of them are being eaten. As the krill population increases, the food supply for the whales grows and, consequently, so does the baleen whale population. And more baleen whales are eating more and more krill again. *In the pristine environment, does this cycle continue indefinitely or does one of the species eventually die out?* The baleen whales in the Southern Ocean have been overexploited to the extent that their current population is about one-sixth its estimated pristine level. Thus there appears to be a surplus of Antarctic krill. (Already something like 100,000 tons of krill are being harvested annually.) What effect does exploitation of the whales have on the balance between the whale and krill populations? What are the implications a krill fishery can hold for the depleted stocks of baleen whales and for other species such as seals, seabirds, penguins, and fish that depend on krill for their main source of food? The ability to answer such questions is important to management of multispecies fisheries. Let's see what answers can be obtained from a graphical modeling approach.

Assumptions   Let $x(t)$ denote the Antarctic krill population at any time t, and let $y(t)$ denote the population of baleen whales in the Southern Ocean. The level of the krill population depends on a number of factors, including the ability of the ocean to support them, the existence of competitors for the plankton they ingest, and the presence and levels of predators. As a rough first model, let's start by assuming that the ocean can support an unlimited number of krill so that

$$\frac{dx}{dt} = ax \quad \text{for } a > 0$$

(Later we may want to refine the model with a limited growth assumption. This refinement is presented in UMAP 610 listed in the 11.1 Projects section.) Second, assume that the krill are eaten primarily by the baleen whales (so neglect any other predators). Then the growth rate of the krill is diminished in a way that is proportional to the number of interactions between them and the baleen whales. One interaction assumption leads to the differential equation

$$\frac{dx}{dt} = ax - bxy = (a - by)x \tag{11.8}$$

Notice that the intrinsic growth rate $k = a - by$ decreases as the level of the baleen whale population increases. The constants a and b indicate the degrees of self-regulation of the krill population and the predatoriness of the baleen whales, respectively. These coefficients must be determined experimentally or from historical data. So far, Equation (11.8) governing the growth of the krill population looks just like either of the equations in the competitive hunter model presented in the preceding section.

Next, consider the baleen whale population $y(t)$. In the absence of krill the whales have no food, so we will assume their population declines at a rate proportional to their numbers. This assumption produces the exponential decay equation

$$\frac{dy}{dt} = -my \quad \text{for } m > 0$$

However, in the presence of krill the baleen whale population increases at a rate proportional to the interactions between the whales and their krill food supply. Thus the preceding equation is modified to give

$$\frac{dy}{dt} = -my + nxy = (-m + nx)y \tag{11.9}$$

Notice from Equation (11.9) that the intrinsic growth rate $r = -m + nx$ of the whales increases as the level of the krill population increases. The positive coefficients m and n would be determined experimentally or from historical data. Putting results (11.8) and (11.9) together gives the following autonomous system of differential equations for our predator–prey model:

$$\frac{dx}{dt} = (a - by)x \tag{11.10}$$

$$\frac{dy}{dt} = (-m + nx)y$$

where $x(0) = x_0$, $y(0) = y_0$, a, b, m, and n are all positive constants. The System (11.10) governs the interaction of the baleen whales and Antarctic krill populations under our unlimited growth assumptions and in the absence of other competitors and predators.

Graphical analysis of the model   Let's determine whether the krill and whale populations reach equilibrium levels. The rest points or equilibrium levels occur when $dx/dt = dy/dt = 0$. Setting the right sides of (11.10) equal to zero and solving for x and y simultaneously gives the rest points $(x, y) = (0, 0)$ and $(x, y) = (m/n, a/b)$. Along the vertical line $x = m/n$ and the x-axis in the phase plane, the growth dy/dt in the baleen whale population is zero; along the horizontal line $y = a/b$ and the y-axis, the growth dx/dt in the krill population is zero. These features are depicted in Figure 11.15.

Because the values for the constants a, b, m, and n in the System (11.10) will only be estimates, we need to investigate the behavior of the solution trajectories near the two rest points $(0, 0)$ and $(m/n, a/b)$. Thus we analyze the directions of dx/dt and dy/dt in the phase plane. In the System (11.10), the vertical line $x = m/n$ divides the phase plane into two half-planes. In the left half-plane dy/dt is negative, and in the right half-plane it is positive. In a similar way, the horizontal line $y = a/b$ determines two half-planes. In the upper half-plane dx/dt is negative, and in the lower half-plane it is positive.

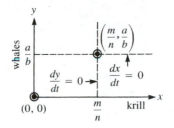

FIGURE 11.15 Rest points of the predator–prey model given by the System (11.10)

The directions of the associated trajectories are indicated in Figure 11.16. Along the y-axis, motion must be vertical and toward the rest point $(0, 0)$; along the x-axis, motion must be horizontal and away from the rest point $(0, 0)$.

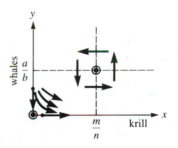

FIGURE 11.16 Trajectory directions in the predator–prey model

From Figure 11.16 we can see that the rest point $(0, 0)$ is unstable. Entrance to the rest point is along the line $x = 0$, where there is no krill population. Thus the whale population declines to zero in the absence of its primary food supply. All other trajectories recede from the rest point.

The rest point $(m/n, a/b)$ is more complicated to analyze. The information given by the figure is insufficient to distinguish between the three possible motions shown in Figure 11.13 of the preceding section. We cannot tell whether the motion is periodic, asymptotically stable, or unstable. Thus we must perform a further analysis.

An analytical solution of the model Because the number of baleen whales will depend on the number of Antarctic krill available for food, we assume that y is a function of x. Then from the chain rule for derivatives, we have

$$\frac{dy}{dx} = \frac{dy/dt}{dx/dt}$$

or

$$\frac{dy}{dx} = \frac{(-m + nx)y}{(a - by)x} \tag{11.11}$$

Equation (11.11) is a separable first-order differential equation and can be rewritten as

$$\left(\frac{a}{y} - b\right) dy = \left(n - \frac{m}{x}\right) dx \tag{11.12}$$

Integration of each side of (11.12) yields

$$a \ln y - by = nx - m \ln x + k_1$$

or

$$a \ln y + m \ln x - by - nx = k_1$$

where k_1 is a constant.

Using properties of the natural logarithm and exponential functions, this last equation can be rewritten as

$$\frac{y^a x^m}{e^{by+nx}} = K \tag{11.13}$$

where K is a constant. Equation (11.13) defines the solution trajectories in the phase plane. We now show that these trajectories are closed and represent periodic motion.

Periodic predator–prey trajectories   Equation (11.13) can be rewritten as

$$\left(\frac{y^a}{e^{by}}\right) = K\left(\frac{e^{nx}}{x^m}\right) \tag{11.14}$$

Let's determine the behavior of the function $f(y) = y^a/e^{by}$. Using the first derivative test (see Problem 1 in the 11.3 problem set), we can easily show that $f(y)$ has a relative maximum at $y = a/b$ and no other critical points. For simplicity of notation, call this maximum value M_y. Moreover, $f(0) = 0$ and from l'Hôpital's rule $f(y)$ approaches 0 as y tends to infinity. Similar arguments apply to the function $g(x) = x^m/e^{nx}$, which achieves its maximum value M_x at $x = m/n$. The graphs of the functions f and g are depicted in Figure 11.17.

From Figure 11.17, the largest value for $y^a e^{-by} x^m e^{-nx}$ is $M_y M_x$. That is, Equation (11.14) has no solutions if $K > M_y M_x$ and exactly one solution, $x = m/n$ and $y = a/b$, when $K = M_y M_x$. Let's consider what happens when $K < M_y M_x$.

Suppose $K = sM_y$, where $s < M_x$ is a positive constant. Then the equation

$$x^m e^{-nx} = s$$

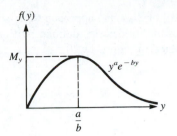
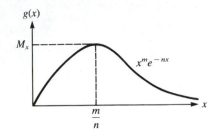

FIGURE 11.17   Graphs of the functions $f(y) = y^a/e^{by}$ and $g(x) = x^m/e^{nx}$

has exactly two solutions: $x_m < m/n$ and $x_M > m/n$ (see Figure 11.18). Now if $x < x_m$, then $x^m e^{-nx} < s$ so that $se^{nx}x^{-m} > 1$ and

$$f(y) = y^a e^{-by} = Ke^{nx}x^{-m} = sM_y e^{nx}x^{-m} > M_y$$

Therefore there is no solution for y in Equation (11.14) when $x < x_m$. Likewise, there is no solution when $x > x_M$. If $x = x_m$ or $x = x_M$, Equation (11.14) has exactly the one solution $y = a/b$.

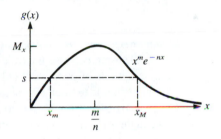

FIGURE 11.18   The equation $x^m e^{-nx} = s$ has exactly two solutions for $s < M_x$.

Finally, if x lies between x_m and x_M, Equation (11.14) has exactly two solutions. The smaller solution $y_1(x)$ is less than a/b, and the larger solution $y_2(x)$ is greater than a/b. This situation is depicted in Figure 11.19. Moreover, as x

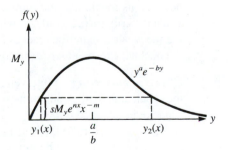

FIGURE 11.19   When $x_m < x < x_M$, there are exactly two solutions for y in Equation (11.14)

approaches either x_m or x_M, $f(y)$ approaches M_y so both $y_1(x)$ and $y_2(x)$ approach
a/b. It follows that the trajectories defined by Equation (11.14) are periodic and
have the form depicted in Figure 11.20.

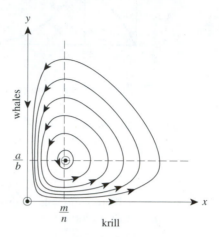

FIGURE 11.20   Trajectories in the vicinity of the rest point $(m/n, a/b)$ are
periodic

Model interpretation   What conclusions can be drawn from the trajectories in Figure 11.20? First, because the trajectories are closed curves, they predict that, under
the assumptions of our model (11.10), neither the baleen whales nor the Antarctic
krill will become extinct. (Remember, the model is based on the pristine situation.)
The second observation is that along a single trajectory the two populations fluctuate
between their maximum and minimum values. That is, starting with populations in
the region where $x > m/n$ and $y > a/b$, the krill population will decline and the
whale population increase until the krill population reaches the level $x = m/n$, at
which point the whale population also begins to decline. Both populations continue
to decline until the whale population reaches the level $y = a/b$ and the krill population begins to increase, and so on, around the trajectory. Recall from our discussion
in Section 11.2 that the trajectories never cross. A sketch of the population curves
is shown in Figure 11.21. In that figure we can see the krill population fluctuates
between its maximum and minimum values over one complete cycle. Notice that
when the krill are plentiful, the whale population has its maximum rate of increase,
but the whale population reaches its maximum value after the krill population is on
the decline. The predator *lags* behind the prey in a cyclic fashion.

Effects of harvesting   For given initial population levels $x(0) = x_0$ and $y(0) = y_0$,
the whale and krill populations will fluctuate with time around one of the closed
trajectories depicted in Figure 11.20. Let T denote the time it takes to complete one
full cycle and return to the starting point. The average levels of the krill and baleen
whale populations over the time cycle are defined, respectively, by the integrals

$$\bar{x} = \frac{1}{T} \int_0^T x(t)\, dt \quad \text{and} \quad \bar{y} = \frac{1}{T} \int_0^T y(t)\, dt$$

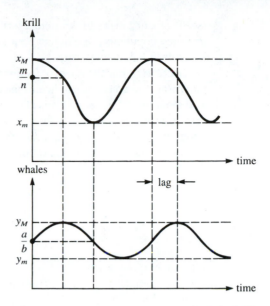

FIGURE 11.21 The whale population lags behind the krill population as both populations fluctuate cyclically between their maximum and minimum values

Now, from Equation (11.8)

$$\left(\frac{1}{x}\right)\left(\frac{dx}{dt}\right) = a - by$$

so that integration of both sides from $t = 0$ to $t = T$ leads to

$$\int_0^T \left(\frac{1}{x}\right)\left(\frac{dx}{dt}\right) dt = \int_0^T (a - by)\, dt$$

or

$$\ln x(T) - \ln x(0) = aT - b \int_0^T y(t)\, dt$$

Because of the periodicity of the trajectory, $x(T) = x(0)$ and this last equation gives the average value

$$\bar{y} = \frac{a}{b}$$

In an analogous manner, it can be shown that

$$\bar{x} = \frac{m}{n}$$

(see Problem 2 in the 11.3 problem set). Therefore, the *average* levels of the predator and prey populations are in fact their *equilibrium* levels. Let's see what this means in terms of harvesting krill.

Let's assume the effect of fishing for krill is to decrease its population level at a rate $rx(t)$. The constant r indicates the intensity of fishing and includes such factors as the number of fishing vessels at sea and the number of fishermen casting nets for krill. Because less food is now available for the baleen whales, assume the whale population also decreases at a rate $ry(t)$. Incorporating these fishing assumptions into our model, we obtain the refined model

$$\frac{dx}{dt} = (a - by)x - rx = \left[(a - r) - by\right] x \tag{11.15}$$

$$\frac{dy}{dt} = (-m + nx)y - ry = \left[-(m + r) + nx\right] y$$

The autonomous system (11.15) is of the same form as (11.10) (provided $a - r > 0$) with a replaced by $a - r$ and m replaced by $m + r$. Thus the new average population levels will be

$$\overline{x} = \frac{m + r}{n} \quad \text{and} \quad \overline{y} = \frac{a - r}{b}$$

Consequently, a moderate amount of harvesting krill (so $r < a$) actually *increases* the average level of krill and *decreases* the average baleen whale population (under our assumptions for the model). The increase in krill population is beneficial to other species in the Southern Ocean that depend on the krill for their main food source (seals, seabirds, penguins, and fish). The fact that some fishing increases the number of krill is known as **Volterra's principle**. The autonomous system (11.10) was first proposed by Lotka (1925) and Volterra (1931) as a simple model of predator–prey interaction.

The Lotka–Volterra model can be modified to reflect the situation in which both the predator and the prey are diminished by some kind of depleting forces, such as when applying insecticide treatments that destroy both the insect predator and its insect prey. An example is given in Problem 3 in the 11.3 problem set.

Some biologists and ecologists argue that the Lotka–Volterra model (11.10) is unrealistic because the system is not asymptotically stable, whereas most observable natural predator–prey systems tend to equilibrium levels as time evolves. Nevertheless, regular population cycles, as suggested by the trajectories in Figure 11.20, do occur in nature. Some scientists have proposed models other than the Lotka–Volterra model that do exhibit oscillations that are asymptotically stable (so that the trajectories approach equilibrium solutions). One such model is given by

$$\frac{dx}{dt} = ax + bxy - rx^2$$

$$\frac{dy}{dt} = -my + nxy - sy^2$$

In this last autonomous system, the term rx^2 indicates the degree of internal competition of the prey for their limited resource (such as food and space), and the term sy^2 indicates the degree of competition among the predators for the finite amount of available prey. An analysis of this model is more difficult than that presented for the Lotka–Volterra model, but it can be shown that the trajectories of the model are not periodic and tend to equilibrium levels. The constants r and s are positive and would be determined by experimentation or historical data.

11.3 PROBLEMS

1. Apply the first- and second-derivative tests to the function $f(y) = y^a/e^{by}$ to show that $y = a/b$ is a unique critical point that yields the relative maximum $f(a/b)$. Show also that $f(y)$ approaches 0 as y tends to infinity.

2. Derive the result that the average value $\bar{x}$ of the prey population modeled by the Lotka–Volterra system (11.10) is given by the population level m/n.

3. In 1868 the accidental introduction into the United States of the cottony cushion insect (*Icerya purchasi*) from Australia threatened to destroy the American citrus industry. To counteract this situation, a natural Australian predator, a ladybird beetle (*Novius cardinalis*) was imported. The beetles kept the scale insects down to a relatively low level. When DDT was discovered to kill scale insects, farmers applied it in the hopes of reducing even further the scale insect population. However, DDT turned out to be fatal to the beetle as well and the overall effect of using the insecticide was to increase the numbers of the scale insect.

   Modify the Lotka–Volterra model to reflect a predator–prey system of two insect species in which farmers apply (on a continuing basis) an insecticide that destroys both the insect predator and the insect prey at a common rate proportional to the numbers present. What conclusions do you reach concerning the effects of the application of the insecticide? Use graphical analysis to determine the effect of using the insecticide once.

4. In a 1969 study, E. R. Leigh concluded that the fluctuations in the numbers of Canadian lynx and its primary food source, the hare, trapped by the Hudson's Bay Company between 1847 and 1903 were periodic. The actual population levels of both species differed greatly from the predicted population levels obtained from the Lotka–Volterra predator–prey model.

   Use the entire model-building process to modify the Lotka–Volterra model to arrive at a more realistic model for the growth rates of both species. Answer the following questions in the appropriate places of the model-building process:

   a. How have you modified the basic assumptions of the predator–prey model?

   b. Why are your modifications an improvement to the basic model?

   c. What are the equilibrium points for your model?

   d. Is it possible to classify each equilibrium point as either stable or unstable? If so, classify them.

e. Based on your equilibrium analysis, what values will the population levels of lynx and hare approach as t tends to infinity?

f. How would you use your revised model to suggest hunting policies for Canadian lynx and hare? *Hint:* You are introducing a second predator—human beings—into the system.

5. Consider two species whose survival depends on their mutual cooperation. An example would be a species of bee that feeds primarily on the nectar of one plant species and simultaneously pollinates that plant. One simple model of this mutualism is given by the autonomous system

$$\frac{dx}{dt} = -ax + bxy$$

$$\frac{dy}{dt} = -my + nxy$$

a. What assumptions are implicitly being made about the growth of each species in the absence of cooperation?

b. Interpret the constants a, b, m, and n in terms of the physical problem.

c. What are the equilibrium levels?

d. Perform a graphical analysis and indicate the trajectory directions in the phase plane.

e. Find an analytical solution and sketch typical trajectories in the phase plane.

f. Interpret the outcomes predicted by your graphical analysis. Do you believe the model is realistic? Why?

11.3 PROJECTS

1. Complete the requirements of the UMAP module, "Graphical Analysis of Some Difference Equations in Biology," by Martin Eisen, UMAP 553. The growth of many biological populations can be modeled by difference equations. This module shows how the behavior of the solutions to certain equations can be predicted by graphical techniques.

2. Prepare a summary of one of the papers by May et al. listed in the further readings for this section.

11.3 Further Reading

CLARK, Colin W. *Mathematical Bioeconomics: The Optimal Management of Renewable Resources.* New York: Wiley, 1976.

MAY, R. M. *Stability and Complexity in Model Ecosystems.* Monographs in Population Biology VI. Princeton, NJ: Princeton University Press, 1973.

MAY, R. M., ed. *Theoretical Ecology: Principles and Applications.* Philadelphia: Saunders, 1976.

MAY, R. M., J. R. Beddington, C. W. Clark, S. J. Holt & R. M. Lewis. "Management of Multispecies Fisheries." *Science 205* (July 1979): 267–277.

11.4	**TWO MILITARY EXAMPLES**[4]

Example 1   *An Arms Race Revisited*

In Chapter 1 we investigated an arms race to ascertain the effects a particular strategy might have on the arms race, such as what is likely to occur if one of the two countries introduces mobile launching pads. Let's examine the economic aspects of the arms race.

Problem identification   Consider again two countries engaged in an arms race. Let's attempt to assess qualitatively the effect of the arms race on the level of defense spending. *Specifically, we are interested in knowing whether the arms race will lead to uncontrolled spending and eventually be dominated by the country with the greatest economic assets. Or will an equilibrium level of spending be reached in which each country spends a steady-state amount on defense?*

Assumptions   Define the variable x as the annual defense expenditure for Country 1 and the variable y as the annual defense expenditure for Country 2. We assume that each nation is ready to defend itself and considers a defense budget to be necessary. Let's examine the situation from the point of view of Country 1. Its rate of spending depends on several factors. In the absence of any spending on the part of Country 2 or any grievance with that country, it is reasonable to assume that the defense spending would decrease at a rate proportional to the amount being spent. The proportionality constant is indicative of the requirement to maintain the current arsenal (a percentage of the current spending) and acts as an economic restraint on defense spending (in the sense that government monies need to be spent in other areas such as health and education). Thus

$$\frac{dx}{dt} = -ax \quad \text{for } a > 0$$

Qualitatively, this last equation says there will be no growth in defense spending. Now what happens when Country 1 perceives Country 2 engaging in defense spending? Country 1 will feel compelled to increase its defense budget to offset the defense buildup of its adversary and shore up its own security. Let's assume the rate of increase for Country 1 is proportional to the amount Country 2 spends, where the proportionality coefficient is a measure of the perceived effectiveness of Country 2's weapons. This assumption seems reasonable, at least up to a point. As Country 2 adds weapons to its arsenal, Country 1 will perceive a need to add weapons to its own arsenal, where the numbers added are based on the assessment of the effectiveness of Country 2's weapons. Thus the previous equation is modified to become

$$\frac{dx}{dt} = -ax + by$$

[4]Optional section

We are assuming b is a constant, although this assumption is somewhat unrealistic. We would expect some kind of diminishing return of perceived effectiveness as Country 2 continues to add weapons. That is, if Country 2 has 100 weapons, Country 1 might perceive the need to add 40 weapons. But if Country 2 adds 200 weapons to its arsenal, Country 1 might perceive the need to add only 75 weapons. Thus, realistically, the constant b is more likely to be a decreasing function of y (and, in some instances, might be an increasing function). Later we may wish to refine our model to account for this probable diminishing return.

Finally, let's add a constant term to reflect any underlying grievance felt by Country 1 toward Country 2. That is, even if the level of spending by both countries were zero, Country 1 would still feel compelled to be armed against Country 2, perhaps based on a fear of future aggressive action on the part of its adversary. If $y = 0$, then $c - ax$ represents a growth in defense spending for Country 1, and, until $c = ax$, growth will continue in order to achieve deterrence. These assumptions lead to the differential equation

$$\frac{dx}{dt} = -ax + by + c$$

where a, b, and c are nonnegative constants. The constant a indicates an economic restraint on defense spending, b indicates the intensity of rivalry with Country 2, and c indicates the deterrent or grievance factor. Although we are assuming c to be constant, it is more likely to be a function of both the variables x and y.

An entirely similar argument for Country 2 yields the differential equation

$$\frac{dy}{dt} = mx - ny + p$$

where m, n, and p are nonnegative constants and are interpreted the same way as b, a, and c, respectively. The preceding equations constitute our model for arms expenditures.

Graphical analysis of the model We are interested in whether the defense expenditures reach equilibrium levels. If so, we will know the arms race will not lead to uncontrolled spending. This situation means that both defense budgets must stop growing so $dx/dt = 0$ and $dy/dt = 0$. Thus we seek the rest points or equilibrium points of our model.

First, consider the case in which neither country has a grievance against the other nor perceives any need for deterrence. Then $c = 0$ and $p = 0$, so our model becomes

$$\frac{dx}{dt} = -ax + by \qquad (11.16)$$

$$\frac{dy}{dt} = mx - ny$$

For the autonomous system (11.16), $(x, y) = (0, 0)$ is a rest point. In this state there are no defense expenditures on either side and the two countries live in permanent peace with all conflicts resolved through nonmilitary means. (Such a peaceful state has existed between the United States and Canada since 1817.) However, if grievances that are not resolved to the mutual satisfaction of both sides do arise, the two countries will feel compelled to arm, leading to the equations

$$\frac{dx}{dt} = c \quad \text{and} \quad \frac{dy}{dt} = p$$

Thus (x, y) will not remain at the rest point $(0, 0)$ if c and p are positive, so the rest point is unstable.

Now consider our general model

$$\frac{dx}{dt} = -ax + by + c \tag{11.17}$$

$$\frac{dy}{dt} = mx - ny + p$$

Setting the right sides equal to zero yields the linear system

$$ax - by = c \tag{11.18}$$

$$mx - ny = -p$$

Each of the equations in (11.18) represents a straight line in the phase plane. If the determinant of the coefficients, $bm - an$, is not equal to zero, then these two straight lines intersect at a unique rest point, denoted by (X, Y). It is easy to solve (11.18) to obtain this rest point:

$$X = \frac{bp + cn}{an - bm}$$

$$Y = \frac{ap + cm}{an - bm}$$

Assume that $an - bm > 0$, so the rest point (X, Y) lies in the first quadrant of the phase plane. The situation is depicted in Figure 11.22. Shown in the figure are

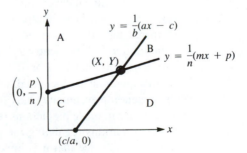

FIGURE 11.22 If $an - bm > 0$, the Model (11.17) has a unique rest point (X, Y) in the first quadrant. Along the line $by = ax - c$, $dx/dt = 0$; along the line $ny = mx + p$, $dy/dt = 0$

four regions labeled A, B, C, and D determined by the two intersecting lines. Let's examine the trajectory directions in each of these regions.

Any point (x, y) in region A lies above both of the lines represented by (11.18) so $ny - mx - p > 0$ and $by - ax + c > 0$. It follows from (11.17) that for region A, $dy/dt < 0$ and $dx/dt > 0$. For a point (x, y) in region B, $ny - mx - p > 0$ and $by - ax + c < 0$, so that $dy/dt < 0$ and $dx/dt < 0$. Similarly, in region C, $dy/dt > 0$ and $dx/dt > 0$; and in region D, $dy/dt > 0$ and $dx/dt < 0$. These features are suggested in Figure 11.23.

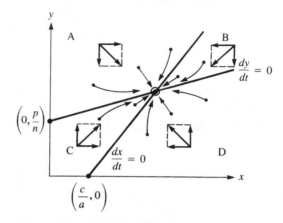

FIGURE 11.23   Composite graphical analysis of the trajectory directions in the four regions determined by the intersecting lines (Equation 11.18)

An analysis of the trajectory motion in the vicinity of the rest point (X, Y) in Figure 11.23 reveals that every trajectory in the phase plane approaches the rest point. Thus under our assumptions that $a/b > m/n$ so the intersection point (X, Y) lies in the first quadrant, the rest point (X, Y) is stable. Therefore, we conclude that defense spending for both countries will approach equilibrium (steady-state) levels of $x = X$ and $y = Y$. (In the problems for this section you are asked to investigate the case in which $a/b < m/n$; you will see that uncontrolled spending occurs.)

Let's examine the meaning of the inequality $a/b > m/n$ (which is equivalent to $an - bm > 0$). From the point of view of Country 1, the economic restraint a compared with the perceived intensity of rivalry b must be high against some specific ratio m/n for Country 2. The constant b has to do with perception and is, in part, a psychological factor. If b is lowered, the chances for the rest point to lie in the first quadrant are increased, with the beneficial result of steady-state levels of defense spending eventually likely for both countries. The value of our model is not in its capacity to make predictions but in its clarification of what can happen under different conditions regulating the parameters a, b, c, m, n, and p in the model. Certainly we can see that it is important for nations to reduce tension levels and perceived threats through mutual cooperation, respect, and disarmament policies if a runaway arms race is to be avoided. Moreover, a permanent peace can be achieved only if grievances are resolved to the mutual satisfaction of each country. However, note the effect of Taylor's deterrence strategy in Chapter 1. In that strategy $(0, 0)$ is no longer an equilibrium point so the possibility of total disarmament is ruled out. We have a stable equilibrium in this model because of the economic constraint $(c - ax)$. Even in the absence of defense spending on the part of Country 2 ($y = 0$), Country 1 will increase its defense budget until the cost of maintaining its armaments becomes exorbitant ($c = ax$).

Example 2   *Lanchester Combat Models*

Consider the situation of combat between two homogeneous forces: a homogeneous X force (e.g., tanks) opposed by another homogeneous Y force (e.g., antitank weapons). We want to know whether one force will eventually win out over the other or whether the combat will end in a draw. Other questions of interest include the following: How do the force levels decrease over time in battle? How many survivors will the winner have? How long will the battle last? How do changes in the initial force levels and weapon-system parameters affect the battle's outcome? In this example we will consider one basic combat model and several of its refinements.

Assumptions   Let $x(t)$ and $y(t)$ denote the strengths of the forces X and Y at time t, respectively. Usually t is measured in hours or days from the beginning of the combat. Let's examine what is meant by the *strengths* of the two forces X and Y. The strength $x(t)$, for example, includes a number of factors. If X is a homogeneous tank force, its strength depends on the number of tanks in operation, the level of technology used in the tank design, the quality of the manufacturing process, the level of training, the skills of the individuals operating the tanks, and so forth. For our purposes, let's assume that the strength $x(t)$ is simply the number of tanks in operation at time t. Likewise, the strength $y(t)$ is the number of antitank weapons operational at time t.

In the actual state of affairs, the numbers $x(t)$ and $y(t)$ are nonnegative integers. However, it is convenient to idealize the situation and assume that $x(t)$ and $y(t)$ are continuous functions of time. For instance, if there are 500 tanks at 1400 hours (2 o'clock in the afternoon in military time) and 487 tanks at 1500 hours, then it is reasonable to assume there are 497.4 tanks at 1412 hours (if we perform a linear interpolation between the data points). That is, 2.6 of the 13 tanks lost in the 1 hr of combat were lost in the first 12 min. We also assume that $x(t)$ and $y(t)$ are differentiable functions of t, so that they are smooth functions without any corners or cusps on their graphs. These idealizations enable us to model the strength functions by differential equations.

What can we assume about how the force levels change as a result of combat? For the basic combat model, let's assume the combat *casualty rate* for the X force is proportional to the strength of the Y force. Other factors affect the change in the X force over time, such as reinforcements of the force by bringing in additional tanks (or troops if it is a troop force), or tank losses caused by mechanical or electronic failures, or losses resulting from operator errors or desertions. (Can you name some additional factors?) We will ignore all these other factors in our initial model so the rate of change in $x(t)$ is given by

$$\frac{dx}{dt} = -ay, \quad a > 0 \qquad (11.19)$$

The positive constant a in Equation (11.19) is called the *antitank weapon kill rate* or *attrition-rate coefficient* and reflects the degree to which a single antitank weapon can destroy tanks. Thus we begin our analysis with the simplest assumption; namely, that the loss rate is proportional to the number of firers. Later we assume an interaction is necessary between firer and target (the firer must locate the target before firing), and we refine our model accordingly. In the refined model the attrition-rate coefficient is proportional to the number of targets.

Under similar assumptions, the rate of change in the Y force is given by

$$\frac{dy}{dt} = -bx, \quad b > 0 \qquad (11.20)$$

Lanchester-type combat model

Here the constant b indicates the degree to which a single tank can destroy antitank weapons. The autonomous system given by (11.19) and (11.20) together with the initial strength levels $x(0) = x_0$ and $y(0) = y_0$ is called a **Lanchester-type combat model**, named after F. W. Lanchester who investigated air combat situations during World War I. Equations (11.19) and (11.20) constitute our basic model subject to the assumptions we have made. We assume throughout that $x \geq 0$ and $y \geq 0$, because negative force levels have no physical meaning.

Analysis of the model Setting the right sides of (11.19) and (11.20) equal to zero, we see that $(0, 0)$ is a rest point for the basic combat model. The trajectory directions in the phase plane are determined from the observations that $dx/dt < 0$ and $dy/dt < 0$ when $x > 0$ and $y > 0$. Moreover, if $x = 0$, then $dy/dt = 0$ (and we assume also that $dx/dt = 0$ because $x < 0$ has no physical meaning). These considerations lead to the trajectory directions depicted in Figure 11.24. Note that our assumptions imply that a trajectory terminates when it reaches either coordinate axis. (Otherwise the rest point $(0, 0)$ would be unstable.)

It is easy to find an analytical solution to the basic model. From the chain rule

$$\frac{dy}{dx} = \frac{dy/dt}{dx/dt}$$

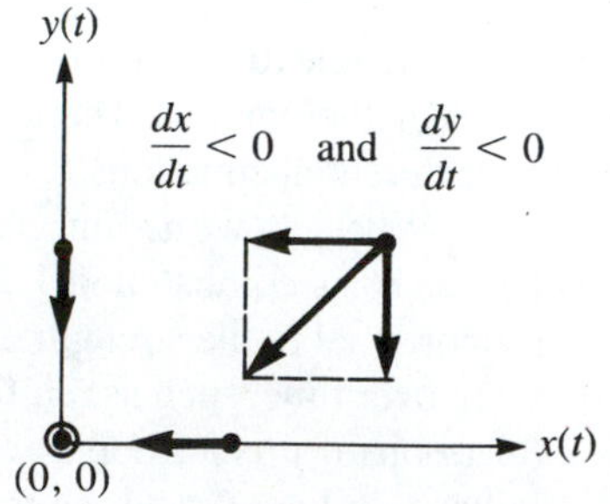

FIGURE 11.24 The rest point $(0, 0)$ for the basic Lanchester combat model

and substitution of (11.19) and (11.20), we get

$$\frac{dy}{dx} = \frac{-bx}{-ay}$$

Separating the variables in this last equation yields

$$-ay\,dy = -bx\,dx \tag{11.21}$$

Lanchester square law model

Integration of each side of (11.21) and use of the initial force levels $x(0) = x_0$ and $y(0) = y_0$ produces the **Lanchester square law model**

$$a(y^2 - y_0^2) = b(x^2 - x_0^2) \tag{11.22}$$

Setting $C = ay_0^2 - bx_0^2$, we obtain the equation

$$ay^2 - bx^2 = C \tag{11.23}$$

Typical trajectories in the phase plane represented by Equation (11.23) are depicted in Figure 11.25. The trajectories for $C \neq 0$ are hyperbolas, and when $C = 0$ the trajectory is the straight line $y = \sqrt{b/a}\,x$. When $C < 0$ the trajectory intersects the x-axis at $x = \sqrt{-C/b}$; then the X (tank) force wins because the Y force has been totally eliminated. On the other hand, if $C > 0$ the Y force wins with a final strength level of $y = \sqrt{C/a}$. These considerations are shown in Figure 11.25.

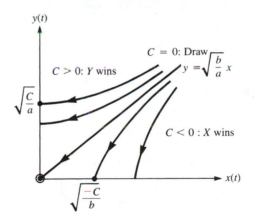

FIGURE 11.25   Trajectories of the basic Lanchester combat model; the trajectories are hyperbolas satisfying the Lanchester square law (11.22)

Let's investigate the situation in which the Y (antitank) force wins. The constant C must be positive, so

$$\left(\frac{y_0}{x_0}\right)^2 > \frac{b}{a} \tag{11.24}$$

The inequality (11.24) gives a necessary and sufficient condition that the Y force wins under the assumptions of our model (so reinforcements are not permitted, for example). From the inequality we can see that a doubling of the initial Y force level results in a fourfold advantage for that force, assuming the X force remains at the same initial level x_0 for constant a and b. This means Country X must increase b (its technology) by a factor of 4 to keep pace with the increase in the size of Country Y's force if the strength x_0 is kept at the same level. Figure 11.26 depicts a typical graph showing the force level curves $x(t)$ and $y(t)$ when the inequality (11.24) is satisfied. Observe from the figure that it is not necessary for the initial Y force level y_0 to exceed the level x_0 of the X force to ensure victory for Y. The crucial relationship is given by the inequality (11.24).

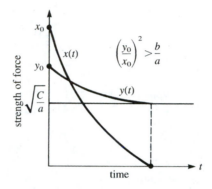

FIGURE 11.26 Force level curves $x(t)$ and $y(t)$ for the basic Lanchester combat model when $C > 0$ and the Y force wins

The model given by Equations (11.19) and (11.20) can be converted to a single second-order differential equation in the dependent variable y as follows: Differentiate (11.20) to obtain

$$\frac{d^2 y}{dt^2} = -b \frac{dx}{dt}$$

Then substitute (11.19) into this last equation to get

$$\frac{d^2 y}{dt^2} = aby$$

or

$$\frac{d^2 y}{dt^2} - aby = 0 \qquad \textbf{(11.25)}$$

In the problem set you are asked to verify that the function

$$y(t) = y_0 \cosh \sqrt{ab}\, t - x_0 \sqrt{a/b} \sinh \sqrt{ab}\, t \qquad \textbf{(11.26)}$$

satisfies the differential equation (11.25) subject to the condition that $y(0) = y_0$. Similarly, the solution for the X force level is

$$x(t) = x_0 \cosh \sqrt{ab}\, t - y_0 \sqrt{a/b} \sinh \sqrt{ab}\, t \qquad (11.27)$$

subject to the initial level $x(0) = x_0$.

Equation (11.26) can be written in the more revealing form

$$\frac{y(t)}{y_0} = \cosh \sqrt{ab}\, t - \left(\frac{x_0}{y_0}\right) \sqrt{\frac{a}{b}} \sinh \sqrt{ab}\, t \qquad (11.28)$$

Expression (11.28) says that Y's current force level divided by the initial one—that is, the *normalized force level*—depends on just two parameters: (1) a dimensionless engagement parameter $E = (x_0/y_0)\sqrt{a/b}$ and (2) a time parameter $T = \sqrt{ab}\, t$. The constant $\sqrt{ab}$ represents the intensity of battle and controls how quickly the battle is driven to its conclusion. The ratio a/b represents the relative effectiveness of individual combatants on the two opposing sides.

Refinements of the Lanchester combat model   In the basic Lanchester combat model

$$\frac{dx}{dt} = -ay \qquad (11.29)$$

$$\frac{dy}{dt} = -bx$$

it has been assumed that the single-weapon attrition rates a and b are constant over time. In many circumstances, however, the X force and the Y force are changing positions and the weapons' effectiveness depends on the distance between firer and target. Thus $a = a(t)$ and $b = b(t)$ are time dependent. In that case Model (11.29) is no longer an autonomous system and it is much more difficult to extract information from the model analytically.

In some situations the single-weapon attrition rate a depends not only on time but also on the number of targets x. This situation occurs, for example, when target detection depends on the number of targets. In this case $a = a(t, x)$ is a function of time as well as of the number of targets. The model now becomes analytically intractable, but numerical methods can be used to generate force-level results.

We can go even further enriching the basic model (11.29). For example, if $a = a(t, x/y)$, then the single-weapon attrition rate depends on time and also the force ratio x/y. In still another operational circumstance it is possible that $a = a(t, x, y)$ so the attrition rate coefficient depends on time, the number of targets, and the number of firers.

When a weapon system uses area fire and enemy targets defend a constant area, the corresponding Lanchester attrition-rate coefficients depend on the number of targets. Then the basic model becomes

$$\frac{dx}{dt} = -gxy \qquad (11.30)$$

$$\frac{dy}{dt} = -hyx$$

where g and h are positive constants, and $x(0) = x_0$ and $y(0) = y_0$ are the initial force levels. Assuming there are no operational losses and no reinforcements on either side, Model (11.30) reflects combat between two guerilla forces.

System (11.30) is easily solvable. From the chain rule we obtain

$$\frac{dy}{dx} = \frac{h}{g}$$

and separation of variables leads to the equation

$$g\,dy = h\,dx$$

Integration then yields the *linear combat law*

$$g(y - y_0) = h(x - x_0) \qquad (11.31)$$

Setting $K = gy_0 - hx_0$, we obtain the equation

$$gy - hx = K \qquad (11.32)$$

If $K > 0$, the Y force wins; if $K < 0$, the X force wins. The trajectories for Model (11.30) are depicted in Figure 11.27. Notice from (11.32) that the Y force wins

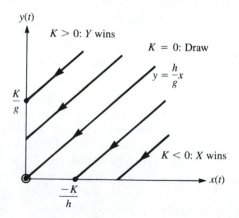

FIGURE 11.27 Trajectories when the attrition-rate coefficients are directly proportional to the number of targets

provided

$$\frac{y_0}{x_0} > \frac{h}{g} \tag{11.33}$$

In this case a doubling of the initial Y force simply doubles the advantage of that force, assuming the X force retains its same initial level x_0.

It can be shown (see Problem 3 in the 11.4 problem set) that the X force level satisfying Model (11.30) is given by

$$x(t) = \begin{cases} x_0 \left[\dfrac{hx_0 - gy_0}{hx_0 - gy_0 e^{-(hx_0 - gy_0)t}} \right] & \text{for } hx_0 \neq gy_0 \\[2em] \dfrac{x_0}{1 + hx_0 t} & \text{for } hx_0 = gy_0 \end{cases} \tag{11.34}$$

A similar result holds for the Y force level.

Let's consider a few more simple, yet natural, enrichments to the basic homogeneous-force combat model (11.29). For example, if the X force were to continuously commit more combatants to the battle at the rate $q(t) \geq 0$, then the rate of change of the X force level would become

$$\frac{dx}{dt} = -ay + q(t) \tag{11.35}$$

Here $q(t) < 0$ would mean the continuous withdrawal of forces. Now what might the function $q(t)$ look like? First, we would observe that replacements are drawn from a limited pool of combatants and weapon stocks. Then we could assume that these resources are committed to the battle at a constant rate m for as long as they last. If R denotes the total number of reserves that can commit to battle, these considerations translate to the submodel

$$q(t) = \begin{cases} m & \text{if } 0 \leq t \leq \dfrac{R}{m} \\[1.5em] 0 & \text{if } t > \dfrac{R}{m} \end{cases} \tag{11.36}$$

Another consideration is that operational losses may occur. By operational losses we mean those resulting from noncombat mishaps such as disease, desertions, and breakdown of machinery (such as a tank). Many of the factors involved in the operational loss rate, such as psychological factors inherent in desertions, would be difficult to identify precisely. However, we might assume the operational loss rate is proportional to the strength of the force. In that case, also assuming there are reinforcements, the rate of change of the X force level would be

$$\frac{dx}{dt} = -cx - ay + q(t) \tag{11.37}$$

where a and c are positive constants and $q(t) \geq 0$. Similar considerations apply to the Y force.

11.4 PROBLEMS

1. In our model (11.17) for the nuclear arms race, assume $an - bm < 0$, so the rest point lies in a quadrant other than the first one in the phase plane. Sketch the lines $dx/dt = 0$ and $dy/dt = 0$ in the phase plane and label them and their intercepts on the coordinate axes. Perform a graphical stability analysis to respond to the following:

   a. Do any potential equilibrium levels for defense spending exist? List any such points and classify them as stable or unstable.

   b. Pick at least four starting points in the first quadrant and sketch their trajectories in the phase plane.

   c. What outcome for defense spending is predicted by your graphical analysis?

   d. From the point of view of Country 1, interpret qualitatively the outcome predicted by your graphical analysis in terms of the relative values of the parameters in the Model (11.17).

2. Verify that the function (11.26) satisfies the differential equation (11.25).

3. a. Solve Equation (11.32) for y and substitute the results into the differential equation $dx/dt = -gxy$ from Model (11.30).

   b. Separate the variables in the differential equation resulting from part **a** and integrate using partial fractions to obtain the force level given by (11.34).

4. In the basic Lanchester model (11.29), assume the two forces are of equal effectiveness so $a = b$. The Y force initially has 50,000 soldiers. There are two geographically separate units of 40,000 and 30,000 soldiers that make up the X force. Use the basic model and result (11.23) to show that if the commander of the Y force fights each of the X units separately, he can bring about a draw.

5. Let X denote a guerilla force and Y denote a conventional force. The autonomous system

$$\frac{dx}{dt} = -gxy$$

$$\frac{dy}{dt} = -bx$$

   is a Lanchestrian model for conventional–guerilla combat in which there are no operational loss rates and no reinforcements.

   a. Discuss the assumptions and relationships necessary to justify the model. Does the model seem reasonable?

   b. Solve the system and obtain the *parabolic law*

$$gy^2 = 2bx + M$$

   where $M = gy_0^2 - 2bx_0$.

   c. What condition must be satisfied by the initial force levels x_0 and y_0 for the conventional Y force to win? If the Y force does win, how many survivors will there be?

6. a. Assuming that the single-weapon attrition rates a and b in (11.29) are constant over time, discuss the submodels

$$a = r_y p_y \qquad \text{and} \qquad b = r_x p_x$$

where r_y and r_x are the respective *firing rates* (shots/combatant/day) of the Y and the X forces, and p_y and p_x are the respective probabilities that a single shot kills an opponent.

   b. How would you model the attrition-rate coefficients g and h in Model (11.30)? It may be helpful to think of (11.30) as modeling guerilla–guerilla combat.

11.4 PROJECTS

Complete the requirements of the referenced UMAP module.

1. "The Richardson Arms Race Model," by Dina A. Zinnes, John V. Gillespie, and G. S. Tahim, UMAP 308. A model is constructed based on the classical assumptions of Lewis Fry Richardson. Difference equations are introduced. Students gain experience in analyzing the equilibrium, stability, and sensitivity properties of an interactive model.

11.4 Further Reading

CALLAHAN, L. G. "Do We Need a Science of War?" *Armed Forces Journal 106*, no. 36 (1969): 16–20.

ENGEL, J. H. "A Verification of Lanchester's Law." *Operations Research 2* (1954): 163–171.

LANCHESTER, F. W. "Mathematics in Warfare." In *The World of Mathematics*, vol. 4, edited by J. R. Newman, pp. 2138–2157. New York: Simon & Schuster, 1956.

McQUIE, R. "Military History and Mathematical Analysis." *Military Review 50*, no. 5 (1970): 8–17.

RICHARDSON, L. R. "Mathematics of War and Foreign Politics." In *The World of Mathematics*, vol. 4, edited by J. R. Newman, pp. 1240–1253. New York: Simon & Schuster, 1956.

U.S. General Accounting Office (GAO). "Models, Data, and War: A Critique of the Foundation Defense Analysis." PAD-80-21. Washington, DC, March 1980.

Chapter Twelve

PROBABILISTIC MODELING

INTRODUCTION

We have been developing models using proportionality and determining the constants of proportionality. But what if variation actually takes place, as in the demand of gasoline at the gasoline station. In this chapter, we will allow the constants of proportionality to vary in a *random* fashion rather than be fixed. We begin by revisiting discrete dynamical systems from Chapter 3 and introducing scenarios that have probabilistic parameters.

12.1 PROBABILISTIC MODELING WITH DISCRETE SYSTEMS

In this section we revisit the systems of difference equations studied in Section 3.4, but now we allow the coefficients of the systems to vary in a probabilistic manner.

Markov chain A special case, called a **Markov chain**, is a process in which there are the same
states finite number of **states** or outcomes that can be occupied at any given time. The states do not overlap and cover all possible outcomes. In a Markov process, the system may move from one state to another one at each time step, and there is a probability associated with this transition for each possible outcome. The sum of the probabilities for transitioning from the *present state* to the *next state* is equal to 1 for each state at each time step. A Markov process with two states is illustrated in Figure 12.1.

Example 1 Consider a rental car company with branches in Orlando and in Tampa. Each branch rents cars to Florida tourists. The company specializes in catering to travel agencies that want to arrange tourist activities in both Orlando and Tampa. Consequently, a traveler will rent a car in one city and drop off the car in either city. Travelers begin their itinerary in either city. Cars can be returned to either location, which can cause an imbalance in available cars to rent. Historical data on the percentages of cars

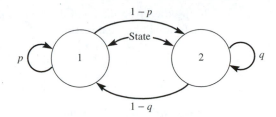

FIGURE 12.1   A Markov chain with two states; the sum of the probabilities for the transition from a present state is 1 for each state (e.g., $p + (1 - p) = 1$ for state 1)

rented and returned to these companies is collected from the previous few years as shown here.

	Next state	
	Orlando	Tampa
Present state Orlando	0.6	0.4
Tampa	0.3	0.7

transition matrix   The recorded data form a **transition matrix** and show that the probability for returning a car to Orlando that was also rented in Orlando is 0.6, whereas the probability it will be returned in Tampa is 0.4. Likewise, if a car is rented in Tampa it has a 0.3 likelihood of being returned to Orlando and a 0.7 probability it will be returned to Tampa. This represents a Markov process with two states: Orlando and Tampa. Notice that the sum of the probabilities for transitioning from a present state to the next state, which is the sum of the probabilities in each row, equals 1 because all possible outcomes are taken into account. The process is illustrated in Figure 12.2.

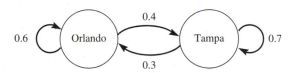

FIGURE 12.2   Two-state Markov chain for the rental car example

Model formulation   Let's define the following variables:

p_n = the percentage of cars available to rent in Orlando at the end of period n

q_n = the percentage of cars available to rent in Tampa at the end of period n

Using the given data and discrete modeling ideas from Section 3.5, we construct the following probabilistic model:

$$\left.\begin{array}{l} p_{n+1} = 0.6p_n + 0.3q_n \\ q_{n+1} = 0.4p_n + 0.7q_n \end{array}\right\} \qquad \textbf{(12.1)}$$

Model solution Assuming that all the cars are originally in Orlando, numerical solutions to System (12.1) give the long-term behavior of the percentages of cars available at each location. The sum of these long-term percentages or probabilities also equals 1. Table 12.1 and Figure 12.3 show the results tabularly and graphically.

TABLE 12.1 Iterated solution to the rental car example

n	Orlando	Tampa
0	1	0
1	0.6	0.4
2	0.48	0.52
3	0.444	0.556
4	0.4332	0.5668
5	0.42996	0.57004
6	0.428988	0.571012
7	0.428696	0.571304
8	0.428609	0.571391
9	0.428583	0.571417
10	0.428575	0.571425
11	0.428572	0.571428
12	0.428572	0.571428
13	0.428572	0.571428
14	0.428571	0.571429

Notice that

$$p_k \rightarrow 3/7 = 0.428571$$
$$q_k \rightarrow 4/7 = 0.571429$$

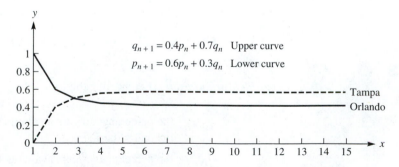

FIGURE 12.3 Graphical solution to the rental car example

Model interpretation If the two branches begin the year with a total of n cars, after 14 time periods or days approximately 57% of the cars will be in Tampa and 43% will be in Orlando. Thus starting with 100 cars in each location, about 114 cars will be based out of Tampa and 86 will be based out of Orlando in the steady state (and it would take only about 5 days to reach this state).

Example 2 Presidential voting tendencies are of interest every 4 years. Over the last decade the Independent party has emerged as a viable alternative for voters in the presidential race. Let's consider a three-party system with Republicans, Democrats, and Independents.

Problem identification *Can we find the long-term behavior of voters in a presidential election?*

Assumptions Over the last decade the trends have been to vote less strictly along party lines. We provide hypothetical historical data for voting trends in the last 10 years of statewide voting. The data are presented in this hypothetical transition matrix and shown in Figure 12.4.

		Next state		
		Republicans	Democrats	Independents
Present state	Republicans	0.75	0.05	0.20
	Democrats	0.20	0.60	0.20
	Independents	0.40	0.20	0.40

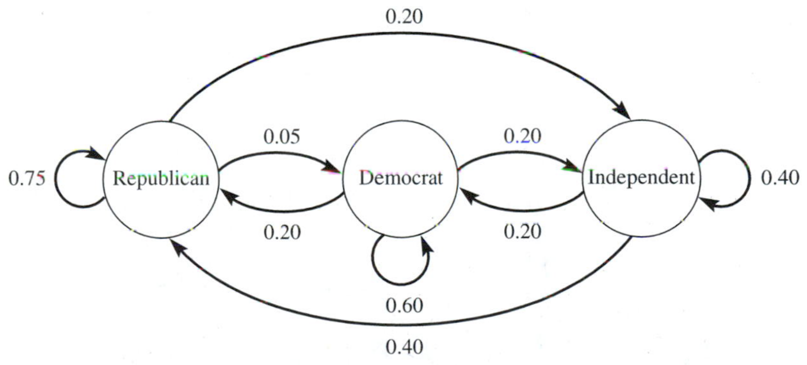

FIGURE 12.4 Three-state Markov chain for presidential voting tendencies

Model formulation Let's define the following variables:

$$R_n = \text{the percentage of voters to vote Republican in period } n$$
$$D_n = \text{the percentage of voters to vote Democratic in period } n$$
$$I_n = \text{the percentage of voters to vote Independent in period } n$$

Using the preceding data and the ideas on discrete dynamical systems from Chapter 3, we can formulate the following system of equations giving the percentage of voters who vote Republican, Democratic, or Independent at each time period:

$$R_{n+1} = 0.75R_n + 0.20D_n + 0.40I_n$$
$$D_{n+1} = 0.05R_n + 0.60D_n + 0.20I_n \qquad \textbf{(12.2)}$$
$$I_{n+1} = 0.20R_n + 0.20D_n + 0.40I_n$$

Model solution   Assume that initially $1/3$ of the voters are Republican, $1/3$ are Democrats, and $1/3$ are Independent. We then obtain the numerical results shown in Table 12.2 for the percentage of voters in each group at each period n. The table

TABLE 12.2 Iterated solution to the presidental voting problem

n	Republican	Democrat	Independent
0	0.33333	0.33333	0.33333
1	0.449996	0.283331	0.266664
2	0.500828	0.245831	0.253331
3	0.52612	0.223206	0.250664
4	0.539497	0.210362	0.250131
5	0.546747	0.203218	0.250024
6	0.550714	0.199273	0.250003
7	0.552891	0.1971	0.249999
8	0.554088	0.195904	0.249998
9	0.554746	0.195247	0.249998
10	0.555108	0.194885	0.249998
11	0.555307	0.194686	0.249998
12	0.555416	0.194576	0.249998
13	0.555476	0.194516	0.249998
14	0.55551	0.194483	0.249998

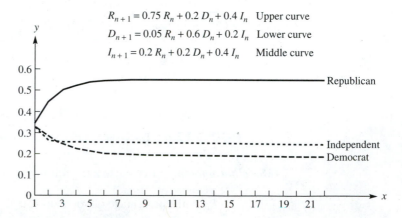

$R_{n+1} = 0.75\,R_n + 0.2\,D_n + 0.4\,I_n$ Upper curve
$D_{n+1} = 0.05\,R_n + 0.6\,D_n + 0.2\,I_n$ Lower curve
$I_{n+1} = 0.2\,R_n + 0.2\,D_n + 0.4\,I_n$ Middle curve

FIGURE 12.5   Graphical solution to the presidential voting tendencies example

shows that in the long run (and after about 10 time periods), about 56% of the voters cast their ballots for the Republican candidate, 19% vote Democratic, and 25% vote Independent. Figure 12.5 shows these results graphically.

Summary of Markov Chains

Markov chain   Let's summarize the ideas of a Markov chain. A **Markov chain** is a process consisting of a sequence of events with the following properties:

states   1. An event has a finite number of outcomes, called **states**. The process is always in one of these states.

2. At each stage or period of the process, a particular outcome can transition from its present state to any other state or remain in the same state.

3. The probability of going from one state to another in a single stage is **transition matrix**   represented by a **transition matrix** for which the entries in each row lie between 0 and 1; each row sums to 1. These probabilities depend only on the present state and not on past states.

12.1 PROBLEMS

1. Consider a model for the long-term dining behavior of the students at College USA. It is found that 25% of the students who eat at the college's Grease Dining Hall return to eat there again, whereas those who eat at Sweet Dining Hall have a 93% return rate. These are the only two dining halls available on campus, and we can assume that all students eat at one of these halls. Formulate a model to solve for the long-term percentage of students eating at each hall.

2. Consider adding a pizza delivery service as an option to the dining halls in Problem 1. Table 12.3 gives the transition percentages based on a student survey. Determine the long-term percentages eating at each place.

TABLE 12.3 Survey of dining at College USA

		Next state	
		Grease Dining Hall	Sweet Dining Hall	Pizza Delivery
Present state	Grease Dining Hall	0.25	0.25	0.50
	Sweet Dining Hall	0.10	0.30	0.60
	Pizza Delivery	0.05	0.15	0.80

3. Example 1 assumed all the cars were initially in Orlando. Try several different starting values. Is equilibrium achieved in each case? If so, what is the final distribution of cars in each case?

4. Example 2 assumed that initially the voters were equally divided among the three parties. Try several different starting values. Is equilibrium achieved in each case? If so, what is the final distribution of voters in each case?

12.1 Projects

1. Consider the pollution in two adjoining lakes where the only flow is between the lakes as illustrated in Figure 12.6. To simplify matters, assume that 100% of the water from Lake *A* comes from Lake *B*. Let a_n and b_n be the total amount of pollution in Lake *A* and Lake *B*, respectively, after *n* years. It has also been determined that we can measure the amount of pollutants given the lake in which they originated. Figure 12.7 shows a diagram reflecting the situation. Formulate and solve the model of the pollution flow as a Markov chain using dynamical systems.

FIGURE 12.6 Pollution of the Great Lakes

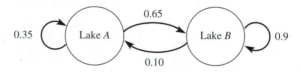

FIGURE 12.7 Two-state Markov chain for the Great Lakes pollution project

12.2 MODELING COMPONENT AND SYSTEM RELIABILITY

Do your personal computer and your automobile perform well for a reasonably long period of time? If they do, we say that these systems are *reliable*. The **reliability** of a component or system is the probability that it will not fail over a specific time period *n*. Let's define $f(t)$ to be the failure rate of an item, component, or system over time *t*, so $f(t)$ is a probability distribution. Let $F(t)$ be the cumulative distribution function corresponding to $f(t)$ as we discussed in Section 7.3. We define the reliability of the item, component, or system by

$$R(t) = 1 - F(t) \tag{12.3}$$

Thus the reliability at any time t is 1 minus the expected cumulative failure rate at time t.

Human–machine systems, whether electronic or mechanical, consist of components some of which may be combined to form subsystems. (Consider systems such as your personal computer, your stereo, or your automobile.) We want to build simple models to examine the reliability of complex systems. We consider design relationships in series, parallel, or combinations of these. Although individual item failure rates can follow a wide variety of distributions, we consider only a few elementary examples.

Example 1

series system

system reliability

A **series system** is one that performs well as long as *all* the components are fully functional. Consider a NASA space shuttle's rocket propulsion system as displayed in Figure 12.8. This is an example of a series system because failure of any one of the *independent* booster rockets will result in a failed mission. If the component reliabilities of each of the three components are given by $R_1(t) = 0.90$, $R_2(t) = 0.95$, and $R_3(t) = 0.96$, then the **system reliability** is defined to be the product

$$R_s(t) = R_1(t)R_2(t)R_3(t) = (0.90)(0.95)(0.96) = 0.8208$$

Notice that in a series relationship the reliability of the whole system is less than any single component's reliability because each component has a reliability that is less than 1.

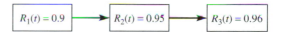

FIGURE 12.8 A NASA rocket propulsion system for a space shuttle showing the booster rockets in series (for three stages)

Example 2

parallel system

A **parallel system** is one that performs as long as a single one of its components remains operational. Consider the communication system of a NASA space shuttle, as displayed in Figure 12.9. Note that there are two separate and *independent* communication systems, either of which can operate to provide satisfactory com-

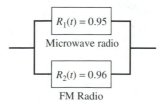

FIGURE 12.9 Two NASA space shuttle communication systems operating in parallel

munications with NASA control. If the independent component reliabilities for each communication system are $R_1(t) = 0.95$ and $R_2(t) = 0.96$, then we define the *system reliability* to be

$$R_s(t) = R_1(t) + R_2(t) - R_1(t)R_2(t) = 0.95 + 0.96 - (0.95) \cdot (0.96) = 0.998$$

Note that in parallel relationships, the system reliability is higher than any of the individual component reliabilities.

Example 3   Let's now consider a system combining series and parallel relationships as we join together the preceding two subsystems for a controlled propulsion ignition system (see Figure 12.10). We examine each subsystem. Subsystem 1 (the communication system) is a parallel relationship and we found its reliability to be 0.998. Subsystem 2 (the propulsion system) is in a series relationship and we found its system reliability to be 0.8208. These two systems are in a series relationship, so the reliability for the entire system is the product of the two subsystem reliabilities:

$$R_s(t) = R_{s_1}(t) \cdot R_{s_2}(t) = (0.998)(0.8208) = 0.8192$$

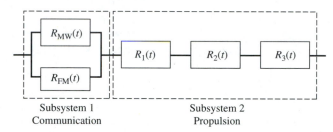

FIGURE 12.10   A NASA-controlled space shuttle propulsion ignition system

12.2 Problems

1. Consider a stereo system with CD player, AM-FM radio tuner, speakers (dual), and power amplifier components, as displayed with their reliabilities in Figure 12.11. Determine the system's reliability. What assumptions are required in your model?

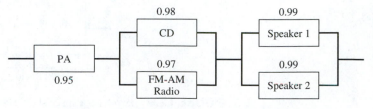

FIGURE 12.11   Reliability of stereo components

2. Consider a personal computer with each item's reliability as shown in Figure 12.12. Determine the reliability of the computer system. What assumptions are required?

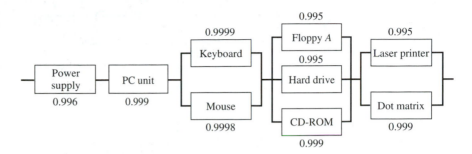

FIGURE 12.12 Personal computer reliabilities

3. Consider a more advanced stereo system with component reliabilities as displayed in Figure 12.13. Determine the system's reliability. What assumptions are required?

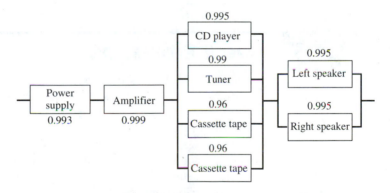

FIGURE 12.13 Advanced stereo system

12.2 PROJECTS

1. Two alternative designs are submitted for a landing module to enable the transport of astronauts to the surface of Mars. The mission is to land safely on Mars, collect several hundred pounds of samples from the planet's surface, and then return to the shuttle in orbit around Mars. The alternative designs are displayed together with their reliabilities in Figure 12.14. Which design would you recommend to NASA? What assumptions are required? Are the assumptions reasonable?

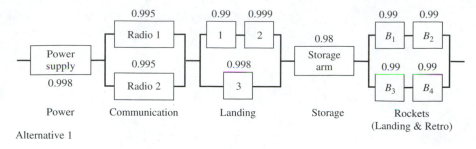

Alternative 1

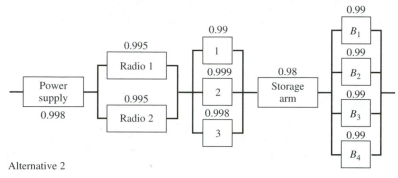

Alternative 2

FIGURE 12.14 Alternative designs for the Mars module

12.3   LINEAR REGRESSION

In Chapter 5 we discussed various criteria for fitting models to collected data. In particular, the least-squares criterion, which minimizes the sum of the squared deviations, was presented. In Chapter 8, we then showed that the formulation of minimizing the sum of the squared deviations is an optimization problem. Until now we have considered only a single observation y_i for each value of the independent variable x_i. But what if we have multiple observations? In this section we explore a statistical methodology for minimizing the sum of the squared deviations, called **linear regression**. Our objectives are

linear regression

1. To illustrate the basic linear regression model and its assumptions.
2. To define and interpret the statistic R^2.
3. To illustrate a graphical interpretation for the fit of the linear regression model by examining and interpreting the residual scatterplots.

   We introduce only basic concepts and interpretations here. More in-depth studies of the subject of linear regression are given in advanced statistics courses.

The Linear Regression Model

The basic linear regression model is defined by

$$y_i = ax_i + b \quad \text{for } i = 1, 2, \ldots, m \text{ data points} \tag{12.4}$$

In Section 5.3 we derived the *normal equations*

$$a \sum_{i=1}^{m} x_i^2 + b \sum_{i=1}^{m} x_i = \sum_{i=1}^{m} x_i y_i \tag{12.5}$$

$$a \sum_{i=1}^{m} x_i + mb = \sum_{i=1}^{m} y_i$$

and solved them to obtain the slope a and y-intercept b for the least-squares best-fitting line:

$$a = \frac{m \sum x_i y_i - \sum x_i \sum y_i}{m \sum x_i^2 - (\sum x_i)^2}, \text{ the slope} \tag{12.6}$$

and

$$b = \frac{\sum x_i^2 \sum y_i - \sum x_i y_i \sum x_i}{m \sum x_i^2 - (\sum x_i)^2}, \text{ the intercept} \tag{12.7}$$

We now add additional equations to aid in the statistical analysis of the basic model (12.4).

error sum of squares   The first of these is the **error sum of squares** given by

$$\text{SSE} = \sum_{i=1}^{m} \left[y_i - (ax_i + b) \right]^2 \tag{12.8}$$

total corrected sum of squares   which reflects variation about the regression line. The second concept is the **total corrected sum of squares** of y defined by

$$\text{SST} = \sum_{i=1}^{m} (y_i - \bar{y})^2 \tag{12.9}$$

where $\bar{y}$ is the average of the y-values for the data points (x_i, y_i), $i = 1, \ldots, m$. (The number $\bar{y}$ is also the average value of the linear regression line $y = ax + b$ over the range of data.) Equations (12.8) and (12.9) then produce the **regression sum of squares** given by the equation

regression sum of squares

$$\text{SSR} = \text{SST} - \text{SSE} \tag{12.10}$$

The quantity SSR reflects the amount of variation in the y-values explained by the linear regression line $y = ax + b$ when compared with the variation in the y-values about the line $y = \bar{y}$.

From Equation 12.10, SST is always at least as large as SSE. This fact prompts the following definition of the coefficient of determination R^2, which is a measure of fit for the regression line:

$$R^2 = 1 - \frac{\text{SSE}}{\text{SST}} \qquad (12.11)$$

The number R^2 expresses the proportion of the total variation in the y variable of the actual data (when compared with the line $y = \bar{y}$) that can be accounted for by the straight-line model whose values are given by $ax + b$ (the predicted values) and calculated in terms of the x variable. If $R^2 = 0.81$, for instance, then 81% of the total variation of the y-values (from the line $y = \bar{y}$) is accounted for by a linear relationship with the values of x. Thus the closer the value of R^2 is to 1, the better the fit of the regression line model to the actual data. If $R^2 = 1$, then the data lie perfectly along the regression line. (Note that $R^2 \leq 1$ always holds.) The following are additional properties of R^2:

1. The value of R^2 does not depend on which of the two variables is labeled x or labeled y.
2. The value of R^2 is independent of the units of x and y.

residuals   Another indicator of the reasonableness of the fit is a plot of the **residuals** versus the independent variable. Recall the residuals are the *errors* between the actual and predicted values:

$$r_i = y_i - f(x_i) = y_i - (ax_i - b) \qquad (12.12)$$

If we plot the residuals versus the independent variable, we obtain valuable information about them:

1. The residuals should be randomly distributed and be contained in a reasonably small band that is commensurate with the accuracy of the data.
2. An extremely large residual warrants further investigation of the associated data point to discover the cause of the large residual.
3. A pattern or trend in the residuals indicates that a predicable effect remains to be modeled. The nature of the pattern often provides clues on how to refine the model, if a refinement is called for. These ideas are illustrated in Figure 12.15.

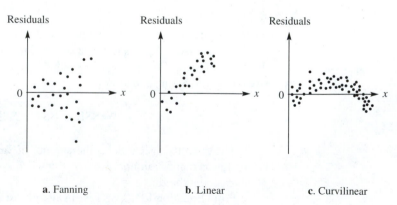

FIGURE 12.15 Possible patterns of residuals plots

Example 1 *Ponderosa Pines*

Recall the ponderosa pine data from Chapter 4, provided in Table 12.4. Figure 12.16 shows a scatterplot of the data and suggests a trend. The plot is concave upward and increasing, which could suggest a power function model (or an exponential model).

TABLE 12.4 Ponderosa pine data

Diameter (inches)	Board-feet
36	192
28	113
28	88
41	294
19	28
32	123
22	51
38	252
25	56
17	16
31	141
20	32
25	86
19	21
39	231
33	187
17	22
37	205
23	57
39	265

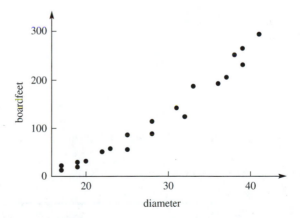

FIGURE 12.16 Scatterplot of the ponderosa pine data

Problem identification *Predict the number of board-feet as a function of the diameter of the ponderosa pine.*

Assumptions We assume that the ponderosa pines are geometrically similar and have the shape of a right circular cylinder. This allows us to use the diameter as a characteristic dimension to predict the volume. We can reasonably assume the height of a tree is proportional to its diameter.

Model formulation Geometric similarity gives the proportionality

$$V \propto d^3 \tag{12.13}$$

where d is the diameter of a tree (measured at ground level). If we further assume that the ponderosa pines have constant height (rather than assuming height is proportional to the diameter), then we obtain

$$V \propto d^2 \tag{12.14}$$

Assuming there is a constant volume associated with the underground root system would then suggest the following refinements of these proportionality models:

$$V = ad^3 + b \tag{12.15}$$

and

$$V = \alpha d^2 + \beta \tag{12.16}$$

Let's use linear regression to find the constant parameters in the four model types, and then compare the results.

Model solution The following solutions were obtained using a computer and applying linear regression on the four model types:

$$V = 0.00431d^3$$
$$V = 0.00426d^3 + 2.08$$
$$V = 0.152d^2$$
$$V = 0.194d^2 - 45.7$$

Table 12.5 displays the results of these regression models.

TABLE 12.5 Key information from regression models using the ponderosa pine data

Model	SSE	SSR	SST	R^2
$V = 0.00431d^3$	3742	458536	462278	0.9919
$V = 0.00426d^3 + 2.08$	3712	155986	159698	0.977
$V = 0.152d^2$	12895	449383	462278	0.9721
$V = 0.194d^2 - 45.7$	3910	155788	159698	0.976

Notice that the R^2 values are all quite large (close to 1), which indicates a strong linear relationship. The residuals are calculated using Equation (12.12), and their plots are shown in Figure 12.17. (Recall we are looking for a random distribution of the residuals having no apparent pattern.) Note there is an apparent trend in the errors corresponding to the model $V = 0.152d^2$. We would probably *reject* (or

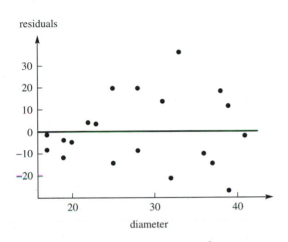

a. Residual plot for board-feet = $0.00431d^3$; the model appears adequate because no apparent trend appears in the residual plot

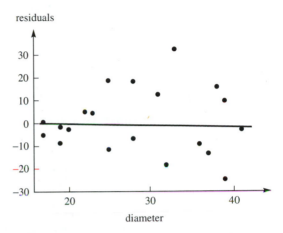

b. Residual plot for board-feet = $2.08 + 0.0042d^3$; the model appears adequate because no trend appears in the plot

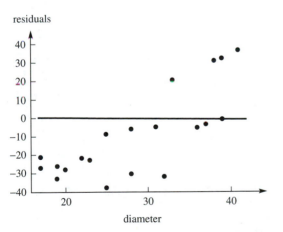

c. Residual plot for the model board-feet = $0.152d^2$; the model does not appear to be adequate because there is a linear trend in the residual plot

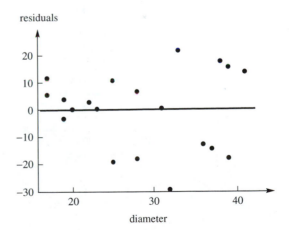

d. Residual plot for the model board-feet = $0.194d^2 - 45.7$; the model appears to be adequate because there is no apparent trend in the residual plot

FIGURE 12.17 Residual plots for various models for board-feet = f(diameter) for the ponderosa pine data

refine) this model based on this plot, while accepting the other models that appear reasonable.

Example 2   The Bass Fishing Derby Revisited

Let's revisit the bass fishing problem from Section 4.6. We collected much more data and now have 100 data points to use for fitting the model. The data are given in Table 12.6 and plotted in Figure 12.18. Based on the analysis in Section 4.5, we assume the following model type:

$$W = al^3 + b \qquad (12.17)$$

where W is the weight of the fish and l is its length.

TABLE 12.6 Data for bass fish with weight (W) in oz and length (l) in inches

W	13	13	13	13	13	14	14	15	15	15
l	12	12.25	12	12.25	14.875	12	12	12.125	12.125	12.25
W	15	15	15	16	16	16	16	16	16	16
l	12	12.5	12.25	12.675	12.5	12.75	12.75	12	12.75	12.25
W	16	16	16	16	16	17	17	17	17	17
l	12	13	12.5	12.5	12.25	12.675	12.25	12.75	12.75	13.125
W	17	17	17	18	18	18	18	18	18	18
l	15.25	12.5	13.5	12.5	13	13.125	13	13.375	16.675	13
W	18	19	19	19	19	19	19	20	20	20
l	13.375	13.25	13.25	13.5	13.5	13.5	13	13.75	13.125	13.75
W	20	20	20	20	20	20	21	21	21	22
l	13.5	13.75	13.5	13.75	17	14.5	13.75	13.5	13.25	13.765
W	22	22	23	23	23	24	24	24	24	24
l	14	14	14.25	14.375	14	14.75	13.5	13.5	14.5	14
W	24	25	25	26	26	27	27	28	28	28
l	17	14.25	14.25	14.375	14.675	16.75	14.25	14.75	13	14.75
W	28	29	29	30	35	36	40	41	41	44
l	14.875	14.5	13.125	14.5	12.5	15.75	16.25	17.375	14.5	13.25
W	45	46	47	47	48	49	53	56	62	78
l	17.25	17	18	16.5	18	17.5	18	18.375	19.25	20

The linear regression solution is

$$W = 0.00800l^3 + 0.95 \qquad (12.18)$$

and the analysis of variance is illustrated in Table 12.7.

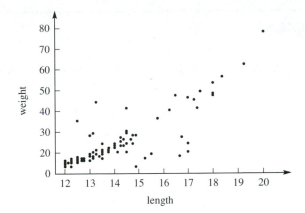

FIGURE 12.18   Scatterplot of the bass fish data

TABLE 12.7   Key information from a regression model for the bass fish example with 100 data points

Model	SSE	SSR	SST	R^2
$W = 0.00800l^3 + 0.95$	3758	10401	14159	0.735

The R^2 value for our model is 0.735, which is reasonably close to 1 considering the nature of the behavior being modeled. The residuals are found from Equation (12.12) and plotted in Figure 12.19. We see no apparent trend in the plot so there is no suggestion on how to refine the model for possible improvement.

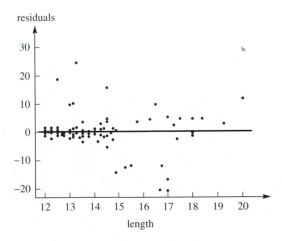

FIGURE 12.19   Residual plot for $W = 0.008l^3 + 0.95$; the model appears adequate because no trends appear in the residual plot

12.3 PROBLEMS

Use the basic linear model $y = ax + b$ to fit the following data sets. Provide the model; the values of SSE, SSR, SST, and R^2; and a residual plot.

1. Problem 2 in Section 4.1.
2. Problem 2 in Section 4.2.

12.3 PROJECTS

Use linear regression to formulate and analyze Projects 1–5 in Section 4.5.

12.3 Further Reading

MENDENHALL, William & Terry Sincich. *A Second Course in Statistics: Regression Analysis*, 5th ed. Upper Saddle River, NJ: Prentice-Hall, 1996.

Chapter Thirteen

DIMENSIONAL ANALYSIS AND SIMILITUDE

INTRODUCTION

In the process of constructing a mathematical model, we have seen that the variables influencing the behavior must be identified and classified, and then appropriate relationships need to be determined. In the case of a single dependent variable, this procedure gives rise to some unknown function

$$y = f(x_1, x_2, \ldots, x_n)$$

where the x_i measure the factors influencing the phenomenon under investigation. In some situations the discovery of the nature of the function f for the chosen factors comes about by making some reasonable assumption based on a law of nature or previous experience and construction of a mathematical model. We were able to use this methodology in constructing our model on vehicular stopping distance (see Section 4.2). On the other hand, especially for those models designed to predict some physical phenomenon, we may find it difficult or impossible to construct a solvable or tractable explicative model because of the inherent complexity of the problem. In certain instances we might conduct a series of experiments to see how the dependent variable y is related to various values of the independent variable(s). In such cases we usually prepare a figure or table and apply an appropriate curve-fitting or interpolation method that can be used to predict the value of y for suitable ranges of the independent variable(s). We used this technique in modeling the elapsed time of a tape recorder in Sections 6.2 and 6.3.

Dimensional analysis is a method for helping determine how the selected variables are related and for reducing significantly the amount of experimental data that must be collected. It is based on the premise that physical quantities have dimensions and that physical laws are not altered by changing the units measuring dimensions. Thus the phenomenon under investigation can be described by a dimensionally correct equation among the variables. A dimensional analysis provides qualitative information about the model. It is especially important when it is necessary to conduct experiments in the modeling process, because the method is helpful in testing the validity of including or neglecting a particular factor, in reducing the number of experiments to be conducted to make predictions, and in improving the usefulness of the results by providing alternatives for the parameters used to present them. Dimensional analysis has proved useful in physics and engineering for many years and now even plays a role in the study of the life sciences, economics, and operations research. Let's consider an example illustrating how dimensional analysis can be used in the modeling process to increase the efficiency of an experimental design.

Introductory Example: A Simple Pendulum

Consider the situation of a simple pendulum as suggested in Figure 13.1. Let r denote the length of the pendulum, m its mass, and θ the initial angle of displacement from the vertical. One characteristic that is vital in understanding the behavior of the **period** pendulum is the **period**, which is the time required for the pendulum bob to swing through one complete cycle and return to its original position (as at the beginning of the cycle). We represent the period of the pendulum by the dependent variable t.

Problem identification   *For a given pendulum system, determine its period.*

Assumptions   First we list the factors that influence the period. Some of these factors are the length r, the mass m, the initial angle of displacement θ, the acceleration caused by gravity g, and frictional forces such as the friction at the hinge and the drag on the pendulum. Assume initially that the hinge is frictionless, that the mass of the pendulum is concentrated at one end of the pendulum, and that the drag force is negligible. Other assumptions about the frictional forces will be examined in Section 13.3. Thus the problem is to determine or approximate the function

$$t = f(r, m, \theta, g)$$

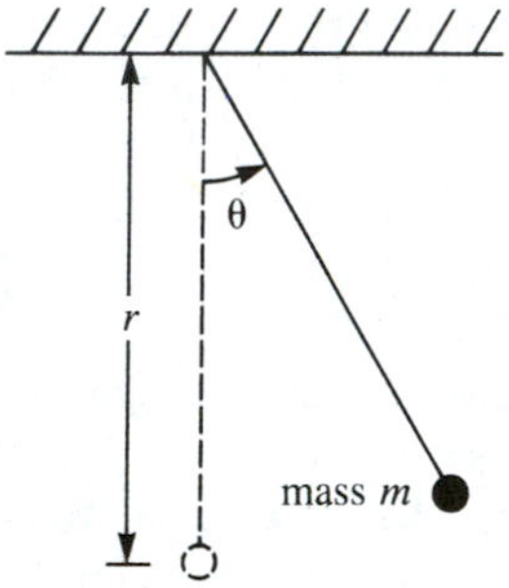

FIGURE 13.1   A simple pendulum

and test its worthiness as a predictor.

Experimental determination of the model   Because gravity is essentially constant under the assumptions, the period t is a function of the three variables length r, mass m, and initial angle of displacement θ. At this point we could systematically conduct experiments to determine how t varies with these three variables. We would want to choose enough values of the independent variables to feel confident in predicting the period t over that range. How many experiments will be necessary?

For the sake of illustration, consider a function of one independent variable $y = f(x)$ and assume that four points have been deemed necessary to predict y over a suitable domain for x. The situation is depicted in Figure 13.2. An appropriate curve-fitting or interpolation method could be used to predict y within the domain for x.

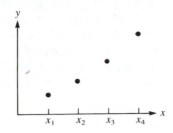

FIGURE 13.2   Four points have been deemed necessary to predict y for this function of one variable x

Next consider what happens when a second independent variable affects the situation under investigation. We then have a function

$$y = f(x, z)$$

For each data value of x in Figure 13.3, experiments must be conducted to obtain y for four values of z. Thus 16 (i.e., 4^2) experiments are required. These observations are illustrated in Figure 13.3. Likewise, a function of three variables requires 64 (i.e., 4^3) experiments. In general, 4^n experiments are required to predict y when n is the number of arguments of the function, assuming four points for the domain of each argument. Thus a procedure that reduces the number of arguments of the function f will dramatically reduce the total number of required experiments. Dimensional analysis is one such procedure.

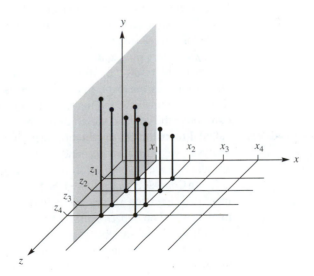

FIGURE 13.3   Sixteen points are necessary to predict y for this function of the two variables x and z

The power of dimensional analysis can also be appreciated when we examine the interpolation curves that would be determined after collecting the data represented in Figures 13.2 and 13.3. Let's assume it is decided to pass a cubic polynomial through the four points shown in Figure 13.2. That is, the four points are used to determine the four constants C_1, C_2, C_3, C_4 in the interpolating curve

$$y = C_1 x^3 + C_2 x^2 + C_3 x + C_4$$

Now consider interpolating from Figure 13.3. If for a fixed value of x, say $x = x_1$, we decide to connect our points using a cubic polynomial in z, the equation of the interpolating surface is

$$\begin{aligned}
y = {} & D_1 x^3 + D_2 x^2 + D_3 x + D_4 \\
& + (D_5 x^3 + D_6 x^2 + D_7 x + D_8)z \\
& + (D_9 x^3 + D_{10} x^2 + D_{11} x + D_{12})z^2 \\
& + (D_{13} x^3 + D_{14} x^2 + D_{15} x + D_{16})z^3
\end{aligned}$$

Note from the equation that there are 16 constants—$D_1, D_2, \ldots, D_{16}$—to determine rather than 4 as in the two-dimensional case. This procedure again illustrates the dramatic reduction in effort required when we reduce the number of arguments of the function we will finally investigate.

At this point we make the important observation that the experimental effort required depends more heavily on the number of arguments of the function to be investigated than on the true number of independent variables originally selected by the modeler. For example, consider a function of two arguments, say $y = f(x, z)$. The discussion concerning the number of experiments necessary would not be altered if x were some particular combination of several variables. That is, x could be uv/w, where u, v, and w are the variables originally selected in the model.

Consider now the following preview of dimensional analysis, which describes how it reduces our experimental effort. Beginning with a function of n variables (hence n arguments), the number of arguments is reduced (ordinarily by three) by combining the original variables into products. These resulting $(n-3)$ products are

dimensionless products   called **dimensionless products** of the original variables. After applying dimensional analysis, we still need to conduct experiments to make our predictions, but the amount of experimental effort required will have been reduced exponentially.

In Chapter 2 we discussed the trade-offs of considering additional variables for increased precision versus neglecting variables for simplification. In constructing models based on experimental data, the preceding discussion suggests that the cost of each additional variable is an exponential increase in the number of experimental trials that must be conducted. In the next two sections we present the main ideas underlying the dimensional analysis process. You may find that some of these ideas are a little more difficult than previous ones we have been investigating, but the methodology is powerful when modeling physical behavior.

13.1 DIMENSIONS AS PRODUCTS

The study of physics is based on abstract concepts such as mass, length, time, velocity, acceleration, force, energy, work, and pressure. To each such concept there is assigned a unit of measurement. A physical law such as $F = ma$ is true provided the units of measurement are consistent. Thus if mass is measured in kilograms and acceleration in meters per second squared, then the force must be measured in newtons. These units of measurement belong to the MKS (meter-kilogram-second) mass system. It would be inconsistent with the equation $F = ma$ to measure mass in slugs, acceleration in feet per second squared, and force in newtons. In this illustration, force must be measured in pounds, giving the American Engineering System of measurement. There are other systems of measurement, but all are prescribed by international standards to be consistent with the laws of physics.

The three primary physical quantities we consider in this chapter are mass, length, and time. We associate with these quantities the *dimensions M, L*, and *T*, respectively. The dimensions are symbols that reveal how the numerical value of a quantity changes when the units of measurement change in certain ways. The dimensions of other quantities follow from definitions or physical laws and are expressed in terms of M, L, and T. For example, velocity v is defined as the ratio of distance s (dimension L) traveled to time t (dimension T) of travel—that is, $v = st^{-1}$, so the dimension of velocity is LT^{-1}. Similarly, because area is fundamentally a product of two lengths, its dimension is L^2. These dimension expressions hold true regardless of the particular system of measurement, and they show, for example, that velocity may be expressed in meters per second, feet per second, miles per hour, and so forth. Likewise, area can be measured in terms of square meters, square feet, square miles, and so on.

There are other entities in physics that are more complex in the sense that they are not usually defined directly in terms of mass, length, and time alone, but their definitions include other quantities such as velocity. We associate dimensions with these more complex quantities in accordance with the algebraic operations involved in the definitions. For example, because momentum is the product of mass and velocity, its dimension is $M(LT^{-1})$ or simply MLT^{-1}.

The basic definition of a quantity may also involve dimensionless constants; these are ignored in finding dimensions. Thus the dimension of kinetic energy, which is one-half (a dimensionless constant) the product of mass with velocity squared, is $M(LT^{-1})^2$ or simply ML^2T^{-2}. As we will see in Example 2, some constants (dimensional constants), such as gravity g, do have an associated dimension, and these must be considered in a dimensional analysis.

These examples illustrate important concepts regarding dimensions of physical quantities:

1. We have based the concept of dimension on three physical quantities: mass m, length s, and time t. These quantities are measured in some appropriate system of units whose choice does not affect the assignment of dimensions. (This underlying system must be linear. A dimensional analysis will not work if the scale is logarithmic, for example.)

product

2. There are other physical quantities, such as area and velocity, that are defined as simple products involving only mass, length, or time. Here we use the term **product** to include any quotient, because we may indicate division by negative exponents.

3. There are still other, more complex physical entities, such as momentum and kinetic energy, whose definitions involve quantities other than mass, length, and time. Because the simpler quantities from (1) and (2) are themselves products, these more complex quantities can also be expressed as products involving mass, length, and time by algebraic simplification. We use the term *product* to refer to any physical quantity from item (1), (2), or (3); a product from (1) is trivial because it has only one factor.

dimension

4. To each product, there is assigned a **dimension**—that is, an expression of the form

$$M^n L^p T^q \tag{13.1}$$

where n, p, and q are real numbers that can be positive, negative, or zero.

When a basic dimension is missing from a product, the corresponding exponent is understood to be zero. Thus the dimension $M^2 L^0 T^{-1}$ can also appear as $M^2 T^{-1}$. When n, p, and q are all zero in an expression of the form (13.1), so the dimension reduces to

$$M^0 L^0 T^0 \tag{13.2}$$

the quantity, or product, is said to be *dimensionless*.

Special care must be taken in forming sums of products because just as we cannot add apples and oranges in an equation, we cannot add products that have unlike dimensions. For example, if F denotes force, m mass, and v velocity, we know immediately that the equation

$$F = mv + v^2$$

cannot be correct because mv has dimension MLT^{-1}, whereas v^2 has dimension $L^2 T^{-2}$. These dimensions are unlike, and hence the products mv and v^2 cannot be added. An equation such as this—that is, one that contains among its terms two products having unlike dimensions—is said to be *dimensionally incompatible*. Equations that involve only sums of products having the same dimension are *dimensionally compatible*.

dimensional homogeneity

The concept of dimensional compatibility is related to another important concept called **dimensional homogeneity**. In general, an equation that is true regardless of the system of units in which the variables are measured is said to be dimensionally homogeneous. For example, $t = \sqrt{2s/g}$ giving the time a body falls a distance s under gravity (neglecting air resistance) is dimensionally homogeneous (true in all systems), whereas the equation $t = \sqrt{s/16.1}$ is not dimensionally homogeneous (because it depends on a particular system). In particular, if an equation involves only sums of dimensionless products (i.e., it is a dimensionless equation), then the equation is dimensionally homogeneous. Because the products are dimensionless, the factors used for conversion from one system of units to another would simply cancel.

The application of dimensional analysis to a real-world problem is based on the assumption that the solution to the problem is given by a dimensionally homogeneous equation in terms of the appropriate variables. Thus the task is to determine the form of the desired equation by finding an appropriate dimensionless equation, and then solving for the dependent variable. To accomplish this task, we must decide which variables enter into the physical problem under investigation and determine all the dimensionless products among them. In general, there may be infinitely many such products, so they will have to be described rather than actually written out. Certain subsets of these dimensionless products are then used to construct dimensionally homogeneous equations. In Section 13.2 we investigate how the dimensionless products are used to find all dimensionally homogeneous equations. The following example illustrates how the dimensionless products can be found.

Example 1   **A Simple Pendulum Revisited**

Consider again the simple pendulum discussed in the introduction. Analyzing the dimensions of the variables for the pendulum problem, we have

Variable	m	g	t	r	θ
Dimension	M	LT^{-2}	T	L	$M^0 L^0 T^0$

Next we find all the dimensionless products among the variables. Any product of these variables must be of the form

$$m^a g^b t^c r^d \theta^e \qquad (13.3)$$

and hence must have dimension

$$(M)^a (LT^{-2})^b (T)^c (L)^d (M^0 L^0 T^0)^e$$

Therefore, a product of the form (13.3) is dimensionless if and only if

$$M^a L^{b+d} T^{c-2b} = M^0 L^0 T^0 \qquad (13.4)$$

Equating the exponents on both sides of this last equation leads to the system of linear equations

$$\left. \begin{array}{l} a \qquad\qquad\quad + 0e = 0 \\ \quad b \quad + d + 0e = 0 \\ -2b + c \quad\; + 0e = 0 \end{array} \right\} \qquad (13.5)$$

Solution of the system (13.5) gives $a = 0$, $c = 2b$, and $d = -b$, where b is arbitrary. Thus there are infinitely many solutions. Here are some general rules for selecting arbitrary variables: (1) choose the dependent variable so it will appear only once; (2) select any variable that expedites the solution of the other equations (i.e., a variable that appears in all equations); (3) choose a variable that always has a zero coefficient, if possible. Notice that the exponent e does not really appear in (13.4) (because it has a zero coefficient in each equation) so that it is arbitrary also. One dimensionless product is obtained by setting $b = 0$ and $e = 1$, yielding $a = c = d = 0$. A second, independent dimensionless product is obtained when $b = 1$ and $e = 0$, yielding $a = 0$, $c = 2$, and $d = -1$. These solutions give the dimensionless products

$$\prod_1 = m^0 g^0 t^0 r^0 \theta^1 = \theta$$

$$\prod_2 = m^0 g^1 t^2 r^{-1} \theta^0 = \frac{gt^2}{r}$$

In Section 13.2 we will learn a methodology for relating these products to carry the modeling process to completion. For now, we will develop a relationship in an intuitive manner.

Assuming $t = f(r, m, g, \theta)$, to determine more about the function f we observe that if the units in which we measure mass are made smaller by some factor (say 10), then the measure of the period t will not change because it is measured in units (T) of time. Because m is the only factor whose dimension contains M, it cannot appear in the model. Similarly, if the scale of the units (L) for measuring length is altered, it cannot change the measure of the period. For this to happen, the factors r and g must appear in the model as r/g, g/r, or more generally $(g/r)^k$. This ensures that any linear change in the way length is measured will be canceled. Finally, if we make the units (T) that measure time smaller by a factor of, say, 10 the measure of the period will directly increase by this same factor 10. Thus to have the dimension of T on the right side of the equation $t = f(r, m, g, \theta)$, g and r must appear as $\sqrt{r/g}$, because T appears to the power -2 in the dimension of g. Note that none of the preceding conditions places any restrictions on the angle θ. Thus the equation of the period should be of the form

$$t = \sqrt{\frac{r}{g}} h(\theta)$$

where the function h must be determined or approximated by experimentation.

We note two things in this analysis that are characteristic of a dimensional analysis. First, in the *MLT* system, three conditions are placed on the model so we should generally expect to reduce the number of arguments of the function relating the variables by three. In the pendulum problem we reduced the number of arguments from four to one. Second, all arguments of the function present at the end of a dimensional analysis (in this case, θ) are dimensionless products.

In the problem of the undamped pendulum we assumed that friction and drag were negligible. Before proceeding with experiments (which might be costly), we would like to know if that assumption is reasonable. Consider the model obtained so far:

$$t = \sqrt{\frac{r}{g}}\, h(\theta)$$

Keeping θ constant while allowing r to vary, form the ratio

$$\frac{t_1}{t_2} = \frac{\sqrt{r_1/g}\, h(\theta_0)}{\sqrt{r_2/g}\, h(\theta_0)} = \sqrt{\frac{r_1}{r_2}}$$

Hence the model predicts that t will vary as $\sqrt{r}$ for constant θ. Thus if we plot t versus $\sqrt{r}$ with fixed θ for some observations, we would expect to get a straight line (see Figure 13.4). If we do not obtain a reasonably straight line, we need to reexamine the assumptions. Note that our judgment here is qualitative. The final measure of the adequacy of any model is always how well it predicts or explains the phenomenon under investigation. Nevertheless, this initial test is useful for eliminating obviously bad assumptions and for choosing among competing sets of assumptions.

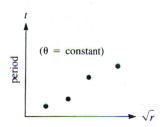

FIGURE 13.4 Testing the assumptions of the simple pendulum model by plotting the period t versus the square root of the length r for constant displacement θ

Dimensional analysis has helped construct a model $t = f(r, m, g, \theta)$ for the undamped pendulum as $t = \sqrt{r/g}\, h(\theta)$. If we are interested in predicting the behavior of the pendulum, we could isolate the effect of h by holding r constant and varying θ. This provides the ratio

$$\frac{t_1}{t_2} = \frac{\sqrt{r_0/g}\, h(\theta_1)}{\sqrt{r_0/g}\, h(\theta_2)} = \frac{h(\theta_1)}{h(\theta_2)}$$

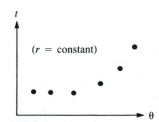

FIGURE 13.5

Determining the unknown function h

Hence a plot of t versus θ for several observations would reveal the nature of h. This plot is illustrated in Figure 13.5. We may never discover the true function h relating the variables. In such cases an empirical model might be constructed from the experimental data, as discussed in Chapter 6. When we are interested in using our model to predict t based on experimental results, it is convenient to use the equation $t\sqrt{g/r} = h(\theta)$ and to plot $t\sqrt{g/r}$ versus θ as in Figure 13.6. Then, for a given value of θ, we would determine $t\sqrt{g/r}$, multiply it by $\sqrt{r/g}$ for a specific r, and finally determine t.

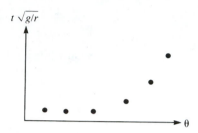

FIGURE 13.6 Presenting the results for the simple pendulum

Example 2   Wind Force on a Van[1]

Suppose you are driving a van down a highway with gusty winds. How does the speed of your vehicle affect the wind force you are experiencing?

The force F of the wind on the van is certainly affected by the speed v of the van and the surface area A of the van directly exposed to the wind's direction. Thus we might hypothesize that the force is proportional to some power of the speed times some power of the surface area; that is,

$$F = kv^a A^b \tag{13.6}$$

for some (dimensionless) constant k. Analyzing the dimensions of the variables gives

Variable	F	k	v	A
Dimension	MLT^{-2}	$M^0 L^0 T^0$	LT^{-1}	L^2

Hence, dimensionally, Equation (13.6) becomes

$$MLT^{-2} = (M^0 L^0 T^0)(LT^{-1})^a (L^2)^b$$

This last equation cannot be correct because the dimension M for mass does not enter into the right-hand side with nonzero exponent.

So consider again Equation (13.6). What is missing in our assumption concerning the wind force? Wouldn't the strength of the wind be affected by its density? After some reflection we would probably agree that density does have an effect. If we include the density ρ as a factor, our refined model becomes

$$F = kv^a A^b \rho^c \tag{13.7}$$

---

[1]This example was suggested from a preprint titled "Dimensional Analysis," by Ralph Mansfield.

Because density is mass per unit volume, the dimension of density is ML^{-3}. Therefore, dimensionally, Equation (13.7) becomes

$$MLT^{-2} = (M^0L^0T^0)(LT^{-1})^a(L^2)^b(ML^{-3})^c$$

Equating the exponents on both sides of this last equation leads to the system of linear equations

$$\left.\begin{array}{rcr} c = & 1 \\ a + 2b - 3c = & 1 \\ -a \qquad\quad = & -2 \end{array}\right\} \qquad \textbf{(13.8)}$$

Solution of the system (13.8) gives $a = 2$, $b = 1$, and $c = 1$. When substituted into (13.7), these values give the model

$$F = kv^2A\rho$$

At this point we make an important observation. When it was assumed that $F = kv^aA^b$, the constant was assumed to be dimensionless. Subsequently, our analysis revealed that for a particular medium (so ρ is constant)

$$F \propto Av^2$$

dimensional constant

giving $F = k_1Av^2$. However, k_1 does have a dimension associated with it and is called a **dimensional constant**. In particular, the dimension of k_1 is

$$\frac{MLT^{-2}}{L^2(L^2T^{-2})} = ML^{-3}$$

Dimensional constants contain important information and must be considered when performing a dimensional analysis. We consider dimensional constants again in Section 13.3 when we investigate a damped pendulum.

If we assume the density ρ is constant, our model shows that the force of the wind is proportional to the square of the speed of the van times its surface area

TABLE 13.1 Dimensions of physical entities in the *MLT* system

Mass	M	Momentum	MLT^{-1}
Length	L	Work	ML^2T^{-2}
Time	T	Density	ML^{-3}
Velocity	LT^{-1}	Viscosity	$ML^{-1}T^{-1}$
Acceleration	LT^{-2}	Pressure	$ML^{-1}T^{-2}$
Specific weight	$ML^{-2}T^{-2}$	Surface tension	MT^{-2}
Force	MLT^{-2}	Power	ML^2T^{-3}
Frequency	T^{-1}	Rotational inertia	ML^2
Angular velocity	T^{-1}	Torque	ML^2T^{-2}
Angular acceleration	T^{-2}	Entropy	ML^2T^{-2}
Angular momentum	ML^2T^{-1}	Heat	ML^2T^{-2}
Energy	ML^2T^{-2}		

directly exposed to the wind. We can test the model by collecting data and plotting the wind force F versus $v^2 A$ to see if the graph approximates a straight line through the origin. This example illustrates one of the ways dimensional analysis can be used to test our assumptions and check whether we have a faulty list of variables identifying the problem. Table 13.1 gives a summary of the dimensions of some common physical entities.

13.1 PROBLEMS

1. Determine whether the equation

$$s = s_0 + v_0 t - 0.5gt^2$$

is dimensionally compatible if s is the position (measured vertically from a fixed reference point) of a body at time t, s_0 the position at $t = 0$, v_0 the initial velocity, and g the acceleration caused by gravity.

2. Find a dimensionless product relating the torque $\tau (ML^2T^{-2})$ produced by an automobile engine, the engine's rotation rate $\psi (T^{-1})$, the volume V of air displaced by the engine, and the air density ρ.

3. The various constants of physics often have physical dimensions (dimensional constants) because their values depend on the system in which they are expressed. For example, Newton's law of gravitation asserts that the attractive force between two bodies is proportional to the product of their masses divided by the square of the distance between them, or symbolically,

$$F = \frac{Gm_1 m_2}{r^2}$$

where G is the gravitational constant. Find the dimension of G so Newton's law is dimensionally compatible.

4. Certain stars, whose light and radial velocities undergo periodic vibrations, are thought to be pulsating. It is hypothesized that the period t of pulsation depends on the star's radius r, its mass m, and the gravitational constant G. (See Problem 3 for the dimension of G.) Express t as a product of m, r, and G, so the equation

$$t = m^a r^b G^c$$

is dimensionally compatible.

5. In checking the dimensions of an equation, note that derivatives also possess dimensions. For example, the dimension of ds/dt is LT^{-1} and the dimension of d^2s/dt^2 is LT^{-2}, where s denotes distance and t denotes time. Determine

whether the equation

$$\frac{dE}{dt} = \left[mr^2 \left(\frac{d^2\theta}{dt^2} \right) mgr \sin\theta \right] \frac{d\theta}{dt}$$

for the time rate of change of total energy E in a pendulum system with damping force is dimensionally compatible.

6. For a body moving along a straight-line path, if the mass of the body is changing over time, then an equation governing its motion is given by

$$m\frac{dv}{dt} = F + u\frac{dm}{dt}$$

where m is the mass of the body, v is the velocity of the body, F is the total force acting on the body, dm is the mass joining or leaving the body in the time interval dt, and u is the velocity of dm at the moment it joins or leaves the body (relative to an observer stationed on the body). Show that the preceding equation is dimensionally compatible.

7. In humans, the hydrostatic pressure of blood contributes to the total blood pressure. The hydrostatic pressure P is a product of blood density ρ, height h of the blood column between the heart and some lower point in the body, and gravity g. Determine

$$P = k\rho^a h^b g^c$$

where k is a dimensionless constant.

8. Assume the force F opposing the fall of a raindrop through air is a product of viscosity μ, velocity v, and the diameter r of the drop. Assume that density is neglected. Find

$$F = k\mu^a v^b r^c$$

where k is a dimensionless constant.

13.1 PROJECTS

1. Complete the requirements of "Keeping Dimension Straight," by George E. Strecker, UMAP 564. This module is a basic introduction to the distinction between dimensions and units and provides the student with practice in using dimensional arguments to set up solutions to elementary problems and to recognize errors.

13.2   THE PROCESS OF DIMENSIONAL ANALYSIS

In the preceding section we learned how to determine all dimensionless products among the variables selected in the problem under investigation. Now we investigate how to use the dimensionless products to find all possible dimensionally homogeneous equations among the variables. The key result is Buckingham's Theorem, which summarizes the entire theory of dimensional analysis.

Example 1 in the preceding section shows that in general many dimensionless products may be formed from the variables of a given system. In that example we determined every dimensionless product to be of the form

$$g^b t^{2b} r^{-b} \theta^e \tag{13.9}$$

where b and e are arbitrary real numbers. Each one of these products corresponds to a solution of the homogeneous system of linear algebraic equations given by (13.5). The two products

$$\prod_1 = \theta \quad \text{and} \quad \prod_2 = \frac{gt^2}{r}$$

obtained when $b = 0, e = 1$ and $b = 1, e = 0$, respectively, are special in the sense that any of the dimensionless products (13.9) can be given as a product of some power of $\prod_1$ times some power of $\prod_2$. Thus, for instance,

$$g^3 t^6 r^{-3} \theta^{1/2} = \prod_1^{1/2} \prod_2^3$$

This observation follows from the fact that $b = 0, e = 1$ and $b = 1, e = 0$ and represent, in some sense, independent solutions of the system (13.5). Let's explore these ideas further.

Consider the following system of m linear algebraic equations in the n unknowns $x_1, x_2, \ldots, x_n$:

$$a_{11}x_1 a_{12}x_2 + \cdots + a_{1n}x_n = b_1$$
$$a_{21}x_1 a_{22}x_2 + \cdots + a_{2n}x_n = b_2$$
$$\vdots \tag{13.10}$$
$$a_{m1}x_1 a_{m2}x_2 + \cdots + a_{mn}x_n = b_m$$

coefficients

constants

The symbols a_{ij} and b_i denote real numbers for each $i = 1, 2, \ldots, m$ and $j = 1, 2, \ldots, n$. The numbers a_{ij} are called the **coefficients** of the system and b_i are referred to as the **constants**. The subscript i in the symbol a_{ij} refers to the ith equation of the system (13.10) and the subscript j refers to the jth unknown x_j to which a_{ij} belongs. Thus the subscripts serve to locate a_{ij}. It is customary to read

a_{13} as "*a*, one, three" and a_{42} as "*a*, four, two," for example, rather than "*a*, thirteen" and "*a*, forty-two."

solution

A **solution** to the system (13.10) is a sequence of numbers $s_1, s_2, \ldots, s_n$ for which $x_1 = s_1, x_2 = s_2, \ldots, x_n = s_n$ solves each equation in the system. If

homogeneous

$b_1 = b_2 = \cdots = b_m = 0$, the system (13.10) is said to be **homogeneous**. The solution $s_1 = s_2 = \cdots = s_n = 0$ always solves the homogeneous system and

trivial solution

is called the **trivial solution**. For a homogeneous system there are two solution possibilities: Either the trivial solution is the only solution or there are infinitely many solutions.

Whenever $s_1, s_2, \ldots, s_n$ and $s_1', s_2', \ldots, s_n'$ are solutions to the homogeneous system, the sequences $s_1 + s_1', s_2 + s_2', \ldots, s_n + s_n'$ and $cs_1, cs_2, \ldots, cs_n$ are also

sum

solutions for any constant c. These solutions are called the **sum** and **scalar multiple**

scalar multiple

of the original solutions, respectively. If S and S' refer to the original solutions, we use the notations $S + S'$ to refer to their sum and cS to refer to a scalar multiple of the first solution. If $S_1, S_2, \ldots, S_k$ is a collection of k solutions to the homogeneous system, then the solution

$$c_1 S_1 + c_2 S_2 + \cdots + c_k S_k$$

linear combination

is called a **linear combination** of the k solutions, where $c_1, c_2, \ldots, c_k$ are arbitrary real numbers. It is an easy exercise to show that any linear combination of solutions to the homogeneous system is still another solution to the system.

independent

A set of solutions to a homogeneous system is said to be **independent** if no solution in the set is a linear combination of the remaining solutions in the set. A

complete

set of solutions is **complete** if it is independent and every solution is expressible as a linear combination of solutions in the set. For a specific homogeneous system, we seek some complete set of solutions because all other solutions are produced from them using linear combinations. For example, the two solutions corresponding to the two choices $b = 0, e = 1$ and $b = 1, e = 0$ form a complete set of solutions to the homogeneous system (13.5).

It is not our intent to present the theory of linear algebraic equations. Such a study is appropriate for a course in linear algebra. We do point out that there is an elementary algorithm known as Gaussian elimination for producing a complete set of solutions to a given system of linear equations. Moreover, Gaussian elimination is readily implemented on computers and handheld programmable calculators. The systems of equations we will encounter in this book are simple enough to be solved by the elimination methods learned in intermediate algebra.

How does our discussion relate to dimensional analysis? Our basic goal thus far has been to find all possible dimensionless products among the variables that influence the physical phenomenon under investigation. We developed a homogeneous system of linear algebraic equations to help us determine these dimensionless products. This system of equations usually has infinitely many solutions. Each solution gives the values of the exponents that result in a dimensionless product among the variables. If we sum two solutions, we produce another solution that yields the same dimensionless product as does multiplication of the dimensionless products corresponding to the original two solutions. For example, the sum of the solutions corresponding to $b = 0$, $e = 1$ and $b = 1$, $e = 0$ for (13.5) yields the solution corresponding to $b = 1$, $e = 1$ with the corresponding dimensionless product from (13.9) given by

$$gt^2 r^{-1} \theta = \Pi_1 \Pi_2$$

The reason for this result is that the unknowns in the system of equations are the exponents in the dimensionless products, and addition of exponents algebraically corresponds to multiplication of numbers having the same base: $x^{m+n} = x^m x^n$. Moreover, multiplication of a solution by a constant produces a solution that yields the same dimensionless product as does raising the product corresponding to the original solution to the power of the constant. For example, -1 times the solution corresponding to $b = 1$, $e = 0$ yields the solution corresponding to $b = -1$, $e = 0$ with the corresponding dimensionless product

$$g^{-1} t^{-2} r = \Pi_2^{-1}$$

The reason for this last result is that algebraic multiplication of an exponent by a constant corresponds to raising a power to a power: $x^{mn} = (x^m)^n$.

In summary, addition of solutions to the homogeneous system of equations results in multiplication of their corresponding dimensionless products, and multiplication of a solution by a constant results in raising the corresponding product to the power given by that constant. Thus if S_1 and S_2 are two solutions corresponding to the dimensionless products Π_1 and Π_2, respectively, then the linear combination $aS_1 + bS_2$ corresponds to the dimensionless product

$$\Pi_1^a \Pi_2^b$$

It follows from our preceding discussion that a complete set of solutions to the homogeneous system of equations produces all possible solutions through linear combination. The dimensionless products corresponding to a complete set of solutions is therefore called a *complete set of dimensionless products*. All dimensionless products can be obtained by forming powers and products of the members of a complete set.

Next, let's investigate how these dimensionless products can be used to produce all possible dimensionally homogeneous equations among the variables. In Section 13.1 we defined an equation to be dimensionally homogeneous if it remains true regardless of the system of units in which the variables are measured. The fundamental result in dimensional analysis that provides for the construction of all dimensionally homogeneous equations from complete sets of dimensionless products is the following theorem.

THEOREM 13.1   **Buckingham's Theorem**   An equation is dimensionally homogeneous if and only if it can be put into the form

$$f\left(\prod_1, \prod_2, \ldots, \prod_n\right) = 0 \qquad (13.11)$$

where f is some function of n arguments and $\{\prod_1, \prod_2, \ldots, \prod_n\}$ is a complete set of dimensionless products.

Let's apply Buckingham's theorem to the pendulum discussed in the preceding sections. The two dimensionless products

$$\prod_1 = \theta \quad \text{and} \quad \prod_2 = \frac{gt^2}{r}$$

form a complete set for the pendulum problem. Thus, according to Buckingham's theorem, there is a function f such that

$$f\left(\theta, \frac{gt^2}{r}\right) = 0$$

Assuming we can solve this last equation for gt^2/r as a function of θ, it follows that

$$t = \sqrt{\frac{r}{g}} \, h(\theta) \qquad (13.12)$$

where h is some function of the single variable θ. Notice that this last result agrees with our intuitive formulation for the simple pendulum presented in Section 13.1. Observe that Equation (13.12) represents only a general form for the relationship among the variables m, g, t, r, and θ. However, it can be concluded from this expression that t does not depend on the mass m and is related to $r^{1/2}$ and $g^{-1/2}$ by some function of the initial angle of displacement θ. Knowing this much, we can determine the nature of the function h experimentally or approximate it, as discussed in Section 13.1.

Consider Equation (13.11) in Buckingham's theorem. For the case in which a complete set consists of a single dimensionless product, say $\prod_1$, the equation reduces to the form

$$f\left(\prod_1\right) = 0$$

In this case we assume the function f has one real root at k (to assume otherwise has little physical meaning). Hence the solution $\prod_1 = k$ is obtained.

Using Buckingham's theorem, let's reconsider Example 2, on the wind force on a van driving down a highway. Because the four variables F, v, A, and ρ were selected and all three equations in (13.8) are independent, a complete set of dimensionless products consists of a single product

$$\prod_1 = \frac{F}{v^2 A \rho}$$

Application of Buckingham's theorem gives

$$f\left(\prod_1\right) = 0$$

which implies from the preceding discussion that $\prod_1 = k$, or

$$F = k v^2 A \rho$$

where k is a dimensionless constant as before. Thus when a complete set consists of a *single dimensionless product*, as is generally the case when we begin with four variables, the application of Buckingham's theorem yields the desired relationship *up to a constant of proportionality*. Of course, the predicted proportionality must be tested to determine the adequacy of our list of variables. If the list does prove to be adequate, then the constant of proportionality can be determined by experimentation, thereby completely defining the relationship.

For the case $n = 2$, Equation (13.11) in Buckingham's theorem takes the form

$$f\left(\prod_1, \prod_2\right) = 0 \tag{13.13}$$

If we choose the products in the complete set $\{\prod_1, \prod_2\}$ so the dependent variable appears in only one of them, say $\prod_2$, we can proceed under the assumption that Equation (13.13) can be solved for that chosen product $\prod_2$ in terms of the remaining product $\prod_1$. Such a solution takes the form

$$\prod_2 = H\left(\prod_1\right)$$

This latter equation can then be solved for the dependent variable. Note that when a complete set consists of more than one dimensionless product, the application of Buckingham's theorem determines the desired relationship *up to an arbitrary function*. After verifying the adequacy of the list of variables, we may be lucky enough to recognize the underlying functional relationship. In general we can expect to construct an empirical model, although the task has been eased considerably.

For the general case of n dimensionless products in the complete set for Buckingham's theorem, we again choose the products in the complete set $\{\prod_1, \prod_2, \ldots, \prod_n\}$ so the dependent variable appears in only one of them, say $\prod_n$ for definiteness. Assuming we can solve Equation (13.11) for that product $\prod_n$ in terms of the remaining ones, we have the form

$$\prod_n = H\left(\prod_1, \prod_2, \ldots, \prod_{n-1}\right)$$

We then solve this last equation for the dependent variable.

Summary of Dimensional Analysis Methodology

We now summarize the steps in the dimensional analysis process:

Step 1. Decide which variables enter the problem under investigation.
Step 2. Determine a complete set of dimensionless products $\{\prod_1, \prod_2, \ldots, \prod_n\}$ among the variables. Make sure the dependent variable of the problem appears in only one of the dimensionless products.
Step 3. Check to ensure that the products found in the previous step are dimensionless and independent. Otherwise we have an algebra error.
Step 4. Apply Buckingham's theorem to produce all possible dimensionally homogeneous equations among the variables. This procedure yields an equation of the form (13.11).
Step 5. Solve the equation in Step 4 for the dependent variable.
Step 6. Test to ensure that the assumptions made in Step 1 are reasonable. Otherwise, the list of variables is faulty.
Step 7. Conduct the necessary experiments and present the results in a useful format.

Let's illustrate the first five steps of the preceding procedure.

Example 1   *Terminal Velocity of a Raindrop*

Consider the problem of determining the terminal velocity v of a raindrop falling from a motionless cloud. We looked at this problem from a simplistic point of view in Chapter 4, but let's take another look using dimensional analysis.

What are the variables influencing the behavior of the raindrop? Certainly the terminal velocity will depend on the size of the raindrop given by, say, its radius r. The density ρ of the air and the viscosity μ of the air will also affect the behavior. (Viscosity measures resistance to motion—a sort of internal molecular friction. In gases this resistance is caused by collisions between fast-moving molecules.) The acceleration caused by gravity g is another variable to be considered. Although the surface tension of the raindrop is a factor that does influence the behavior of the fall, we will ignore this factor. If necessary, surface tension can be taken into account in a later, refined model. These considerations give the following table relating the selected variables to their dimensions:

Variable	v	r	g	ρ	μ
Dimension	LT^{-1}	L	LT^{-2}	ML^{-3}	$ML^{-1}T^{-1}$

Next we find all the dimensionless products among the variables. Any such product must be of the form

$$v^a r^b g^c \rho^d \mu^e \tag{13.14}$$

and hence must have dimension

$$(LT^{-1})^a (L)^b (LT^{-2})^c (ML^{-3})^d (ML^{-1}T^{-1})^e$$

Therefore, a product of the form (13.14) is dimensionless if and only if the following system of equations in the exponents is satisfied:

$$\left. \begin{array}{r} d + e = 0 \\ a + b + c - 3d - e = 0 \\ -a \quad - 2c \quad - e = 0 \end{array} \right\} \tag{13.15}$$

Solution of the system (13.15) gives $b = (3/2)d - (1/2)a$, $c = (1/2)d - (1/2)a$, and $e = -d$, where a and d are arbitrary. One dimensionless product $\prod_1$ is obtained by setting $a = 1, d = 0$; another independent dimensionless product $\prod_2$ is obtained when $a = 0, d = 1$. These solutions give

$$\prod_1 = vr^{-1/2}g^{-1/2} \quad \text{and} \quad \prod_2 = r^{3/2}g^{1/2}\rho\mu^{-1}$$

Next we check the results to ensure that the products are indeed dimensionless:

$$\frac{LT^{-1}}{L^{1/2}(LT^{-2})^{1/2}} = M^0 L^0 T^0$$

and

$$\frac{L^{3/2}(LT^{-2})^{1/2}(ML^{-3})}{ML^{-1}T^{-1}} = M^0 L^0 T^0$$

Thus according to Buckingham's theorem, there is a function f such that

$$f\left(vr^{-1/2}g^{-1/2}, \frac{r^{3/2}g^{1/2}\rho}{\mu} \right) = 0$$

Assuming we can solve this last equation for $vr^{-1/2}g^{-1/2}$ as a function of the second product $\prod_2$, it follows that

$$v = \sqrt{rg}\, h\left(\frac{r^{3/2}g^{1/2}\rho}{\mu}\right)$$

where h is some function of the single product $\prod_2$.

The preceding example illustrates a characteristic feature of dimensional analysis. Normally the modeler studying a given physical system has an intuitive idea of the variables involved and has a working knowledge of general principles and laws (such as Newton's Second Law) but lacks the precise laws governing the interaction of the variables. Of course, the modeler can always experiment with each independent variable separately, holding the others constant and measuring the effect on the system. Often, however, the efficiency of the experimental work can be improved through an application of dimensional analysis. Although we did not illustrate Steps 6 and 7 of the dimensional analysis process for the preceding example, we will illustrate these steps in Section 13.3.

We now make some observations concerning the dimensional analysis process. Suppose n variables have been identified in the physical problem under investigation. When determining a complete set of dimensionless products, we form a system of three linear algebraic equations by equating the exponents for M, L, and T to zero. That is, a system of three equations in n unknowns (the exponents) is obtained. If the three equations are independent, we can solve the system for three of the unknowns in terms of the remaining $n - 3$ unknowns (declared to be arbitrary). In this case, we find $n - 3$ independent dimensionless products that make up the complete set we seek. For instance, in the preceding example there are five unknowns a, b, c, d, e, and we determined three of them ($b, c,$ and e) in terms of the remaining $(5 - 3)$ two arbitrary ones (a and d). Thus we obtained a complete set of two dimensionless products. When choosing the $n - 3$ dimensionless products, we must be sure the dependent variable appears in only one of them. We can then solve Equation (13.11) guaranteed by Buckingham's theorem for the dependent variable, at least under suitable assumptions on the function f in that equation. (The full story telling when such a solution is possible is the content of an important result in advanced calculus known as the Implicit Function Theorem.)

We acknowledge that we have been rather sketchy in our presentation for solving the system of linear algebraic equations that results in the process of determining all dimensionless products. We simply assumed that you remember how to solve simple linear systems by the method of elimination of variables. We conclude this section with another example.

Example 2   *Automobile Gas Mileage Revisited*

Consider again the automobile gasoline mileage problem presented in Section 4.3. One of our submodels in that problem was for the force of propulsion F_p. The variables we identified that affect the propulsion force are C_r, the amount of fuel burned per unit time, the amount K of energy contained in each gallon of gasoline, and the speed v. Let's perform a dimensional analysis. The following table relates the various variables to their dimensions:

Variable	F_p	C_r	K	v
Dimension	MLT^{-2}	L^3T^{-1}	$ML^{-1}T^{-2}$	LT^{-1}

Thus the product

$$F_p^a C_r^b K^c v^d \tag{13.16}$$

must have the dimension

$$(MLT^{-2})^a (L^3T^{-1})^b (ML^{-1}T^{-2})^c (LT^{-1})^d$$

The requirement for a dimensionless product leads to the system

$$\left.\begin{array}{r} a \quad + c \quad = 0 \\ a + 3b - \; c + d = 0 \\ -2a - \; b - 2c - d = 0 \end{array}\right\} \tag{13.17}$$

Solution of the system (13.17) gives $b = -a$, $c = -a$, and $d = a$, where a is arbitrary. Choosing $a = 1$, we obtain the dimensionless product

$$\prod_1 = F_p C_r^{-1} K^{-1} v$$

From Buckingham's theorem there is a function f with $f(\prod_1) = 0$, so $\prod_1$ equals a constant. Therefore,

$$F_p \propto \frac{C_r K}{v}$$

in agreement with the conclusion reached in Chapter 4.

13.2 PROBLEMS

1. Predict the time of revolution for two bodies of mass m_1 and m_2 in empty space revolving about each other under their mutual gravitational attraction.

2. A projectile is fired with initial velocity v at an angle θ with the horizon. Predict the range R.

3. Consider an object falling under the influence of gravity. Assume air resistance is negligible. Using dimensional analysis, find the speed v of the object after it

has fallen a distance *s*. Let $v = f(m, g, s)$, where *m* is the mass of the object and *g* is the acceleration caused by gravity. Does your answer agree with your knowledge of the physical situation? Explain.

4. Using dimensional analysis, find a proportionality relationship for the centrifugal force *F* of a particle in terms of its mass *m*, its velocity *v*, and radius *r* of the curvature of its path.

5. We would like to know the nature of the drag forces experienced by a sphere as it passes through a fluid. Assume the sphere has a low speed. Therefore, the drag force is highly dependent on the viscosity of the fluid. Neglect the fluid density. Use the dimensional analysis process to develop a model for drag force *F* as a function of the radius *r* and velocity *v* of the sphere and the viscosity μ of the fluid.

6. The volume flow rate *q* for laminar flow in a pipe depends on the pipe radius *r*, the viscosity μ of the fluid, and the pressure drop per unit length dp/dz. Develop a model for the flow rate *q* as a function of *r*, μ, and dp/dz.

7. In fluid mechanics, the Reynolds number is a dimensionless number involving the fluid velocity *v*, density ρ, viscosity μ, and a characteristic length *r*. Use dimensional analysis to find the Reynolds number.

8. The power *P* delivered to a pump depends on the specific weight *w* of the fluid pumped, the height *h* to which the fluid is pumped, and the fluid flow rate *q* in cubic feet per second. Use dimensional analysis to determine an equation for power.

9. Find the volume flow rate dV/dt of blood flowing in an artery as a function of the pressure drop per unit length of artery *P*, the radius *r* of the artery, the blood density ρ, and the blood viscosity μ.

10. The speed of sound in a gas depends on the pressure and the density. Use dimensional analysis to find the speed of sound in terms of pressure and density.

11. The lift force *F* on a missile depends on its length *r*, velocity *v*, diameter δ, and initial angle θ with the horizon; it also depends on the density ρ, viscosity μ, gravity *g*, and speed of sound *s* in the air. Show that

$$F = \rho v^2 r^2 h \left(\frac{\delta}{r}, \theta, \frac{\mu}{\rho v r}, \frac{s}{v}, \frac{rg}{v^2} \right)$$

12. The height *h* that a fluid will rise in a capillary tube decreases as the diameter *D* of the tube increases. Use dimensional analysis to determine how *h* varies with *D* and the specific weight *w* and surface tension σ of the liquid.

13.3   A DAMPED PENDULUM

In Section 13.1 we investigated the pendulum problem under the assumptions that the hinge is frictionless, the mass is concentrated at one end of the pendulum, and the drag force is negligible. Suppose we are not satisfied with the results predicted by the constructed model. We can refine the model by incorporating drag forces. If F represents the total drag force, the problem now is to determine the function

$$t = f(r, m, g, \theta, F)$$

Let's consider a submodel for the drag force. As we have seen in previous examples, the modeler is usually faced with a trade-off between simplicity and accuracy. For the pendulum it might seem reasonable to expect the drag force to be proportional to some positive power of the velocity. To keep our model simple, we assume F is proportional to either v or v^2, as depicted in Figure 13.7.

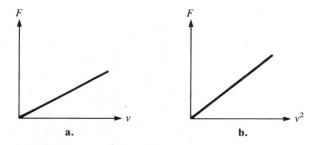

FIGURE 13.7 Possible submodels for the drag force

Now we can experiment to determine directly the nature of the drag force. However, we will first perform a dimensional analysis because we expect it to reduce our experimental effort. Assume F is proportional to v so $F = kv$. For convenience we choose to work with the dimensional constant $k = F/v$, which has dimension MLT^{-2}/LT^{-1}, or simply MT^{-1}. Notice that the dimensional constant captures the assumption about the drag force. Thus we apply dimensional analysis to the model

$$t = f(r, m, g, \theta, k)$$

An analysis of the dimensions of the variables gives

Variable	t	r	m	g	θ	k
Dimension	T	L	M	LT^{-2}	$M^0 L^0 T^0$	MT^{-1}

Any product of the variables must be of the form

$$t^a r^b m^c g^d \theta^e k^f \tag{13.18}$$

and hence must have dimension

$$(T)^a (L)^b (M)^c (LT^{-2})^d (M^0 L^0 T^0)^e (MT^{-1})^f$$

Therefore, a product of the form (13.18) is dimensionless if and only if

$$
\left.\begin{array}{rrrr}
 & c & & + f = 0 \\
b & + d & & = 0 \\
a & & - 2d & - f = 0
\end{array}\right\} \qquad \textbf{(13.19)}
$$

The equations in the system (13.19) are independent, so we know we can solve for three of the variables in terms of the remaining $(6 - 3)$ three variables. We would like to choose the solutions in such a way that t appears in only one of the dimensionless products. Thus we choose a, e, and f as the arbitrary variables with

$$
c = -f, \; b = -d = \frac{-a}{2} + \frac{f}{2}, \; d = \frac{a}{2} - \frac{f}{2}
$$

Setting $a = 1$, $e = 0$, and $f = 0$, we obtain $c = 0$, $b = -1/2$, and $d = 1/2$ with the corresponding dimensionless product $t\sqrt{g/r}$. Similarly, choosing $a = 0$, $e = 1$, and $f = 0$, we get $c = 0$, $b = 0$, and $d = 0$, corresponding to the dimensionless product θ. Finally, choosing $a = 0$, $e = 0$, and $f = 1$, we obtain $c = -1$, $b = 1/2$, and $d = -1/2$, corresponding to the dimensionless product $k\sqrt{r}/m\sqrt{g}$. Notice that t appears in only the first of these products. From Buckingham's theorem there is a function h with

$$
h\left(t\sqrt{g/r}, \theta, \frac{k\sqrt{r}}{m\sqrt{g}}\right) = 0
$$

Assuming we can solve this last equation for $t\sqrt{g/r}$, we obtain

$$
t = \sqrt{r/g}\, H\left(\theta, \frac{k\sqrt{r}}{m\sqrt{g}}\right)
$$

for some function H of two arguments.

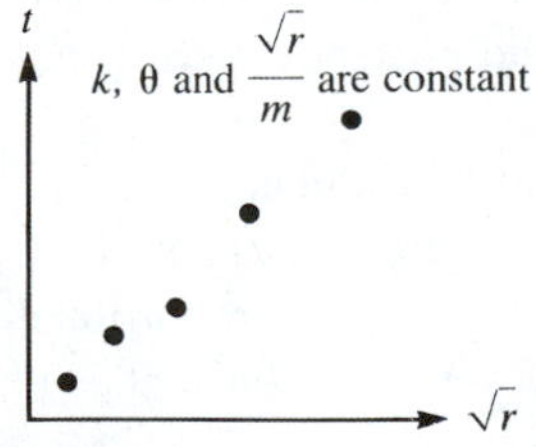

FIGURE 13.8 A plot of t versus $\sqrt{r}$ keeping the variables k, θ, and $\sqrt{r}/m$ constant

Testing the Model (Step 6)

Given $t = \sqrt{r/g}\, H(\theta, k\sqrt{r}/m\sqrt{g})$, our model predicts that $t_1/t_2 = \sqrt{r_1/r_2}$ if the parameters of the function H (namely, θ and $k\sqrt{r}/m\sqrt{g}$) could be held constant. There is no difficulty keeping θ and k constant. Varying r while simultaneously keeping $k\sqrt{r}/m\sqrt{g}$ constant is more complicated. Because g is constant, we could try varying r and m in such a manner that $\sqrt{r}/m$ remains constant. This might be done using a pendulum with a hollow mass to vary m without altering the drag characteristics. Under these conditions we would expect the plot in Figure 13.8.

Presenting the Results (Step 7)

As was suggested in predicting the period of the undamped pendulum, we can plot $t\sqrt{g/r} = H(\theta, k\sqrt{r}/m\sqrt{g})$. However, because H is a function of two arguments, this would yield a three-dimensional figure that is not easy to use. An alternative technique is to plot $t\sqrt{g/r}$ versus $k\sqrt{r}/m\sqrt{g}$ for various values of θ. This is illustrated in Figure 13.9. To be safe in predicting t over the range of interest for representative values of θ, it would be necessary to conduct sufficient experiments at various values of $k\sqrt{r}/m\sqrt{g}$. Note that once data are collected, various empirical models could be constructed using an appropriate interpolating scheme for each value of θ.

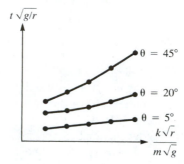

FIGURE 13.9 Presenting the results

Choosing Among Competing Models

Because dimensional analysis involves only algebra, we are tempted to develop several models under different assumptions before proceeding with, perhaps quite costly, experimentation. In the case of the pendulum, under different assumptions we can develop the following three models (see Problem 1 in the 13.3 problem set):

A. $t = \sqrt{r/g}\, h(\theta)$ no drag forces

B. $t = \sqrt{r/g}\, h\left(\theta, \dfrac{k\sqrt{r}}{m\sqrt{g}}\right)$ drag forces proportional to v: $F = kv$

C. $t = \sqrt{r/g}\, h\left(\theta, \dfrac{k_1 r}{m}\right)$ drag forces proportional to v^2: $F = k_1 v^2$

Because all the preceding models are approximations, it is reasonable to ask which, if any, is suitable in a particular situation. We now describe the experimentation necessary to distinguish among these models and present experimental results.

Model A predicts that when the angle of displacement θ is held constant, the period t is proportional to $\sqrt{r}$. Model B predicts that when θ and $\sqrt{r}/m$ are both held constant, while maintaining the same drag characteristics k, t is proportional to $\sqrt{r}$. Finally, Model C predicts that if θ, r/m, and k_1 are held constant, t is proportional to $\sqrt{r}$.

The following discussion describes our experimental results for the pendulum.[2] Various types of balls were suspended from a string in a way to minimize

[2]Data collected by Michael Jaye.

the friction at the hinge. The balls included tennis balls as well as different types and sizes of plastic balls. A hole was made in each ball to permit variations in the mass without appreciably altering the aerodynamic characteristics of the ball nor the location of the center of mass. The models were then compared with one another. In the case of the tennis ball, Model A proved superior. The period was independent of the mass. A plot of t versus $\sqrt{r}$ for constant θ is shown in Figure 13.10.

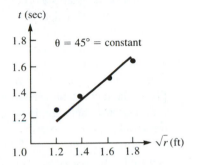

FIGURE 13.10   Model A for a tennis ball

Having decided that $t = \sqrt{r/g}\, h(\theta)$ is the best of the models for the tennis ball, we isolated the effect of θ by holding r constant to gain insight into the nature of the function h. A plot of t versus θ for constant r is shown in Figure 13.11.

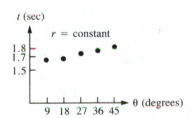

FIGURE 13.11   Isolating the effect of θ

Note from Figure 13.11 that for small angles of initial displacement θ, the period is virtually independent of θ. However, the displacement effect becomes more noticeable as θ is increased. Thus for small angles we might hypothesize that $t = c\sqrt{r/g}$ for some constant c. If we plot t versus $\sqrt{r}$ for small angles, the slope of the resulting straight line should be constant.

For larger angles, the experiment demonstrates that the effect of θ needs to be considered. In such cases, we may desire to estimate the period for various angles. For example, if $\theta = 45°$ and we know a particular value of $\sqrt{r}$, we can estimate t from Figure 13.10. Although not shown, plots for several different angles can be graphed in the same figure.

Dimensional Analysis in the Model-Building Process

Let's summarize how dimensional analysis provides assistance in the model-building process. In the determination of a model, we must first decide which factors to neglect and which to include. A dimensional analysis provides additional information on how the included factors are related. Moreover, in large problems, we often determine one or more submodels before attacking the larger problem. For example, in the pendulum problem we had to develop a submodel for drag forces. A dimensional analysis is helpful in choosing among the various submodels.

A dimensional analysis is also useful for obtaining an initial test of the assumptions in the model. For example, suppose we hypothesize that the dependent variable y is some function of five variables, $y = f(x_1, x_2, x_3, x_4, x_5)$. A dimensional analysis in the *MLT* system in general yields $\prod_1 = h(\prod_2, \prod_3)$, where each $\prod_i$ is a dimensionless product. The model predicts that $\prod_1$ will remain constant if $\prod_2$ and $\prod_3$ are held constant, even though the components of $\prod_2$ and $\prod_3$ may themselves vary. Because there are, in general, an infinite number of ways of choosing the $\prod_i$, we should choose those that can be controlled in laboratory experiments. Having determined that $\prod_1 = h(\prod_2, \prod_3)$, we can isolate the effect of $\prod_2$ by holding $\prod_3$ constant and vice versa. This can help explain the functional relationship among the variables. For instance, we saw in our example that the period of the pendulum did not depend on the initial displacement for small displacements.

Perhaps the greatest contribution of dimensional analysis is that it reduces the number of experiments required to predict the behavior. If we wanted to conduct experiments to predict values of y for the assumed relationship $y = f(x_1, x_2, x_3, x_4, x_5)$, and it was decided that five data points would be necessary over the range of each variable, then 5^5 or 3125 experiments would be necessary. Because a two-dimensional chart is required to interpolate conveniently, y might be plotted against x_1 for five values of x_1, holding x_2, x_3, x_4, x_5 constant. Because x_2, x_3, x_4, and x_5 must vary as well, 5^4 or 625 charts would be necessary. However, after a dimensional analysis yields $\prod_1 = h(\prod_2, \prod_3)$, only 25 data points would be required. Moreover, $\prod_1$ can be plotted versus $\prod_2$, for various values of $\prod_3$, on a single chart. Ultimately an empirical model is usually constructed for purposes of interpolation, but the task is far easier after applying a dimensional analysis.

Finally, dimensional analysis is helpful in presenting the results. It is usually best to present experimental results using those $\prod_i$ that are classical within the field of study. For instance, in the field of fluid mechanics there are eight factors that might be significant in a particular situation. These factors are velocity v, length r, mass density ρ, viscosity μ, acceleration of gravity g, speed of sound c, surface tension σ, and pressure p. Thus a dimensional analysis could require as many as five independent dimensionless products. The five generally used are the Reynolds number, Froude number, Mach number, Weber number, and pressure coefficient. These numbers are defined as follows (and discussed in Section 13.5):

$$\text{Reynolds number} \qquad \frac{vr\rho}{\mu}$$

$$\text{Froude number} \qquad \frac{v^2}{rg}$$

Mach number $\dfrac{v}{c}$

Weber number $\dfrac{\rho v^2 r}{\sigma}$

Pressure coefficient $\dfrac{p}{\rho v^2}$

Thus the application of dimensional analysis becomes quite easy. Depending on which of the eight variables are considered in a particular problem, the following steps are performed:

1. Choose an appropriate subset from the preceding five dimensionless products.
2. Apply Buckingham's theorem.
3. Test the reasonableness of the choice of variables.
4. Conduct the necessary experiments and present the results in a useful format.

We illustrate an application of these steps to a fluid mechanics problem in Section 13.5.

13.3 PROBLEMS

1. For the damped pendulum

    a. Assume F is proportional to v^2 and use dimensional analysis to show that $t = \sqrt{r/g}\, h(\theta, rk_1/m)$.

    b. Assume F is proportional to v^2 and describe an experiment to test the model $t = \sqrt{r/g}\, h(\theta, rk_1/m)$.

2. Under appropriate conditions, all three models for the pendulum imply that t is proportional to $\sqrt{r}$. Explain how the conditions distinguish between the three models by considering how m must vary in each case.

3. Use a model employing a differential equation to predict the period of a simple frictionless pendulum for small initial angles of displacement. (*Hint:* Let $\sin\theta = \theta$.) Under these conditions, what should be the constant of proportionality? Compare your results with those predicted by Model A in the text.[3]

13.4   EXAMPLES ILLUSTRATING DIMENSIONAL ANALYSIS

In this section we present several examples that illustrate the modeling process and dimensional analysis.

---

[3]For students who have studied differential equations.

Example 1   *Explosion Analysis*[4]

In excavation and mining operations it is important to be able to predict the size of a crater resulting from a given explosive such as TNT in some particular soil medium. Direct experimentation is often impossible or too costly. Thus it is desirable to use small laboratory or field tests, and then scale these up in some manner to predict the results for explosions far greater in magnitude.

We may wonder how the modeler determines which variables to include in the initial list. Experience is necessary to determine intelligently which variables can be neglected. Even with experience, however, the task is usually difficult in practice, as this example will illustrate. It also illustrates that the modeler often must change the list of variables to get usable results.

Problem identification   *Predict the crater volume V produced by a spherical explosive located at some depth d in a particular soil medium.*

Assumptions and model formulation   Initially let's assume the craters are geometrically similar (see Chapter 4) and the crater size depends on three variables: the radius r of the crater, the density ρ of the soil, and the mass W of the explosive. These three variables are composed of only two primary dimensions, length L and mass M, and a dimensional analysis results in only one dimensionless product (see Problem 1a in the 13.4 problem set):

$$\prod_r = r \left(\frac{\rho}{W} \right)^{1/3}$$

According to Buckingham's theorem $\prod_r$ must equal a constant. Thus the crater dimensions of radius or depth vary with the cube root of the mass of the explosive. Because the crater volume is proportional to r^3, it follows that the volume of the crater is proportional to the mass of the explosive for constant soil density. Symbolically, we have

$$V \propto \frac{W}{\rho} \tag{13.20}$$

Experiments have shown that the proportionality (13.20) is satisfactory for small explosions (less than 300 lb of TNT) at zero depth in soils, such as moist alluvium, that have good cohesion. For larger explosions, however, the rule proves unsatisfactory. Other experiments suggest that gravity plays a key role in the explosion process, and because we want to consider extraterrestrial craters as well, we need to incorporate gravity as a variable.

If gravity is taken into account, we assume crater size to be dependent on four variables: crater radius r, density of the soil ρ, gravity g, and charge energy E. Here

---

[4]This example is adapted with permission from R. M. Schmidt, "A Centrifuge Cratering Experiment: Development of a Gravity-scaled Yield Parameter," in *Impact and Explosion Cratering*, ed. D. J. Roddy, R. O. Pepin, and R. B. Merrill (New York: Pergamon Press, 1977), pp. 1261–1276.

the charge energy is the mass W of the explosive times its specific energy. Applying a dimensional analysis to these four variables again leads to a single dimensionless product (see Problem 1b in the 13.4 problem set):

$$\prod_{rg} = r\left(\frac{\rho g}{E}\right)^{1/4}$$

Thus $\prod_{rg}$ equals a constant and the linear crater dimensions (radius or depth of the crater) vary with the one-fourth root of the energy (or mass) of the explosive for a constant soil density. This leads to the following proportionality known as quarter-root scaling and is a special case of *gravity scaling*:

$$V \propto \left(\frac{E}{\rho g}\right)^{3/4} \tag{13.21}$$

Experimental evidence indicates that gravity scaling holds for large explosions (greater than 100 tons of TNT) where the stresses in the cratering process are much larger than the material strengths of the soil. The proportionality (13.21) predicts that crater volume decreases with increased gravity. The effect of gravity on crater formation is of interest in the study of extraterrestrial craters. Gravitational effects can be tested experimentally using a centrifuge to increase gravitational accelerations.[5]

A question of interest to explosion analysts is whether the material properties of the soil do become less important with increased charge size as well as with increased gravity. Let's consider the case in which the soil medium is characterized only by its density ρ. Thus the crater volume V depends on the explosive, soil density ρ, gravity g, and depth of burial d of the charge. In addition, the explicit role of material strength or cohesion has been tested and the strength–gravity transition is shown to be a function of charge size and soil strength. (See Holsapple and Schmidt, 1979, cited in Further Reading at the end of this section.)

We now describe our explosive in more detail than in previous models. To characterize an explosive, three independent variables are needed: size, energy yield, and explosive density δ. The size can be given as charge mass W, charge energy E, or the radius α of the spherical explosive. The energy yield can be given as a measure of the specific energy Q_e or the energy density per unit volume Q_V. The following equations relate the variables:

$$W = \frac{E}{Q_e}$$

$$Q_V = \delta Q_e$$

$$\alpha^3 = \left(\frac{3}{4\pi}\right)\left(\frac{W}{\delta}\right)$$

[5] See the papers by R. M. Schmidt (1977, 1980) and by Schmidt and Holsapple (1980), cited in Further Reading, which discuss the effects when a centrifuge is used to perform explosive cratering tests under the influence of gravitational acceleration up to 480 G, where 1 G is the terrestrial gravity field strength of 981 cm/sec^2.

One choice of these variables leads to the model formulation

$$V = f(W, Q_e, \delta, \rho, g, d)$$

Because there are seven variables under consideration and the *MLT* system is being used, a dimensional analysis generally will result in four $(7 - 3)$ dimensionless products. The dimensions of the variables are shown in the following table:

Variable	V	W	Q_e	δ	ρ	g	d
Dimension	L^3	M	$L^2 T^{-2}$	ML^{-3}	ML^{-3}	LT^{-2}	L

Any product of the variables must be of the form:

$$V^a W^b Q_e^c \delta^e \rho^f g^k d^m \qquad (13.22)$$

and hence have dimensions:

$$(L^3)^a (M)^b (L^2 T^{-2})^c (ML^{-3})^{e+f} (LT^{-2})^k (L)^m$$

Therefore, a product of the form (13.22) is dimensionless if and only if the exponents satisfy the following homogeneous system of equations:

$$
\begin{aligned}
M: & \quad b \quad + e + f \quad = 0 \\
L: & \quad 3a \quad + 2c - 3e - 3f + k + m = 0 \\
T: & \quad - 2c \quad - 2k \quad = 0
\end{aligned}
$$

Solution to this system produces

$$b = \frac{k - m}{3} - a, \qquad c = -k, \qquad e = a - f + \frac{m - k}{3}$$

where a, f, k, and m are arbitrary. By setting one of these arbitrary exponents equal to 1 and the other three equal to 0, in succession, we obtain the following set of dimensionless products:

$$\frac{V\delta}{W}, \qquad \left(\frac{g}{Q_e}\right)\left(\frac{W}{\delta}\right)^{1/3}, \qquad d\left(\frac{\delta}{W}\right)^{1/3}, \qquad \frac{\rho}{\delta}$$

(Convince yourself that these are dimensionless.) Because the dimensions of ρ and δ are equal, we can rewrite these dimensionless products as follows:

$$\Pi_1 = \frac{V\rho}{W}$$

$$\Pi_2 = \left(\frac{g}{Q_e}\right)\left(\frac{W}{\delta}\right)^{1/3}$$

$$\Pi_3 = d \left(\frac{\rho}{W} \right)^{1/3}$$

$$\Pi_4 = \frac{\rho}{\delta}$$

so Π_1 is consistent with the dimensionless product implied by (13.20). Then, applying Buckingham's theorem, we obtain the model

$$h\left(\Pi_1, \Pi_2, \Pi_3, \Pi_4 \right) = 0 \tag{13.23}$$

or

$$V = \frac{W}{\rho} H \left(\frac{gW^{1/3}}{Q_e \delta^{1/3}}, \frac{d\rho^{1/3}}{W^{1/3}}, \frac{\rho}{\delta} \right)$$

Presenting the results For oil-base clay the value of ρ is about 1.53 g/cm^3; for wet sand, 1.65; and for desert alluvium, 1.60. For TNT, δ has the value 2.23 g/cm^3. Thus $0.69 < \Pi_4 < 0.74$, so for simplicity we can assume for these soils and TNT that Π_4 is constant. Then (13.23) becomes

$$h\left(\Pi_1, \Pi_2, \Pi_3 \right) = 0 \tag{13.24}$$

R. M. Schmidt gathered experimental data to plot the surface described by (13.24). A plot of the surface is depicted in Figure 13.12, showing the crater and volume parameter Π_1 as a function of the scaled energy charge Π_2 and the depth of the burial parameter Π_3. Cross-sectional data for the surface parallel to the $\Pi_1 \Pi_3$-plane when $\Pi_2 = 1.15 \times 10^{-6}$ are depicted in Figure 13.13.

Experiments have shown that the physical effect of increasing gravity is to reduce crater volume for a given charge yield. This result suggests that increased gravity can be compensated for by increasing the size of the charge to maintain the same cratering efficiency. Note also that both Figures 13.12 and 13.13 can be used for prediction once an empirical interpolating model is constructed from the data. Holsapple and Schmidt (1982) extend these methods to impact cratering, and Housen, Schmidt, and Holsapple (1983) extend them to crater ejecta scaling.

Example 2   How Long Should You Roast a Turkey?

One general rule for roasting a turkey is the following: Set the oven to 400°F and allow 20 min per pound for cooking. How good is this rule?

Assumptions Let t denote the cooking time for the turkey. On what variables does t depend? Certainly the size of the turkey is a factor that must be considered. Let's

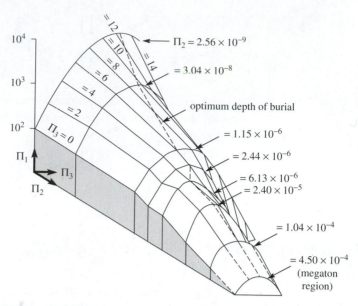

FIGURE 13.12 A plot of the surface $h(\Pi_1, \Pi_2, \Pi_3) = 0$, showing the crater volume parameter Π_1 as a function of gravity-scaled yield Π_2 and depth of burial parameter Π_3 (reprinted by permission of R. M. Schmidt)

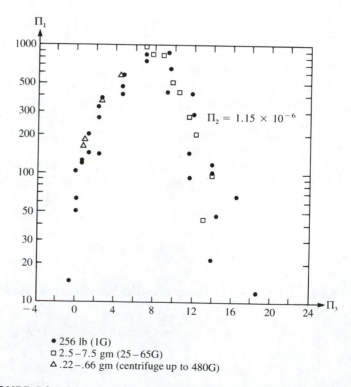

• 256 lb (1G)
□ 2.5 – 7.5 gm (25 – 65G)
△ .22 – .66 gm (centrifuge up to 480G)

FIGURE 13.13 Data values for a cross section of the surface depicted in Figure 13.12 (data reprinted by permission from R. M. Schmidt)

assume the turkeys are geometrically similar and use l to denote some characteristic dimension of the meat to be cooked; specifically, we assume l represents the length of the turkey. Another factor is the difference between the temperature of the raw meat and the oven ΔT_m. (We know from experience that it takes longer to cook a bird that is nearly frozen than it does to cook one that is initially at room temperature.) Because the turkey will have to reach a certain interior temperature before it is considered fully cooked, the difference ΔT_c between the temperature of the cooked meat and the oven is a variable determining the cooking time. Finally, we know that different foods require different cooking times quite independent of size: It takes only 10 min or so to bake cookies, whereas cooking a roast beef or turkey requires several hours. A measure of the factor representing the differences between foods is the *coefficient of heat conduction* for a particular food to be cooked. Let k denote the coefficient of heat conduction for turkey. Thus we have the following model formulation for the cooking time:

$$t = f(\Delta T_m, \Delta T_c, k, l)$$

thermal conductivity

Dimensional analysis   Consider the dimensions of the independent variables. The temperature variables ΔT_m and ΔT_c measure the energy per volume and therefore have the dimension ML^2T^{-2}/L^3, or simply $ML^{-1}T^{-2}$. Now what about the heat conduction variable k? **Thermal conductivity** k is defined as the amount of energy crossing one unit cross-sectional area per second divided by the temperature gradient perpendicular to the area. That is,

$$k = \frac{\text{energy}/(\text{area} \times \text{time})}{\text{temperature}/\text{length}}$$

Accordingly, the dimension of k is $(ML^2T^{-2})(L^{-2}T^{-1})/(ML^{-1}T^{-2})(L^{-1})$, or simply L^2T^{-1}. Our analysis gives the following table:

Variable	ΔT_m	ΔT_c	k	l	t
Dimension	$ML^{-1}T^{-2}$	$ML^{-1}T^{-2}$	L^2T^{-1}	L	T

Any product of the variables must be of the form

$$\Delta T_m^a \Delta T_c^b k^c l^d t^e \tag{13.25}$$

and hence have dimension

$$(ML^{-1}T^{-2})^a(ML^{-1}T^{-2})^b(L^2T^{-1})^c(L)^d(T)^e$$

Therefore a product of the form (13.25) is dimensionless if and only if the exponents satisfy

$$
\begin{aligned}
M: &\quad a + b &&= 0 \\
L: &\quad -a - b + 2c + d &&= 0 \\
T: &\quad -2a - 2b - c + e &&= 0
\end{aligned}
$$

Solution of this system of equations gives

$$a = -b, \qquad c = e, \qquad d = -2e$$

where b and e are arbitrary. If we set $b = 1$, $e = 0$, we obtain $a = -1$, $c = 0$, and $d = 0$; likewise, $b = 0$, $e = 1$ produces $a = 0$, $c = 1$, and $d = -2$. These independent solutions yield the complete set of dimensionless products

$$\prod_1 = \Delta T_m^{-1} \Delta T_c \qquad \text{and} \qquad \prod_2 = kl^{-2}t$$

From Buckingham's theorem we obtain

$$h\left(\prod_1, \prod_2\right) = 0$$

or

$$t = \left(\frac{l^2}{k}\right) H\left(\frac{\Delta T_c}{\Delta T_m}\right) \tag{13.26}$$

The rule stated in our opening remarks gives the roasting time for the turkey in terms of its weight w. Let's assume the turkeys are geometrically similar, or $V \propto l^3$. If we assume the turkey is of constant density (which is not quite correct because the bones and flesh differ in density), then because weight is density times volume and volume is proportional to l^3, we get $w \propto l^3$. Moreover, if we set the oven to a constant baking temperature and specify that the turkey must initially be near room temperature (65°F), then $\Delta T_c / \Delta T_m$ is a dimensionless constant. Combining these results with Equation (13.26), we get the proportionality

$$t \propto w^{2/3} \tag{13.27}$$

because k is constant for turkeys. Thus the required cooking time is proportional to weight raised to the two-thirds power. Therefore, if t_1 hours are required to cook a turkey weighing w_1 pounds and t_2 is the time for a weight of w_2 pounds,

$$\frac{t_1}{t_2} = \left(\frac{w_1}{w_2}\right)^{2/3}$$

it follows that a doubling of the weight of a turkey increases the cooking time by the factor $2^{2/3} \approx 1.59$.

How does our result (13.27) compare with the rule stated previously? Assume ΔT_m, ΔT_c, and k are independent of the length or weight of the turkey, and consider cooking a 23-lb turkey versus an 8-lb bird. According to our general rule, the ratio of cooking times is given by

$$\frac{t_1}{t_2} = \left(\frac{20 \cdot 23}{20 \cdot 8}\right) = 2.875$$

On the other hand, from dimensionless analysis and (13.27),

$$\frac{t_1}{t_2} = \left(\frac{23}{8}\right)^{2/3} \approx 2.02$$

Thus the rule predicts it will take nearly three times as long to cook a 23-lb turkey as it will to cook an 8-lb bird. Dimensional analysis predicts it will take only twice as long. Which rule is correct? Why have so many cooks overcooked a turkey?

Testing the results Suppose various sized turkeys are cooked in an oven preheated to 325°F. The initial temperature of all turkeys is 65°F. All turkeys are removed from the oven when their internal temperature, measured by a meat thermometer, reaches 195°F. The (hypothetical) cooking times for the various turkeys are recorded as follows:

w (lb)	5	10	15	20
t (hr)	2	3.4	4.5	5.4

A plot of t versus $w^{2/3}$ is shown in Figure 13.14. Because the graph approximates a straight line through the origin, we conclude that $t \propto w^{2/3}$, as predicted by our model.

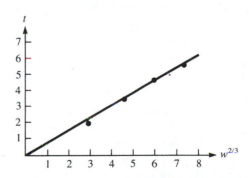

FIGURE 13.14 Plot of cooking times versus weight to the two-thirds power reveals the predicted proportionality

13.4 PROBLEMS

1. a. Use dimensional analysis to establish the cube-root law:

$$r\left(\frac{\rho}{W}\right)^{1/3} = \text{constant}$$

for scaling of explosions, where r is the radius or depth of the crater, ρ is the density of the soil medium, and W the mass of the explosive.

b. Use dimensional analysis to establish the one-fourth root law

$$r\left(\frac{\rho g}{E}\right)^{1/4} = \text{constant}$$

for scaling of explosions, where r is the radius or depth of the crater, ρ is the density of the soil medium, g is gravity, and E the charge energy of the explosive.

2. a. Show that the products $\prod_1$, $\prod_2$, $\prod_3$, $\prod_4$ for the refined explosion model presented in the text are dimensionless products.

b. Assume ρ is essentially constant for the soil being used and restrict the explosive to a specific type, say TNT. Under these conditions ρ/δ is essentially constant yielding

$$\prod_1 = f\left(\prod_2, \prod_3\right)$$

You have collected the following data with $\prod_2 = 1.5 \times 10^{-6}$:

$\prod_3$	0	2	4	6	8	10	12	14
$\prod_1$	15	150	425	750	825	425	250	90

i. Construct a scatterplot of $\prod_1$ versus $\prod_3$. Does a trend exist?

ii. How accurate do you think the data are? Find an empirical model that captures the *trend* of the data with accuracy commensurate with your appraisal of the accuracy of the data.

iii. Use your empirical model to predict the volume of a crater using TNT in desert alluvium with (CGS system) $W = 1500$ g, $\rho = 1.53$ g/cm³, and $\prod_3 = 12.5$.

3. Consider a zero-depth burst, spherical explosive in a soil medium. Assume the value of the crater volume V depends on the explosive size, energy yield, and explosive energy, as well as on the strength Y of the soil (considered a resistance to pressure with dimensions $ML^{-1}T^{-2}$), soil density ρ, and gravity g. In this problem, assume

$$V = f(W, Q_e, \delta, Y, \rho, g)$$

and use dimensional analysis to produce the following *mass set* of dimensionless products:

$$\prod_1 = \frac{V\rho}{W} \qquad \prod_2 = \left(\frac{g}{Q_e}\right)\left(\frac{W}{\delta}\right)^{1/3}$$

$$\prod_3 = \frac{Y}{\delta Q_e} \qquad \prod_4 = \frac{\rho}{\delta}$$

4. For the explosion process and material characteristics discussed in Problem 3, consider

$$V = f(E, Q_v, \delta, Y, \rho, g)$$

and use dimensional analysis to produce the following *energy set* of dimensionless products:

$$\overline{\overline{\Pi}}_1 = \frac{VQ_V}{E} \qquad \overline{\overline{\Pi}}_2 = \frac{\rho g E^{1/3}}{Q_V^{4/3}}$$

$$\overline{\overline{\Pi}}_3 = \frac{Y}{Q_V} \qquad \overline{\overline{\Pi}}_4 = \frac{\rho}{\delta}$$

5. Repeat Problem 4 for

$$V = f(E, Q_e, \delta, Y, \rho, g)$$

and use dimensional analysis to produce the following *gravity set* of dimensionless products:

$$\overline{\overline{\Pi}}_1 = V\left(\frac{\rho g}{E}\right)^{3/4} \qquad \overline{\overline{\Pi}}_2 = \left(\frac{1}{Q_e}\right)\left(\frac{g^3 E}{\delta}\right)^{1/4}$$

$$\overline{\overline{\Pi}}_3 = \frac{Y}{\delta Q_e} \qquad \overline{\overline{\Pi}}_4 = \frac{\rho}{\delta}$$

6. An experiment consists of dropping spheres into a tank of heavy oil and measuring the times of descent. It is desired that a relationship for the time of descent be determined and verified by experimentation. Assume the time of descent is a function of mass m, gravity g, radius r, viscosity μ, and distance traveled d. Fluid density is to be neglected. That is,

$$t = f(m, g, r, \mu, d)$$

a. Use dimensional analysis to find a relationship for the time of descent.

b. How will the spheres be chosen to verify the time of descent relationship is independent of fluid density? Assuming you have verified the assumptions on fluid density, describe how you would determine the nature of your function experimentally.

c. Using differential equations techniques, find the velocity of the sphere as a function of time, radius, mass, viscosity, gravity, and fluid density. Using this result and that found in part **a**, predict under what conditions fluid density may be neglected. (*Hints:* Use the results of Problem 5 of Section 13.2 as a submodel for drag force. Consider the buoyant force.)[6]

7. A windmill is being rotated by air flow to produce power to pump water. It is desired to find the power output *P* of the windmill. Assume that *P* is a function of the density of the air ρ, viscosity of the air μ, diameter of the windmill *d*, wind speed *v*, and the rotational speed of the windmill ω (measured in radians per second). Thus

$$P = f(\rho, \mu, d, v, \omega)$$

   a. Using dimensional analysis, find a relationship for *P*. Be sure to check your products to make sure they are dimensionless.

   b. Does your result make sense? Explain.

   c. Discuss how you would design an experiment to determine the nature of your function.

8. For a sphere traveling through a liquid, assume the drag force F_D is a function of the fluid density ρ, fluid viscosity μ, radius of the sphere *r*, and speed of the sphere *v*. Use dimensional analysis to find a relationship for the drag force

$$F_D = f(\rho, \mu, r, v)$$

Make sure you provide justification that the given independent variables influence the drag force.

13.4 PROJECTS

1. Complete the requirements for the module, "Listening to the Earth: Controlled Source Seismology," by Richard G. Montgomery, UMAP 292, 293. This module develops the elementary theory of wave reflection and refraction and applies it to a model of the Earth's subsurface. The model shows how information on layer depth and sound velocity can be obtained to provide data on width, density, and composition of the subsurface. This module is a good introduction to controlled seismic methods and requires no previous knowledge of physics or geology.

13.4 Further Reading

HOLSAPPLE, K. A. & R. M. Schmidt. "A Material-strength Model for Apparent Crater Volume." *Proc. Lunar Planet Sci. Conf.* no. 10 (1979): 2757–2777.

HOLSAPPLE, K. A. & R. M. Schmidt. "On Scaling of Crater Dimensions-2: Impact Processes." *J. Geophys. Res.* no. 87 (1982): 1849–1870.

[6]For students who have studied differential equations.

HOUSEN, K. R., K. A. Holsapple & and R. M. Schmidt. "Crater Ejecta Scaling Laws 1: Fundamental Forms Based on Dimensional Analysis." *J. Geophys. Res.* no. 88 (1983): 2485–2499.

SCHMIDT, R. M. "A Centrifuge Cratering Experiment: Development of a Gravity-Scaled Yield Parameter," in *Impact and Explosion Cratering*, edited by D. J. Roddy, et al., pp. 1261–1278. New York: Pergamon Press, 1977.

SCHMIDT, R. M. "Meteor Crater: Energy of Formation—Implications of Centrifuge Scaling." *Proc. Lunar Planet*. Sci. Conf. 11 (1980): 2099–2128.

SCHMIDT, R. M. & K. A. Holsapple. "Theory and Experiments on Centrifuge Cratering." *J. Geophys. Res.* no. 85 (1980): 235–252.

13.5   SIMILITUDE

Suppose we are interested in the effects of wave action on a large ship at sea, or heat loss of a submarine and the drag force it experiences in its underwater environment, or the wind effects on an aircraft wing. Quite often because it is physically impossible to duplicate the actual phenomenon in the laboratory, we study a scaled-down model in a simulated environment to predict accurately the performance of the physical system. The actual physical system for which the predictions are to be made is called **prototype** the **prototype**. How do we scale experiments in the laboratory to ensure the effects observed for the model will be the same effects experienced by the prototype?

Although extreme care must be exercised in using simulations, the dimensional products resulting from dimensional analysis of the problem can provide insight into how the scaling for a model should be done. The idea comes from Buckingham's theorem. If the physical system can be described by a dimensionally homogeneous equation in the variables, then it can be put into the form

$$f\left(\prod_1, \prod_2, \ldots, \prod_n\right) = 0$$

for a complete set of dimensionless products. Assume the independent variable of the problem appears only in the product $\prod_n$ and

$$\prod_n = H\left(\prod_1, \prod_2, \ldots, \prod_{n-1}\right)$$

For the solution to the model and the prototype to be the same, it is sufficient that the value of all independent dimensionless products $\prod_1, \prod_2, \ldots, \prod_{n-1}$ be the same for the model and the prototype.

For example, suppose the Reynolds number $vr\rho/\mu$ appears as one of the dimensionless products in a fluid mechanics problem, where v represents fluid velocity, r a characteristic dimension (such as the diameter of a sphere or the length of a ship), ρ the fluid density, and μ the fluid viscosity. These values refer to the prototype. Next let v_m, r_m, ρ_m, and μ_m denote the corresponding values for the scaled-down model. For the effects on the model and the prototype to be the same, we want the two Reynolds numbers to agree so that

$$\frac{v_m r_m \rho_m}{\mu_m} = \frac{v r \rho}{\mu}$$

The last equation is referred to as a *design condition* to be satisfied by the model. If the length of the prototype is too large for the laboratory experiments so we have to scale down the length of the model—say, $r_m = r/10$—then the same Reynolds number for the model and the prototype can be achieved by using the same fluid ($\rho_m = \rho$ and $\mu_m = \mu$) and varying the velocity $v_m = 10v$. If it is impractical to scale the velocity by the factor of 10, we can scale it by a lesser amount $0 < k < 10$ and use a different fluid so the equation

$$\frac{k \rho_m}{10 \mu_m} = \frac{\rho}{\mu}$$

is satisfied. We need to be careful in generalizing the results from the scaled-down model to the prototype. Certain factors (such as surface tension, for instance) that may be negligible for the prototype may become significant for the model. Such factors would have to be taken into account before making any predictions for the prototype.

Example 1   *Drag Force on a Submarine*

We are interested in the drag forces experienced by a submarine to be used for deep-sea oceanographic explorations. We assume the variables affecting the drag D are fluid velocity v, characteristic dimension r (here the length of the submarine), fluid density ρ, fluid viscosity μ, and velocity of sound in the fluid c. We wish to predict the drag force by studying a model of the prototype. How shall the experiments for the model be scaled?

A major stumbling block in our problem is in describing shape factors related to the physical object being modeled, in this case, the submarine. Let's consider submarines that are ellipsoidal in shape. In two dimensions, if a is the length of the major axis and b is the length of the minor axis of an ellipse, we can define $r_1 = a/b$ and assign a characteristic dimension such as r, the length of the submarine (see Figure 4.20). In three dimensions, also define $r_2 = a/b'$, where a is the original major axis and b' is the second minor axis. Then r, r_1, and r_2 describe the shape of the submarine. In a more irregularly shaped object, additional shape factors would be required. The basic idea is that the object can be described using a characteristic dimension and an appropriate collection of shape factors. In the case of our three-dimensional ellipsoidal submarine, the shape factors r_1 and r_2 are needed. These shape factors are dimensionless constants.

Returning to our list of six fluid mechanics variables D, v, r, ρ, μ, and c, notice that we are neglecting surface tension (because it is small) and that gravity is not being considered. Thus it is expected that a dimensional analysis will produce three $(6 - 3)$ independent dimensionless products. We can choose the following three products for convenience:

$$\text{Reynolds number} \qquad R = \frac{vr\rho}{\mu}$$

$$\text{Mach number} \qquad M = \frac{v}{c}$$

$$\text{Pressure coefficient} \qquad P = \frac{p}{\rho v^2}$$

The added shape factors are dimensionless, so Buckingham's theorem gives the equation

$$h(P, R, M, r_1, r_2) = 0$$

Assuming we can solve for P yields

$$P = H(R, M, r_1, r_2)$$

Substituting $P = p/\rho v^2$ and solving for p gives

$$p = \rho v^2 H(R, M, r_1, r_2)$$

Remembering that the total drag force is the pressure (force per unit area) times the area (which is proportional to r^2 for geometrically similar objects) gives the proportionality $D \propto pr^2$, or

$$D = kp v^2 r^2 H(R, M, r_1, r_2) \qquad (13.28)$$

Now a similar equation must hold to give the same proportionality for the model:

$$D_m = kp_m v_m^2 r_m^2 H(R_m, M_m, r_{1m}, r_{2m}) \qquad (13.29)$$

Because the prototype and model equations refer to the same physical system, both equations are identical in form. Therefore the design conditions for the model require that

$$
\begin{array}{ll}
\text{Condition (a)} & R_m = R \\
\text{Condition (b)} & M_m = M \\
\text{Condition (c)} & r_{1m} = r_1 \\
\text{Condition (d)} & r_{2m} = r_2
\end{array}
$$

Note that if conditions (a), (b), (c), and (d) are satisfied, then Equations (13.28) and (13.29) give

$$\frac{D_m}{D} = \frac{\rho_m v_m^2 r_m^2}{\rho v^2 r^2} \qquad (13.30)$$

Thus D can be computed once D_m is measured. Note that the design conditions (c) and (d) imply geometric similarity between the model and the prototype submarine

$$\frac{a_m}{b_m} = \frac{a}{b} \quad \text{and} \quad \frac{a_m}{b_m'} = \frac{a}{b'}$$

If the velocities are small compared with the speed of sound in a fluid, then v/c can be considered constant in accordance with condition (b). If the same fluid is used for both the model and prototype, then condition (a) is satisfied if

$$v_m r_m = vr$$

or

$$\frac{v_m}{v} = \frac{r}{r_m}$$

which states that the velocity of the model must increase inversely as the scaling factor r_m/r. Under these conditions, Equation (13.30) yields

$$\frac{D_m}{D} = \frac{\rho_m v_m^2 r_m^2}{\rho v^2 r^2} = 1$$

If increasing the velocity of the scaled model proves unsatisfactory in the laboratory, then a different fluid may be considered for the scaled model ($\rho_m \neq \rho$ and $\mu_m \neq \mu$). If the ratio v/c is small enough to neglect, then both v_m and r_m can be varied to ensure that

$$\frac{v_m r_m \rho_m}{\mu_m} = \frac{vr\rho}{\mu}$$

in accordance with condition (a). Having chosen values that satisfy design condition (a) and knowing the drag on the scaled model, we can use Equation (13.30) to compute the drag on the prototype. Consider the additional difficulties if the velocities are sufficiently great that we must satisfy condition (b) as well.

A few comments are in order. One distinction between the Reynolds number and the other four numbers in fluid mechanics is that the Reynolds number contains the viscosity of the fluid. Dimensionally, the Reynolds number is proportional to the ratio of the inertia forces of an element of fluid to the viscous force acting on the fluid. In certain problems the numerical value of the Reynolds number may be significant. For example, the flow of a fluid in a pipe is virtually always parallel to the edges of the pipe (giving *laminar flow*) if the Reynolds number is less than 2000. Reynolds numbers in excess of 3000 almost always indicate turbulent flow. Normally, there is a critical Reynolds number between 2000 and 3000 at which the flow becomes turbulent.

The design condition (a) mentioned earlier requires the Reynolds number of the model and the prototype to be the same. This requirement precludes the possibility of laminar flow in the prototype being represented by turbulent flow in the model and vice versa. The equality of the Reynolds number for a model and prototype is important in all problems in which viscosity plays a significant role.

The Mach number is the ratio of fluid velocity to the speed of sound in the fluid. It is generally important for problems involving objects moving with high speed in fluids, such as projectiles, high-speed aircraft, rockets, and submarines. Physically, if the Mach number is the same in model and prototype, the effect of the compressibility force in the fluid relative to the inertia force will be the same for model and prototype. This is the situation that is required by condition (b) in our example on the submarine.

13.5 PROBLEMS

1. A model of an airplane wing is being tested in a wind tunnel. The model wing has an 18-in. chord, and the prototype has a 4-ft chord moving at 250 mph. Assuming the air in the wind tunnel is at atmospheric pressure, at what velocity should wind tunnel tests be conducted so the Reynolds number of the model is the same as that of the prototype?

2. Two smooth balls of equal weight but different diameters are dropped from an airplane. The ratio of their diameters is 5. Neglecting compressibility (assume constant Mach number), what is the ratio of the terminal velocities of the balls? Are the flows similar?

3. Consider predicting the pressure drop Δp between two points along a smooth horizontal pipe under the condition of steady laminar flow. Assume

$$\Delta p = f(s, d, \rho, \mu, v)$$

where s is the control distance between two points in the pipe, d is the diameter of the pipe, ρ is the fluid density, μ is the fluid viscosity, and v is the velocity of the fluid.

a. Determine the design conditions for a scaled model of the prototype.

b. Must the model be geometrically similar to the prototype? Why?

c. May the same fluid be used for model and prototype?

d. Show that if the same fluid is used for the model and prototype, the equation is

$$\Delta p = \frac{\Delta p_m}{n^2}$$

where $n = d/d_m$.

4. It is desired to study the velocity v of a fluid flowing in a smooth open channel. Assume

$$v = f(r, \rho, \mu, \sigma, g)$$

where r is the characteristic length of the channel cross-sectional area divided by the wetted perimeter, ρ is the fluid density, μ is the fluid viscosity, σ is the surface tension, and g is the acceleration of gravity.

a. Describe the appropriate pair of shape factors r_1 and r_2.

b. Show that

$$\frac{v^2}{gr} = H\left(\frac{\rho v r}{\mu}, \frac{\rho v^2 r}{\sigma}, r_1, r_2\right)$$

Discuss the design conditions required of the model.

c. Will it be practical to use the same fluid in the model and the prototype? Why?

d. Suppose the surface tension σ is ignored and the design conditions are satisfied. If $r_m = r/n$, what is the equation for the velocity of the prototype? When is the equation compatible with the design conditions?

e. What is the equation for the velocity v if gravity is ignored? What if viscosity is ignored? What fluid would you use if you were to ignore viscosity?

13.5 Further Reading

LANGHAAR, Henry L. *Dimensional Analysis and Theory of Models*. New York: Wiley, 1951.

MASSEY, Bernard S. *Units, Dimensional Analysis and Physical Similarity*. London: Van Nostrand Reinhold Company, 1971.

MURPHY, Glenn. *Similitude in Engineering*. New York: Ronald Press, 1950.

Appendix A

PROBLEMS FROM THE MATHEMATICS CONTEST IN MODELING, 1985–1996

1985: THE ANIMAL POPULATION PROBLEM

Choose a fish or mammal for which appropriate data are available to model it accurately. Model the animal's natural interactions with its environment by expressing population levels of different groups in terms of the significant parameters of the environment. Then adjust the model to account for harvesting in a form consistent with the actual method by which the animal is harvested. Include any outside constraints imposed by food or space limitations that are supported by the data. Consider the value of the various quantities involved, the number harvested, and the population size itself, in order to devise a numerical quantity that represents the overall value of the harvest. Find a harvesting policy in terms of population size and time that optimizes the value of the harvest over a long period of time. Check that the policy optimizes this value over a realistic range of environmental conditions.

1985: THE STRATEGIC RESERVE PROBLEM

Cobalt, which is not produced in the U.S., is essential to a number of industries. (Defense accounted for 17% of the cobalt production in 1979.) Most cobalt comes from central Africa, a politically unstable region. The Strategic and Critical Materials Stockpiling Act of 1946 requires a cobalt reserve that will carry the U.S.

through a three-year war. The government built up a cobalt stockpile in the 1950s, sold most of it in the early 1970s, and then decided to build it up again in the late 1970s, with a stockpile goal of 85.4 million pounds. About half of this stockpile had been acquired by 1982.

Build a mathematical model for managing a stockpile of the strategic metal cobalt. You will need to consider such questions as:

- How big should the stockpile be?
- At what rate should it be acquired?
- What is a reasonable price to pay for the metal?

You will also want to consider such questions as:

- At what point should the stockpile be drawn down?
- At what rate should it be drawn down?
- What is a reasonable price at which to sell the metal?
- How should sold metal be allocated?

Below we give more information on the sources, cost, demand, and recycling aspects of cobalt.

Useful Information on Cobalt

The government has projected a need of 25 million pounds of cobalt in 1985.

The U.S. has about 100 million pounds of proven cobalt deposits. Production becomes economically feasible when the price reaches $22/lb (as occurred in 1981). It takes four years to get operations rolling, and then six million pounds per year can be produced.

In 1980, 1.2 million pounds of cobalt were recycled, 7% of total consumption.

Please see Figures A.1 to A.3, whose source is *Mineral Facts and Problems*, U.S. Bureau of Mines (Washington, DC: Government Printing Office, 1980).

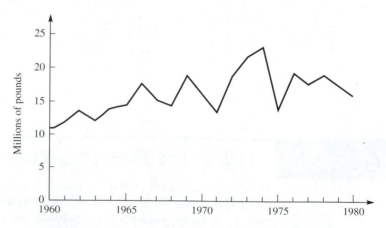

FIGURE A.1 U.S. primary demand for cobalt, 1960–1980

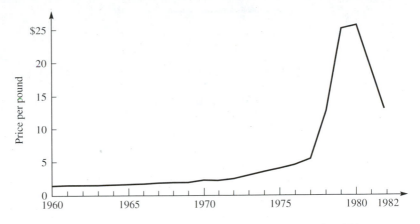

FIGURE A.2   Cobalt prices in the U.S. market, 1960–1982

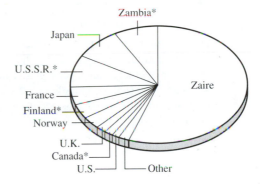

FIGURE A.3   Producers of refind metal and/or oxide 1979; an asterisk denotes a country with domestic production. Source: U.S. Bureau of Mines, *Mineral Facts and Problems* (1980).

1986: THE HYDROGRAPHIC DATA PROBLEM

The following table gives the depth Z of water in feet for surface points with rectangular coordinates X, Y in yards. The depth measurements were taken at low tide. Your ship has a draft of five feet. What region should you avoid within the rectangle $(150, -50) \times (200, 75)$?

X	Y	Z
129.0	7.5	4
140.0	141.5	8
108.5	28.0	6
88.0	147.0	8
185.5	22.5	6
195.0	137.5	8
105.5	85.5	8
157.5	−6.5	9
107.5	−81.0	9
77.0	3.0	8
162.0	−66.5	9
162.0	84.0	4
117.5	−38.5	9

1986: THE EMERGENCY-FACILITIES LOCATION PROBLEM

The township of Rio Rancho has hitherto not had its own emergency facilities. It has secured funds to erect two emergency facilities in 1986, each of which will combine ambulance, fire, and police services. Figure A.4 indicates the demand, or number of emergencies per square block, for 1985. The L region in the north is an obstacle, while the rectangle in the south is a park with a shallow pond. It takes an emergency vehicle an average of 15 seconds to go one block in the N–S direction and 20 seconds in the E–W direction. Your task is to locate the two facilities so as to minimize the total response time.

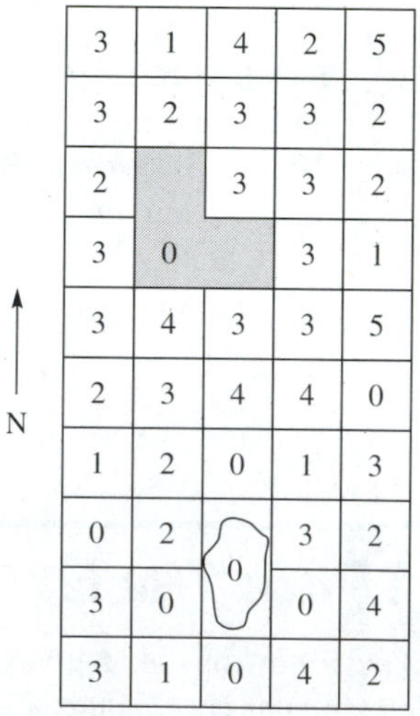

FIGURE A.4   A map of Rio Rancho, with number of emergencies in 1985 indicated for each block

- Assume that the demand is concentrated at the center of the block and that the facilities will be located on corners.
- Assume that the demand is uniformly distributed on the streets bordering each block and that the facilities may be located anywhere on the streets.

1987: THE SALT STORAGE PROBLEM

For approximately 15 years, a Midwestern state has stored salt used on roads in the winter in circular domes. Figure A.5 shows how salt has been stored in the past. The salt is brought into and removed from the domes by driving front-end loaders up ramps of salt leading into the domes. The salt is piled 25 to 30 ft high, using the buckets on the front-end loaders.

Recently, a panel determined that this practice is unsafe. If the front-end loader gets too close to the edge of the salt pile, the salt might shift, and the loader could be thrown against the retaining walls that reinforce the dome. The panel recommended that if the salt is to be piled with the use of loaders, then the piles should be restricted to a maximum height of 15 ft.

Construct a mathematical model for this situation and find a recommended maximum height for salt in the domes.

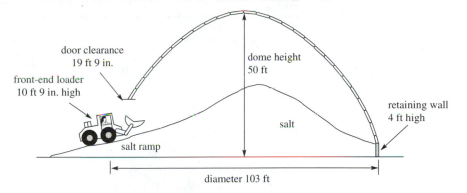

FIGURE A.5   Diagram of a salt storage dome

1987: THE PARKING LOT PROBLEM

The owner of a paved, 100-ft-by-200-ft, corner parking lot in a New England town hires you to design the layout; that is, to design how the lines are to be painted.

You realize that squeezing as many cars into the lot as possible leads to right-angle parking with the cars aligned side by side. However, inexperienced drivers have difficulty parking their cars this way, which can give rise to expensive insurance claims. To reduce the likelihood of damage to parked vehicles, the owner might then have to hire expert drivers for valet parking. On the other hand, most drivers seem to have little difficulty in parking in one attempt if there is a large enough turning radius from the access lane. Of course, the wider the access lane, the fewer cars that can be accommodated in the lot, leading to less revenue for the parking lot owner.

1988: THE RAILROAD FLATCAR PROBLEM

Two railroad flatcars are to be loaded with seven types of packing crates. The crates have the same width and height but varying thickness (t, in cm) and weight (w, in kg). Table A.1 gives, for each crate, the thickness, weight, and number available. Each car has 10.2 meters of length available for packing the crates (like slices of toast) and can carry up to 40 metric tons. There is a special constraint on the total number of C_5, C_6, and C_7 crates because of a subsequent local trucking restriction: The total space (thickness) occupied by these crates must not exceed 302.7 cm. Load the two flatcars (see Figure A.6) so as to minimize the wasted floor space.

TABLE A.1 The thickness, weight, and number of each kind of crate

	C_1	C_2	C_3	C_4	C_5	C_6	C_7	
t	48.7	52.0	61.3	72.0	48.7	52.0	64.0	cm
w	2000	3000	1000	500	4000	2000	1000	kg
	8	7	9	6	6	4	8	

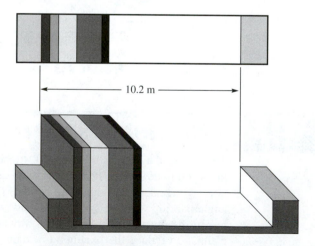

FIGURE A.6 Diagram of loading of a flatcar

1988: THE DRUG RUNNER PROBLEM

Two listening posts 5.43 miles apart pick up a brief radio signal. The sensing devices were oriented at 110° and 119°, respectively, when the signal was detected (see Figure A.7); and they are accurate to within 2°. The signal came from a region of active drug exchange, and it is inferred that there is a powerboat waiting for someone to pick up drugs. It is dusk, the weather is calm, and there are no currents. A small helicopter leaves a pad from Post 1 and is able to fly accurately along the 110° angle direction. The helicopter's speed is three times the speed of the boat. The helicopter will be heard when it gets within 500 ft of the boat. This helicopter has only one detection device, a searchlight. At 200 ft, it can just illuminate a circular region with a radius of 25 ft.

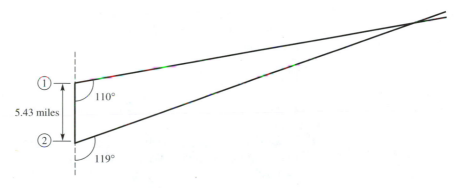

FIGURE A.7   Geometry of the problem

- Describe the (smallest) region where the pilot can expect to find the waiting boat.
- Develop an optimal search method for the helicopter.

Use a 95% confidence level in your calculations.

1989: THE AIRCRAFT QUEUEING PROBLEM

A common procedure at airports is to assign aircraft (A/C) to runways on a first-come-first-served basis. That is, as soon as an A/C is ready to leave the gate (push back), the pilot calls ground control and is added to the queue. Suppose that a control tower has access to a fast online database with the following information for each A/C:

- The time it is scheduled for pushback
- The time it actually pushes back
- The number of passengers on board

- The number of passengers who are scheduled to make a connection at the next stop, as well as the time to make that connection
- The schedule time of arrival at its next stop

Assume that there are seven types of A/C with passenger capacities varying from 100 to 400 in steps of 50. Develop and analyze a mathematical model that takes into account both the travelers' and airlines' satisfaction.

1989: THE MIDGE CLASSIFICATION PROBLEM

Two species of midges, Af and Apf, have been identified by biologists Grogan and Wirth (1981) on the basis of antenna and wing length. (See Figure A.8.) Each of nine Af midges is denoted by □, and each of six Apf midges is denoted by ○. It is important to be able to classify a specimen as Af or Apf, given the antenna and wing length.

1. Given a midge that you know is species Af or Apf, how would you go about classifying it?
2. Apply your method to three specimens with (antenna, wing) lengths (1.24, 1.80), (1.28, 1.84), (1.40, 2.04).
3. Assume that species Af is a valuable pollinator and species Apf is a carrier of a debilitating disease. Would you modify your classification scheme and if so, how?

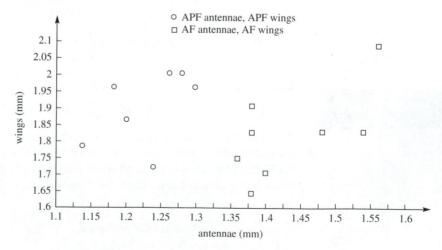

FIGURE A.8   Display of data collected by Grogan and Wirth (1981)

1990: THE BRAIN–DRUG PROBLEM

Researchers on brain disorders test the effects of the new medical drugs—for example, dopamine against Parkinson's disease—with intracerebral injections. To this end, they must estimate the size and the shape of the spatial distribution of the drug after the injection in order to estimate accurately the region of the brain that the drug has affected.

The research data consist of the measurements of the amounts of drug in each of 50 cylindrical tissue samples (see Figure A.9 and Table A.2). Each cylinder has length 0.76 mm and diameter 0.66 mm. The centers of the parallel cylinders lie on a

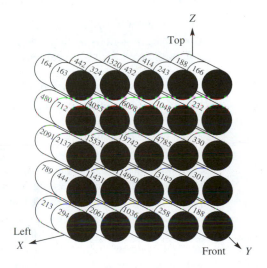

FIGURE A.9   Orientation of the cylinders of tissue

TABLE A.2   Amounts of drug in each of 50 cylindrical tissue samples

Rear vertical section				
164	442	1320	414	188
480	7022	14411	5158	352
2091	23027	28353	13138	681
789	21260	20921	11731	727
213	1303	3765	1715	453
Front vertical section				
163	324	432	243	166
712	4055	6098	1048	232
2137	15531	19742	4785	330
444	11431	14960	3182	301
294	2061	1036	258	188

grid with mesh 1 mm $\times$ 0.76 mm $\times$ 1 mm so that the cylinders touch one another on their circular bases but not along their sides, as shown in the accompanying figure. The injection was made near the center of the cylinder with the highest scintillation count. Naturally, one expects that there is drug also between the cylinders and outside the region covered by the sample.

Estimate the distribution in the region affected by the drug.

One unit represents a scintillation count, or 4.753×10^{-13} mole of dopamine. For example, the table shows that the middle rear cylinder contains 28,353 units.

1991: THE WATER TANK PROBLEM

Some state water-right agencies require from communities data on the rate of water use, in gallons per hour, and the total amount of water used each day. Many communities do not have equipment to measure the flow of water in or out of the municipal tank. Instead, they can measure only the *level* of water in the tank, within 0.5% accuracy, every hour. More importantly, whenever the level in the tank drops below some minimum level L, a pump fills the tank up to the maximum level, H; however, there is no measurement of the pump flow, either. Thus, one cannot readily relate the level in the tank to the amount of water used while the pump is working, which occurs once or twice per day, for a couple of hours each time.

Estimate the flow out of the tank $f(t)$ at all times, even when the pump is working, and estimate the total amount of water used during the day. Table A.3 gives the real data, from an actual small town, for one day.

The table gives the time, in seconds, since the first measurement, and the level of water in the tank, in hundredths of a foot. For example, after 3316 seconds, the depth of water in the tank reached 31.10 ft. The tank is a vertical circular cylinder, with a height of 40 ft and a diameter of 57 ft. Usually, the pump starts filling the tank when the level drops to about 27 ft, and the pump stops when the level rises back to about 35.50 ft.

TABLE A.3 Water-tank levels over a single day for a small town (time is in seconds and level is in 0.01 ft)

Time	Level	Time	Level	Time	Level
0	3175	35932	pump on	68535	2842
3316	3110	39332	pump on	71854	2767
6635	3054	39435	3550	75021	2697
10619	2994	43318	3445	79254	pump on
13937	2947	46636	3350	82649	pump on
17921	2892	49953	3260	85968	3475
21240	2850	53936	3167	89953	3397
25223	2797	57254	3087	93270	3340
28543	2752	60574	3012		
32284	2697	64554	2927		

1991: THE STEINER TREE PROBLEM

The cost for a communication line between two stations is proportional to the length of the line. The cost for conventional minimal spanning trees of a set of stations can often be cut by introducing phantom stations and then constructing a new *Steiner tree*. This device allows costs to be cut by up to 13.4% ($= 1 - \sqrt{3}/2$). Moreover, a network with n stations never requires more than $n - 2$ points to construct the cheapest Steiner tree. Two simple cases are shown in Figure A.10.

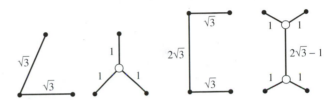

FIGURE A.10 Two simple cases of forming the shortest Steiner tree for a network

For local networks, it is often necessary to use rectilinear or checkerboard distances, instead of straight Euclidean lines. Distances in this metric are computed as shown in Figure A.11.

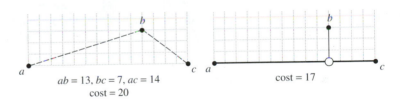

FIGURE A.11 Comparison of distances using straight Euclidean line distances ($ab = 13, bc = 7, ac = 14$; cost $= 20$) versus using rectilinear distances (cost $= 17$)

Suppose you wish to design a minimum cost spanning tree for a local network with 9 stations. Their rectangular coordinates are

$$a(0, 15), \quad b(5, 20), \quad c(16, 24), \quad d(20, 20), \quad e(33, 25),$$
$$f(23, 11), \quad g(35, 7), \quad h(25, 0), \quad i(10, 3)$$

You are restricted to using rectilinear lines. Moreover, all phantom stations must be located at lattice points (i.e., the coordinates must be integers). The cost for each line is its length.

1. Find a minimal cost tree for the network.
2. Suppose each station has a cost $d^{3/2}w$, where d = degree of the station. If $w = 1.2$, find a minimal cost tree.
3. Try to generalize this problem.

1992: THE EMERGENCY POWER-RESTORATION PROBLEM

Power companies serving coastal regions must have emergency-response systems for power outages due to storms. Such systems require the input of data that allow the time and cost required for restoration to be estimated and the value of the outage judged by objective criteria. In the past, Hypothetical Electric Company (HECO) has been criticized in the media for its lack of a prioritization scheme.

You are a consultant to HECO power company. HECO possesses a computerized database with real-time access to service calls that currently require the information:

- Time of report
- Type of requestor
- Estimated number of people affected
- Location (x, y)

Crew sites are located at coordinates $(0, 0)$ and $(40, 40)$, where x and y are in miles. The region serviced by HECO is within $-65 < x < 65$ and $-50 < y < 50$. The region is largely metropolitan with an excellent road network. Crews must return to their dispatch site only at the beginning and end of shift. Company policy requires that no work be initiated until the storm leaves the area, unless the facility is a commuter railroad or hospital, which may be processed immediately if crews are available.

HECO has hired you to develop the objective criteria and schedule the work for the storm restoration requirements listed in Table A.4 using the work force described in Table A.5. Note that the first call was received at 4:20 A.M. and that the storm left the area at 6:00 A.M. Also note that many outages were not reported until much later in the day.

TABLE A.4 Storm restoration requirements

Time (A.M.)	Location	Type	# Affected	Estimated repair time (hrs for crew)
4:20	$(-10, 30)$	Business (cable TV)	?	6
5:30	$(3, 3)$	Residential	20	7
5:35	$(20, 5)$	Business (hospital)	240	8
5:55	$(-10, 5)$	Business (railroad sys.)	25 workers; 75,000 commuters	5

(continues)

TABLE A.4 *(continued)*

Time (A.M.)	Location	Type	# Affected	Estimated repair time (hrs for crew)
6:00	All-clear given; storm leaves area; crews can be dispatched			
6:05	(13, 30)	Residential	45	2
6:06	(5, 20)	Area*	2000	7
6:08	(60, 45)	Residential	?	9
6:09	(1, 10)	Government (city hall)	?	7
6:15	(5, 20)	Business (shopping mall)	200 workers	5
6:20	(5, −25)	Government (fire dept.)	15 workers	3
6:20	(12, 18)	Residential	350	6
6:22	(7, 10)	Area*	400	12
6:25	(−1, 19)	Industry (newspaper co.)	190	10
6:40	(−20, −19)	Industry (factory)	395	7
6:55	(−1, 30)	Area*	?	6
7:00	(−20, 30)	Government (high school)	1200 students	3
7:00	(40, 20)	Government (elementary school)	1700	?
7:00	(7, −20)	Business (restaurant)	25	12
7:00	(8, −23)	Government (police station & jail)	125	7
7:05	(25, 15)	Government (elementary school)	1900	5
7:10	(−10, −10)	Residential	?	9
7:10	(−1, 2)	Government (college)	3000	8
7:10	(8, −25)	Industry (computer manuf.)	450 workers	5
7:10	(18, 55)	Residential	350	10
7:20	(7, 35)	Area*	400	9
7:45	(20, 0)	Residential	800	5
7:50	(−6, 30)	Business (hospital)	300	5
8:15	(0, 40)	Business (several stores)	50	6
8:20	(15, −25)	Government (traffic lights)	?	3
8:35	(−20, −35)	Business (bank)	20	5
8:50	(47, 30)	Residential	40	?
9:50	(55, 50)	Residential	?	12
10:30	(−18, −35)	Residential	10	10
10:30	(−1, 50)	Business (civic center)	150	5
10:35	(−7, −8)	Business (airport)	350 workers	4
10:50	(5, −25)	Government (fire dept.)	15	5
11:30	(8, 20)	Area*	300	12

TABLE A.5 Crew descriptions

- Dispatch locations at $(0, 0)$ and $(40, 40)$.
- Crews consist of three trained workers.
- Crews report to the dispatch location only at the beginning and end of their shifts.
- One crew is scheduled for duty at all times on jobs assigned to each dispatch location. These crews would normally be performing routine assignments. Until the storm leaves the area, they can be dispatched for emergencies only.
- Crews work 8-hr shifts.
- There are six crew teams available at each location.
- Crews can work only one overtime shift in a work day and receive time-and-a-half for overtime.

HECO has asked for a technical report for their purposes and an executive summary in laymen's terms that can be presented to the media. Further, they would like recommendations for the future. To determine your prioritized scheduling system, you will have to make additional assumptions. Detail those assumptions. In the future, you may desire additional data. If so, detail the information desired.

1992: THE AIR-TRAFFIC-CONTROL RADAR PROBLEM

You are to determine the power to be radiated by an air-traffic-control radar at a major metropolitan airport. The airport authority wants to minimize the power of the radar consistent with safety and cost.

The authority is constrained to operate with its existing antennae and receiver circuitry. The only option that they are considering is upgrading the transmitter circuits to make the radar more powerful.

The question that you are to answer is what power (in watts) must be released by the radar to ensure detection of standard passenger aircraft at a distance of 100 kilometers.

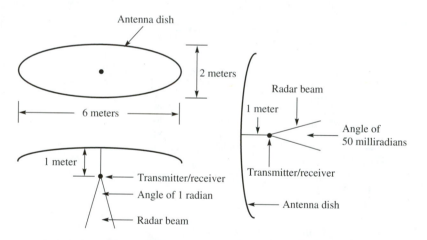

FIGURE A.12   Measurements for the radar system

Technical specifications (see also Figure A.12):

• The radar antenna is a section of a paraboloid of revolution with focal length of 1 meter. Its projection onto a plane tangent to its vertex is an ellipse with a major axis of 6 meters and a minor axis of 2 meters. The main lobe energy beam pattern, located at the focus, is an elliptical cone that has a major axis of 1 radian and a minor axis of 50 milliradians. The antenna and beam are sketched in the figures provided.

• The nominal class of aircraft is one that has an effective radar reflection cross-section of 75 square meters. For the purposes of this problem, this means that in your initial model, the aircraft is equivalent to a 100% reflective circular disc of 75 square meters, which is centered on the axis of the antennae and is perpendicular to it. You may want to consider alternatives or refinements to this initial model.

1993: THE COAL-TIPPLE OPERATIONS PROBLEM

The Aspen-Boulder Coal Company runs a loading facility consisting of a large coal tipple. When the coal trains arrive, they are loaded from the tipple. The standard coal train takes 3 hours to load, and the tipple's capacity is 1.5 standard trainloads of coal. Each day, the railroad sends three standard trains to the loading facility, and they arrive at any time between 5 A.M. and 8 P.M. local time. Each of the trains has three engines. If a train arrives and sits idle while waiting to be loaded, the railroad charges a special fee, called a *demurrage*. The fee is $5,000 per engine per hour. In addition, a high-capacity train arrives once a week every Thursday between 11 A.M. and 1 P.M. This special train has five engines and holds twice as much coal as a standard train. An empty tipple can be loaded directly from the mine to its capacity in six hours by a single loading crew. This crew (and its associated equipment) costs $9,000 per hour. A second crew can be called out to increase the loading rate by conducting an additional tipple-loading operation at the cost of $12,000 per hour. Because of safety requirements, during tipple loading no trains can be loaded. Whenever train loading is interrupted to load the tipple, demurrage charges are in effect.

The management of the Coal Company has asked you to determine the expected annual costs of this tipple's loading operations. Your analysis should include the following considerations:

• How often should the second crew be called out?
• What are the expected monthly demurrage costs?
• If the standard trains could be scheduled to arrive at precise times, what daily schedule would minimize loading costs?
• Would a third tipple-loading crew at $12,000 per hour reduce annual operations costs?
• Can this tipple support a fourth standard train every day?

1993: THE OPTIMAL COMPOSTING PROBLEM

An environmentally conscious institutional cafeteria is recycling customers' uneaten food into compost by means of microorganisms. Each day, the cafeteria blends the leftover food into a slurry, mixes the slurry with crisp salad wastes from the kitchen and a small amount of shredded newspaper, and feeds the resulting mixture to a culture of fungi and soil bacteria, which digest slurry, greens, and paper into usable compost. The crisp greens provide pockets of oxygen for the fungi culture, and the paper absorbs excess humidity. At times, however, the fungi culture appears unable or unwilling to digest as much of the leftovers as customers leave; the cafeteria does not blame the chef for the fungi culture's lack of appetite. Also, the cafeteria has received offers for the purchase of large quantities of its compost. Therefore, the cafeteria is investigating ways to increase its production of compost. Since it cannot yet afford to build a new composting facility, the cafeteria seeks methods to accelerate the fungi culture's activity, for instance, by optimizing the fungi culture's environment (currently held at about 120°F and 100% humidity), or by optimizing the composition of the mixture fed to the fungi culture, or both.

Determine whether any relation exists between the proportions of slurry, greens, and paper in the mixture fed to the fungi culture and the rate at which the fungi culture composts the mixture. If no relation exists, state so. Otherwise, determine what proportions would accelerate the fungi culture's activity.

In addition to the technical report following the format prescribed in the contest instructions, provide a one-page nontechnical recommendation for implementation for the cafeteria manager.

Table A.6 shows the composition of various mixtures in pounds of each ingredient kept in separate bins and the time it took the fungi culture to compost the mixtures, from the date fed to the date completely composted.

TABLE A.6 Composting data

Slurry (pounds)	Greens (pounds)	Paper (pounds)	Fed (date)	Composted (date)
86	31	0	13 Jul 90	10 Aug 90
112	79	0	17 Jul 90	13 Aug 90
71	21	0	24 Jul 90	20 Aug 90
203	82	0	27 Jul 90	22 Aug 90
79	28	0	10 Aug 90	12 Sep 90
105	52	0	13 Aug 90	18 Sep 90
121	15	0	20 Aug 90	24 Sep 90
110	32	0	22 Aug 90	8 Oct 90
82	44	9	30 Apr 91	18 Jun 91
57	60	7	2 May 91	20 Jun 91
77	51	7	7 May 91	25 Jun 91
52	38	6	10 May 91	28 Jun 91

1994: THE CONCRETE SLAB PROBLEM

The U.S. Dept. of Housing and Urban Development (HUD) is considering constructing dwellings of various sizes, ranging from individual houses to large apartment complexes. A principal concern is to minimize recurring costs to occupants, especially the costs of heating and cooling. The region in which the construction is to take place is temperate, with a moderate variation in temperature throughout the year.

With special construction techniques, HUD engineers can build dwellings that do not need to rely on convection—that is, there is no need to rely on opening doors or windows to assist in temperature variation. The dwellings will be single-story, with concrete slab floors as the only foundation. You have been hired as a consultant to analyze the temperature variation in the concrete slab floor to determine if the temperature averaged over the floor surface can be maintained within a prescribed comfort zone throughout the year. If so, what size/shape of slabs will permit this?

Part 1, Floor Temperature

Consider the temperature variation in a concrete slab given that the ambient temperature varies daily within the ranges given in Table A.7. Assume that the high occurs at noon and the low at midnight. Determine if slabs can be designed to maintain a temperature averaged over the floor surface within the prescribed comfort zone considering radiation only. Initially, assume that the heat transfer into the dwelling is through the exposed perimeter of the slab and that the top and bottom of the slabs are insulated. Comment on the appropriateness and sensitivity of these assumptions. If you cannot find a solution that satisfies Table A.7, can you find designs that satisfy a Table A.7 that you propose?

TABLE A.7 Daily variation in temperature

Ambient temperature		Comfort zone	
High:	85°F	High:	76°F
Low:	60°F	High:	65°F

Part 2, Building Temperature

Analyze the practicality of the initial assumptions and extend the analysis to temperature variation within the single-story dwelling. Can the house be kept within the comfort zone?

Part 3, Cost of Construction

Suggest a design that considers HUD's objective of reducing or eliminating heating and cooling costs, considering construction restrictions and costs.

1994: THE COMMUNICATIONS NETWORK PROBLEM

In your company, information is shared among departments on a daily basis. This information includes the previous day's sales statistics and current production guidance. It is important to get this information out as quickly as possible.

Suppose that a communications network is to be used to transfer blocks of data (files) from one computer to another. As an example, consider the graph model in Figure A.13.

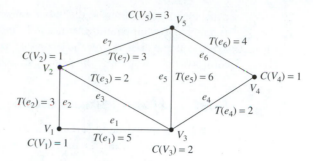

FIGURE A.13   Example of a file transfer network

Vertices $V_1, V_2, \ldots, V_m$ represent computers, and edges $e_1, e_2, \ldots, e_n$ represent files to be transferred (between computers represented by edge endpoints). $T(e_x)$ is the time that it takes to transfer file e_x, and $C(V_y)$ is the capacity of the computer represented by V_y to transfer files simultaneously. A file transfer involves the engagement of both computers for the entire time it takes to transfer the file. For example, $C(V_y) = 1$ means that computer V_y can be involved in only one transfer at a time.

We are interested in scheduling the transfers in an optimal way, to minimize the total time that it takes to complete them all. This minimum total time is called the *makespan*. Consider the following situations for your company:

Situation A

Your corporation has 28 departments. Each department has a computer, each of which is represented by a vertex in Figure A.14. Each day, 27 files must be transferred, represented by the edges in Figure A.14. For this network, $T(e_x) = 1$ and $C(V_y) = 1$ for all x and y. Find an optimal schedule and the makespan for the given network. Can you prove to your supervisor that your makespan is the smallest

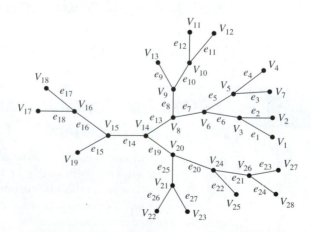

FIGURE A.14   Network for situations A and B

possible (optimal) for the given network? Describe your approach to solving the problem. Does our approach work for the general case; that is, where $T(e_x)$, $C(V_y)$, and the graph structure are arbitrary?

Situation B

Suppose that your company changes the requirements for data transfer. You must now consider the same basic network structure (again, see Figure A.14) with different types and sizes of files. These files take the amount of time to transfer indicated in Table A.8 by the $T(e_x)$ terms for each edge. We still have $C(V_y) = 1$ for all y. Find an optimal schedule and the makespan for the new network. Can you prove that your makespan is the smallest possible for the new network? Describe your approach to solving this problem. Does your approach work for the general case? Comment on any peculiar or unexpected results.

TABLE A.8 File transfer time data for situation B

x	1	2	3	4	5	6	7	8	9
$T(e_x)$	3.0	4.1	4.0	7.0	1.0	8.0	3.2	2.4	5.0
x	10	11	12	13	14	15	16	17	18
$T(e_x)$	8.0	1.0	4.4	9.0	3.2	2.1	8.0	3.6	4.5
x	19	20	21	22	23	24	25	26	27
$T(e_x)$	7.0	7.0	9.0	4.2	4.4	5.0	7.0	9.0	1.2

Situation C

Your corporation is considering expansion. If that happens, there are several new files (edges) that will need to be transferred daily. This expansion will also include an upgrade of the computer system. Some of the 28 departments will get new computers that can handle more than one transfer at a time. All of these changes are indicated in Figure A.15 and Tables A.9 and A.10. What is the best schedule

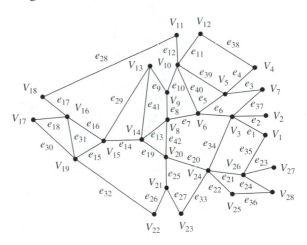

FIGURE A.15 Network for situation C

TABLE A.9 File transfer time data for situation C, for the added transfers

x	28	29	30	31	32	33	34	35
$T(e_x)$	6.0	1.1	5.2	4.1	4.0	7.0	2.4	9.0

x	36	37	38	39	40	41	42
$T(e_x)$	3.7	6.3	6.6	5.1	7.1	3.0	6.1

TABLE A.10 Computer capacity data for situation C

y	1	2	3	4	5	6	7	8	9	10
$C(V_y)$	2	2	1	1	1	1	1	1	2	3

y	11	12	13	14	15	16	17	18	19
$C(V_y)$	1	1	1	2	1	2	1	1	1

y	20	21	22	23	24	25	26	27	28
$C(V_y)$	1	1	2	1	1	1	2	1	1

and makespan that you can find? Can you prove that your makespan is the smallest possible for this network? Describe your approach to solving the problem. Comment on any peculiar or unexpected results.

1995: THE SINGLE HELIX

A small biotechnical company must design, prove, program, and test a mathematical algorithm to locate in real time all the intersections of a helix and a plane in general positions in space.

Computer Aided Geometric Design (CAGD) programs enable engineers to view a plane section of an object they design, such as an automobile suspension or a medical device. Engineers may also display on the plane section quantities such as air flow, stress, or temperature, coded by colors or level curves. Plane sections may be rapidly swept through the entire object to gain a three-dimensional visualization of the object and its reactions to motion, forces, or heat. To achieve such results, the computer programs must quickly and accurately locate all the intersections of the viewed plane and every part of the designed object. General equation solvers may in principle compute such intersections, but for specific problems, specific methods may prove faster and more accurate than general methods. In particular, general CAGD software may prove too slow to complete computations in real time or too large to fit in the company's finished medical devices. These considerations have led the company to the following problem.

Problem

Design, justify, program, and test a method to compute all the intersections of a plane and a helix, both in general positions (at any locations and with any orientations) in space. A segment of the helix may represent, for example, a helicoidal suspension spring or a piece of tubing in a chemical or medical apparatus.

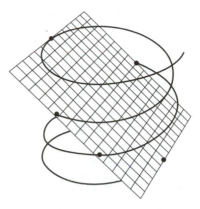

Theoretical justification of the proposed algorithm is necessary to verify the solution from several points of view; for instance, through mathematical proofs of parts of the algorithm and through tests of the final program with known examples. Such documentation and tests will be required by government agencies for medical use.

1995: ALUACHA BALACLAVA COLLEGE

Aluacha Balaclava College, an undergraduate facility, has just hired a new Provost whose first priority is the institution of a fair and reasonable faculty-compensation system. She has hired your consulting team to design a compensation system that reflects the following circumstances and principles:

Faculty are ranked as Instructor, Assistant Professor, Associate Professor, and Professor. Those with Ph.D. degrees are hired at the rank of Assistant Professor. Ph.D. candidates are hired at the rank of Instructor and promoted automatically to Assistant Professor upon completion of their degrees. Faculty may apply for promotion from Associate Professor to Professor after serving at the rank of Associate for seven or more years. Promotions are determined by the Provost with recommendations from a faculty committee.

Faculty salaries are for the 10-month period September through June, with raises effective beginning in September. The total amount of money available for raises varies yearly and is generally disclosed in March for the following year.

The starting salary this year for an Instructor with no prior teaching experience was $27,000; $32,000 for an Assistant Professor. Upon hire, faculty can receive credit for up to 7 years of teaching experience at other institutions.

Principles

• All faculty should get a raise any year that money is available.

• Promotion should incur a substantial benefit; e.g., promotion in the minimum-possble time should result in a benefit roughly equal to 7 years of normal raises.

• Faculty promoted after 7 or 8 years in rank with careers of at least 25 years should make roughly twice as much at retirement as a starting Ph.D.

• Experienced faculty should be paid more than less-experienced in the same rank. The effect of additional years of experience should diminish over time; i.e., if two faculty stay in the same rank, their salaries should equalize over time.

Design a new pay system, first without cost-of-living increases. Incorporate cost-of-living increases, and then finally, design a transition process for current faculty that will move all salaries toward your system without reducing anyone's salary. Existing faculty salaries, ranks, and years of service are in Table A.11. Discuss any refinements you think would improve your system.

The Provost requires a detailed compensation system plan for implementation, as well as a brief, clear, executive summary outlining the model, its assumptions, strengths, weaknesses, and expected results, which she can present to the Board and faculty.

TABLE A.11 Salary data for Aluacha Balaclava College

Case	Years	Rank	Salary	Case	Years	Rank	Salary	Case	Years	Rank	Salary
1	4	ASSO	54,000	2	19	ASST	43,508	3	20	ASST	39,072
4	11	PROF	53,900	5	15	PROF	44,206	6	17	ASST	37,538
7	23	PROF	48,844	8	10	ASST	32,841	9	7	ASSO	49,981
10	20	ASSO	43,549	11	18	ASSO	42,649	12	19	PROF	60,087
13	15	ASSO	38,002	14	4	ASST	30,000	15	34	PROF	60,576
16	28	ASST	44,562	17	9	ASST	30,893	18	22	ASSO	46,351
19	21	ASSO	50,979	20	20	ASST	48,000	21	4	ASST	32,500
22	14	ASSO	38,462	23	23	PROF	53,500	24	21	ASSO	42,488
25	20	ASSO	43,892	26	5	ASST	35,330	27	19	ASSO	41,147
28	15	ASST	34,040	29	18	PROF	48,944	30	7	ASST	30,128
31	5	ASST	35,330	32	6	ASSO	35,942	33	8	PROF	57,295
34	10	ASST	36,991	35	23	PROF	60,576	36	20	ASSO	48,926
37	9	PROF	57,956	38	32	ASSO	52,214	39	15	ASST	39,259
40	22	ASSO	43,672	41	6	INST	45,500	42	5	ASSO	52,262
43	5	ASSO	57,170	44	16	ASST	36,958	45	23	ASST	37,538
46	9	PROF	58,974	47	8	PROF	49,971	48	23	PROF	62,742
49	39	ASSO	52,058	50	4	INST	26,500	51	5	ASST	33,130
52	46	PROF	59,749	53	4	ASSO	37,954	54	19	PROF	45,833
55	6	ASSO	35,270	56	6	ASSO	43,037	57	20	PROF	59,755
58	21	PROF	57,797	59	4	ASSO	53,500	60	6	ASST	32,319
61	17	ASSO	35,668	62	20	PROF	59,333	63	4	ASST	30,500
64	16	ASSO	41,352	65	15	PROF	43,264	66	20	PROF	50,935
67	6	ASST	45,365	68	6	ASSO	35,941	69	6	ASST	49,134
70	4	ASST	29,500	71	4	ASST	30,186	72	7	ASST	32,400

(continues)

TABLE A.11 (continued)

Case	Years	Rank	Salary	Case	Years	Rank	Salary	Case	Years	Rank	Salary
73	12	ASSO	44,501	74	2	ASST	31,900	75	1	ASSO	62,500
76	1	ASST	34,500	77	16	ASSO	40,637	78	4	ASSO	35,500
79	21	PROF	50,521	80	12	ASST	35,158	81	4	INST	28,500
82	16	PROF	46,930	83	24	PROF	55,811	84	6	ASST	30,128
85	16	PROF	46,090	86	5	ASST	28,570	87	19	PROF	44,612
88	17	ASST	36,313	89	6	ASST	33,479	90	14	ASSO	38,624
91	5	ASST	32,210	92	9	ASSO	48,500	93	4	ASST	35,150
94	25	PROF	50,583	95	23	PROF	60,800	96	17	ASST	38,464
97	4	ASST	39,500	98	3	ASST	52,000	99	24	PROF	56,922
100	2	PROF	78,500	101	20	PROF	52,345	102	9	ASST	35,798
103	24	ASST	43,925	104	6	ASSO	35,270	105	14	PROF	49,472
106	19	ASSO	42,215	107	12	ASST	40,427	108	10	ASST	37,021
109	18	ASSO	44,166	110	21	ASSO	46,157	111	8	ASST	32,500
112	19	ASSO	40,785	113	10	ASSO	38,698	114	5	ASST	31,170
115	1	INST	26,161	116	22	PROF	47,974	117	10	ASSO	37,793
118	7	ASST	38,117	119	26	PROF	62,370	120	20	ASSO	51,991
121	1	ASST	31,500	122	8	ASSO	35,941	123	14	ASSO	39,294
124	23	ASSO	51,991	125	1	ASST	30,000	126	15	ASST	34,638
127	20	ASSO	56,836	128	6	INST	35,451	129	10	ASST	32,756
130	14	ASST	32,922	131	12	ASSO	36,451	132	1	ASST	30,000
133	17	PROF	48,134	134	6	ASST	40,436	135	2	ASSO	54,500
136	4	ASSO	55,000	137	5	ASST	32,210	138	21	ASSO	43,160
139	2	ASST	32,000	140	7	ASST	36,300	141	9	ASSO	38,624
142	21	PROF	49,687	143	22	PROF	49,972	144	7	ASSO	46,155
145	12	ASST	37,159	146	9	ASST	32,500	147	3	ASST	31,500
148	13	INST	31,276	149	6	ASST	33,378	150	19	PROF	45,780
151	5	PROF	70,500	152	27	PROF	59,327	153	9	ASSO	37,954
154	5	ASSO	36,612	155	2	ASST	29,500	156	3	PROF	66,500
157	17	ASST	36,378	158	5	ASSO	46,770	159	22	ASST	42,772
160	6	ASST	31,160	161	17	ASST	39,072	162	20	ASST	42,970
163	2	PROF	85,500	164	20	ASST	49,302	165	21	ASSO	43,054
166	21	PROF	49,948	167	5	PROF	50,810	168	19	ASSO	51,378
169	18	ASSO	41,267	170	18	ASST	42,176	171	23	PROF	51,571
172	12	PROF	46,500	173	6	ASST	35,798	174	7	ASST	42,256
175	23	ASSO	46,351	176	22	PROF	48,280	177	3	ASST	55,500
178	15	ASSO	39,265	179	4	ASST	29,500	180	21	ASSO	48,359
181	23	PROF	48,844	182	1	ASST	31,000	183	6	ASST	32,923
184	2	INST	27,700	185	16	PROF	40,748	186	24	ASSO	44,715
187	9	ASSO	37,389	188	28	PROF	51,064	189	19	INST	34,265
190	22	PROF	49,756	191	19	ASST	36,958	192	16	ASST	34,550
193	22	PROF	50,576	194	5	ASST	32,210	195	2	ASST	28,500
196	12	ASSO	41,178	197	22	PROF	53,836	198	19	ASSO	43,519
199	4	ASST	32,000	200	18	ASSO	40,089	201	23	PROF	52,403
202	21	PROF	59,234	203	22	PROF	51,898	204	26	ASSO	47,047

1996: THE SUBMARINE DETECTION PROBLEM

The world's oceans contain an ambient noise field. Seismic disturbances, surface shipping, and marine mammals are sources that, in different frequency ranges, contribute to this field. We wish to consider how this ambient noise might be used to detect large moving objects; e.g., submarines located below the ocean surface. Assuming that a submarine makes no intrinsic noise, develop a method for detecting the presence of a moving submarine, its speed, its size, and its direction of travel using only information obtained by measuring changes to the ambient noise field. Begin with noise at one fixed frequency and amplitude.

1996: THE CONTEST JUDGING PROBLEM

When determining the winner of a competition like the Mathematical Contest in Modeling, there is generally a large number of papers to judge. Let's say there are $P = 100$ papers. A group of J judges is collected to accomplish the judging. Funding for the contest constrains both the number of judges that can be obtained and the amount of time that they can judge. For example, if $P = 100$, then $J = 8$ is typical.

Ideally, each judge would read each paper and rank-order them, but there are too many papers for this. Instead, there will be a number of screening rounds in which each judge will read some number of papers and give them scores. Then some selection scheme is used to reduce the number of papers under consideration: If the papers are rank-ordered, then the bottom 30% that each judge rank-orders could be rejected. Alternatively, if the judges do not rank-order the papers, but instead give them numerical scores (say, from 1 to 100), then all papers falling below some cutoff level could be rejected.

The new pool of papers is then passed back to the judges, and the process is repeated. A concern is that the total number of papers that each judge reads must be substantially less than P. The process is stopped when there are only W papers left. These are the winners. Typically for $P = 100$, $W = 3$.

Your task is to determine a selection scheme, using a combination of rank-ordering, numerical scoring, and other methods, by which the final W papers will include only papers from among the best $2W$ papers. (By "best," we assume that there is an absolute rank-ordering to which all judges would agree.) For example, the top three papers found by your method will consist entirely of papers from among the best six papers. Among all such methods, the one that requires each judge to read the least number of papers is desired.

Note the possibility of systematic bias in a numerical scoring scheme. For example, for a specific collection of papers, one judge could average 70 points, while another could average 80 points. How would you scale your scheme to accommodate for changes in the content parameters (P, J, and W)?

For further information on the Mathematics Contest in Modeling (MCM), write to: COMAP, 57 Bedford St., Lexington, MA 02173.

An Elevator Simulation Algorithm

We will define the terms used in the following algorithm and explain some of its underlying logic. Because the algorithm is rather complex, this approach should be more revealing than using some hypothetical numbers and taking you step by step through the algorithm. (This is a difficult program to write if you are not using GPSS or another simulation language.)

During the simulation there is a TIME clock that keeps track of the time (given in seconds). Initially, the value of TIME is 0 sec (at 7:50 A.M.), and the simulation ends when TIME reaches 4800 sec (at 9:10 A.M.). Each customer is assigned a number according to the order of his or her arrival: The first customer is labeled 1, the second customer 2, and so forth. Whenever another customer arrives at the lobby, the time between the customer's arrival and the time when the immediately preceding customer arrived is added to the TIME clock. This time between successive arrivals of customers i and $i - 1$ is labeled $between_i$ in the algorithm, and the arrival time of customer i is labeled $arrive_i$. Initially, for the customer arrival submodel, we assume that all values between 0 and 30 have an equal likelihood of occurring.

All four elevators have their own availability times, called $return_j$ for elevator j. If elevator j is currently available at the main floor, its time is the current time, so $return_j = $ TIME. If an elevator is in transit, its availability time is the time at which it will return to the main floor. Passengers enter an available elevator in the numerical order of the elevators: first elevator 1 (if it is available), next elevator 2 (if it is available), and so on. Maximum occupancy of an elevator is 12 passengers.

Whenever another customer arrives in the lobby of the building, two possible situations exist. Either an elevator is available for receiving passengers, or no elevator is available and a queue is forming as customers wait for one to become available. Once an elevator becomes available, it is "tagged" for loading and passengers can enter *only* that elevator until either it is fully occupied (with 12 passengers) or the 15-sec time delay is exceeded before the arrival of another customer. After loading, the elevator departs to deliver all of its passengers. Even if fully loaded with 12 passengers, it is assumed the elevator awaits 15 sec to load the last passenger, allow floor selection, and get under way.

To keep track of which floors have been selected during the loading period of an elevator and the number of times a particular floor has been selected, the algorithm sets up two one-dimensional arrays (having a component for each of the floors 1–12). (Although no one selects floor 1, the indexing is simplified with its inclusion.) These arrays are called *selvec$_j$* and *flrvec$_j$* for the tagged elevator j. If a customer selects floor 5, for instance, than a 1 is entered into the 5th component position of selvec$_j$ and also into the 5th component of flrvec$_j$. If another customer selects floor 5, then the 5th component of flrvec$_j$ is updated to a 2, and so forth. For example, suppose the passengers in elevator j have selected floor 3 twice, floor 5 twice, and floors 7, 8, and 12 once. Then for this elevator, selvec$_j$ = $(0, 0, 1, 0, 1, 0, 1, 1, 0, 0, 0, 1)$ and flrvec$_j$ = $(0, 0, 2, 0, 2, 0, 1, 1, 0, 0, 0, 1)$. These arrays are then used to calculate the transport time of elevator j so we can determine when it will return to the main floor. As stated previously, that return time is designated *return$_j$*. The arrays are also used to calculate the delivery times of each passenger in elevator j. Initially, we assume that a customer chooses a floor with equal likelihood. We assume that it takes 10 sec for an elevator to travel between floors, 10 sec to open *and* close its doors, and 3 sec for each passenger to disembark. We also assume that it takes 3 sec for each passenger in a queue to enter the next available elevator.

Summary of Elevator Simulation Algorithm Terms

between$_i$ time between successive arrivals of customers i and $i - 1$ (a random integer varying between 0 and 30 sec)

arrive$_i$ time of arrival from start of clock at $t = 0$ for customer i (calculated only if customer enters a queue waiting for an elevator)

floor$_i$ floor selected by customer i (a random integer varying between 2 and 12)

elevator$_i$ time customer i spends in an elevator

wait$_i$ time customer i waits before stepping into an elevator (calculated only if customer enters a queue waiting for an elevator)

delivery$_i$ time required to deliver customer i to destination floor from time of arrival, including any waiting time

selvec$_j$ binary 0, 1 one-dimensional array representing the floors selected for elevator j, not counting the number of times a particular floor has been selected

flrvec$_j$ integer one-dimensional array representing the number of times each floor has been selected for elevator j for the group of passengers currently being transported to their respective floors

occup$_j$ number of current occupants of elevator j

return$_j$ time from start of clock at $t = 0$ that elevator j returns to the main floor and is available for receiving passengers

first$_j$ an index, the customer number of the first passenger who enters elevator j after it returns to the main floor

quecust   customer number of the first person waiting in the queue

queue	total length of current queue of customers waiting for an elevator to become available
startque	clock time at which the (possibly updated) current queue commences to form
$stop_j$	total number of stops made by elevator j during the entire simulation
$eldel_j$	total time elevator j spends in delivering its current load of passengers
$operate_j$	total time elevator j operates during the entire simulation
limit	customer number of the last person to enter an available elevator before it commences transport
max	largest index of a nonzero entry in the array $selvec_j$ (highest floor selected)
remain	number of customers left in the queue after loading next available elevator
quetotal	total number of customers who spent time waiting
TIME	current clock time in seconds, starting at $t = 0$
DELTIME	average delivery time of a customer to reach destination floor from time of arrival, including any waiting time
ELEVTIME	average time a person spends in an elevator
MAXDEL	maximum time required for a customer to reach his or her floor of destination from time of arrival
MAXELEV	maximum time a customer spends in an elevator
QUELEN	number of customers waiting in the longest queue
QUETIME	average time a customer who must wait spends in a queue
MAXQUE	longest time a customer spends in a queue

Elevator System Simulation Algorithm

Input	None required.
Output	Number of passengers serviced, DELTIME, ELEVTIME, MAXDEL, MAXELEV, QUELEN, QUETIME, MAXQUE, $stop_j$, and the percent time each elevator is in use.
Step 1	Initially set the following parameters to zero: DELTIME, ELEVTIME, MAXDEL, MAXELEV, QUELEN, QUETIME, MAXQUE, quetotal, remain.
Step 2	For the first customer, generate time between successive arrivals and floor destination, and initialize delivery time:

$$i = 1$$

Generate $between_i$ and $floor_i$

$$delivery_i = 15$$

Step 3	Initialize clock time, elevator available clock times, elevator stops, and elevator operating times. Also, initialize all customer waiting times.

$$\text{TIME} = \text{between}_i$$

For $k = 1$ to 4: $\text{return}_k = \text{TIME}$ and $\text{stop}_k = \text{operate}_k = 0$

For $k = 1$ to 400: $\text{wait}_k = 0$

(* The number 400 is an upper-bound guess for the total number of customers *)

Step 4 While $\text{TIME} \leq 4800$, do Steps 5–32:

Step 5 Select the first available elevator:

If $\text{TIME} \geq \text{return}_1$, then $j = 1$ else

If $\text{TIME} \geq \text{return}_2$, then $j = 2$ else

If $\text{TIME} \geq \text{return}_3$, then $j = 3$ else

If $\text{TIME} \geq \text{return}_4$, then $j = 4$

ELSE (* no elevator is currently available *) GOTO Step 19.

Step 6 Set as an index the customer number of first person to occupy tagged elevator, and initialize the elevator occupancy floor selection vectors:

$$\text{first}_j = i, \qquad \text{occup}_j = 0$$

For $k = 1$ to 12: $\text{selvec}_j[k] = \text{flrvec}_j[k] = 0$

Step 7 Load current customer on elevator j by setting the floor selection vectors and incrementing elevator occupancy:

$$\text{selvec}_j[\text{floor}_i] = 1$$

$$\text{flrvec}_j[\text{floor}_i] = \text{flrvec}_j[\text{floor}_i] + 1$$

$$\text{occup}_j = \text{occup}_j + 1$$

Step 8 Get next customer and update clock time:

$$i = i + 1$$

Generate between_i and floor_i

$$\text{TIME} = \text{TIME} + \text{between}_i$$

$\text{delivery}_i = 15$

Step 9 Set all available elevators to current clock time:

For $k = 1$ to 4:

If $\text{TIME} \geq \text{return}_k$, then $\text{return}_k = \text{TIME}$.

Else leave return_k as is.

Step 10 If $\text{between}_i \leq 15$ and $\text{occup}_j < 12$, then increase the delivery times for each customer on the tagged elevator j:

For $k = \text{first}_j$ to $i - 1$:

$\text{delivery}_k = \text{delivery}_k + \text{between}_i$

and GOTO Step 7 to load current customer on the elevator and get the next customer.

Else (* send off the tagged elevator *):

Set $\text{limit} = i - 1$ and GOTO Step 11.

The sequence of Steps 11–18 implements delivery of all passengers on the currently tagged elevator j:

Step 11   For $k = \text{first}_j$ to limit, do Steps 12–16.

Step 12   Calculate time customer k spends in elevator:

$$N = \text{floor}_k - 1 \text{ (an index)}$$

$\text{elevator}_k = $ travel time up to floor + time to drop off previous
customers + customer k drop off time
+ open/close door times on previous floors
+ open door on current floor

$$= 10N + 3\sum_{m=1}^{N} \text{flrvec}_j[m] + 3 + 10\sum_{m=1}^{N} \text{selvec}_j[m] + 5$$

Step 13   Calculate delivery time for customer k:

$$\text{delivery}_k = \text{delivery}_k + \text{elevator}_k$$

Step 14   Sum to total delivery time for averaging:

$$\text{DELTIME} = \text{DELTIME} + \text{delivery}_k$$

Step 15   If $\text{delivery}_k > \text{MAXDEL}$, then $\text{MAXDEL} = \text{delivery}_k$.
Else leave MAXDEL as is.

Step 16   If $\text{elevator}_k > \text{MAXELEV}$, then $\text{MAXELEV} = \text{elevator}_k$.
Else leave MAXELEV as is.

Step 17   Calculate total number of stops for elevator j, its time in transit, and the time at which it returns to the main floor:

$$\text{stop}_j = \text{stop}_j + \sum_{m=1}^{12} \text{selvec}_j[m]$$

$\text{Max} = $ index of largest nonzero entry in selvec_j
(i.e., the highest floor visited)

$\text{eldel}_j = $ travel time + passenger drop-off time + doors time

$$= 20(\text{Max} - 1) + 3\sum_{m=1}^{12} \text{flrvec}_j[m] + 10\sum_{m=1}^{12} \text{selvec}_j[m]$$

$$\text{return}_j = \text{TIME} + \text{eldel}_j$$

$$\text{operate}_j = \text{operate}_j + \text{eldel}_j$$

Step 18   GOTO Step 5.

The sequence of Steps 19–32 is taken when no elevator is currently available and a queue of customers waiting for elevator service is set up.

Step 19   Initialize queue:

$\text{quecust} = i$ (number for first customer in queue)

$\text{startque} = \text{TIME}$ (starting time of queue)

$\text{queue} = 1$

$\text{arrive}_i = \text{TIME}$

Step 20   Get the next customer and update clock time:

$$i = i + 1$$

Generate between$_i$ and floor$_i$

$$\text{TIME} = \text{TIME} + \text{between}_i$$

$$\text{arrive}_i = \text{TIME}$$

$$\text{queue} = \text{queue} + 1$$

Step 21   Check for elevator availability:

If TIME $\geq$ return$_1$, then $j = 1$ and GOTO Step 22 else

If TIME $\geq$ return$_2$, then $j = 2$ and GOTO Step 22 else

If TIME $\geq$ return$_3$, then $j = 3$ and GOTO Step 22 else

If TIME $\geq$ return$_4$, then $j = 4$ and GOTO Step 22

ELSE ($*$ no elevator is available yet $*$) GOTO Step 20.

Step 22   Elevator j is available. Initialize floor selection vectors and assess the length of the queue:

For $k = 1$ to 12: selvec$_j[k]$ = flrvec$_j[k]$ = 0

$$\text{remain} = \text{queue} - 12$$

Step 23   If remain ≤ 0, then $R = i$ and occup$_j$ = queue.
Else R = quecust + 11 and occup$_j$ = 12.

Step 24   Load customers onto elevator j:

For k = quecust to R:

selvec$_j$[floor$_k$] = 1   and   flrvec$_j$[floor$_k$] = flrvec$_j$[floor$_k$] + 1

Step 25   If queue $\geq$ QUELEN, then QUELEN = queue.
Else leave QUELEN as is.

Step 26   Update queuing totals:

$$\text{quetotal} = \text{quetotal} + \text{occup}_j$$

$$\text{QUETIME} = \text{QUETIME} + \sum_{m=\text{quecust}}^{R} [\text{TIME} - \text{arrive}_m]$$

Step 27   If (TIME − startque) $\geq$ MAXQUE, then MAXQUE = TIME − startque.
Else leave MAXQUE as is.

Step 28   Set index giving number of first customer to occupy tagged elevator:

$$\text{first}_j = \text{quecust}$$

Step 29   Calculate delivery and waiting times for each passenger on the tagged elevator:

For $k = \text{first}_j$ to R:

delivery$_k$ = 15 + (TIME − arrive$_k$)

wait$_k$ = TIME − arrive$_k$

Step 30   If remain ≤ 0, then set queue = 0 and GOTO Step 8 to get next customer.
Else set limit = R and, for $k = \text{first}_j$ to limit, do Steps 12–17. When finished, GOTO Step 31.

Step 31   Update queue length and check for elevator availability:

$$\text{queue} = \text{remain}$$

$$\text{quecust} = R + 1$$

$$\text{startque} = \text{arrive}_{R+1}$$

Step 32   GOTO Step 20.

The sequence of Steps 33–36 calculates output values for the morning rush-hour elevator simulation.

Step 33   Output the following values:
$N = i -$ queue, the total number of customers served
DELTIME = DELTIME/N, average delivery time
MAXDEL, maximum delivery time of a customer

Step 34   Output the average time spent in an elevator and the maximum time spent in an elevator:

$$\text{ELEVTIME} = \sum_{m=1}^{\text{limit}} \frac{\text{elevator}[m]}{\text{limit}} \quad \text{and MAXELEV}$$

Step 35   Output the number of customers waiting in the longest queue, the average time a customer who waits in line spends in a queue, and the longest time spent in a queue:

$$\text{QUELEN}$$

$$\text{QUETIME} = \frac{\text{QUETIME}}{\text{quetotal}}$$

$$\text{MAXQUE}$$

Step 36   Output the total number of stops for each elevator and the percent time each elevator is in transport:

$$\text{For } k = 1 \text{ to } 4\text{: display stop}_k \text{ and } \frac{\text{operate}_k}{4800}$$

STOP

Note   For ease of presentation in the elevator simulation, TIME is updated only as the next customer arrives in the lobby. Therefore TIME is not an actual clock being updated every second. It is possible, when a queue has formed, that an elevator returns to the main floor during Step 20, before the next customer arrives. However, loading of the available elevator does not commence until that customer actually arrives. For this reason, the times spent waiting in a queue are slightly on the high side. In Problem 2 you are asked to modify Steps 20–32 in the algorithm so that loading commences immediately upon the return of the first available elevator.

Table B.1 gives the results of 15 independent simulations, representing three weeks of morning rush hour, according to the preceding algorithm.

TABLE B.1 Results of elevator simulation for 15 consecutive days

Simulation number	Numbers of customers serviced	Average delivery time	Maximum delivery time	Average time in elevator	Maximum time in elevator	Number of customers in longest queue	Average time in a queue	Longest time in a queue	Total number of stops for each elevator				Percent of total time each elevator is in transport			
									1	2	3	4	1	2	3	4
1	328	147	412	89	208	12	40	166	67	76	56	52	84	87	80	75
2	322	146	409	88	211	18	43	176	74	62	52	61	88	80	78	77
3	309	139	385	87	201	12	37	161	62	61	61	62	85	83	80	80
4	331	149	371	89	205	13	42	149	72	69	68	44	85	82	87	73
5	320	146	404	87	208	15	48	178	72	52	58	58	86	78	80	74
6	313	153	405	91	211	13	45	146	69	62	72	47	85	82	82	74
7	328	138	341	88	195	10	35	120	59	66	70	61	86	81	84	82
8	312	147	377	86	198	12	46	163	69	60	61	43	82	82	81	73
9	329	139	352	87	208	11	37	155	58	63	70	57	86	86	82	78
10	314	143	325	88	205	9	35	128	65	68	57	65	83	84	78	83
11	317	137	344	85	202	10	38	129	64	75	64	56	86	85	81	77
12	341	153	396	90	211	18	45	177	83	63	63	53	87	82	78	77
13	318	136	345	80	208	11	36	140	58	64	63	55	91	82	83	73
14	319	140	356	88	208	13	33	135	64	67	58	65	84	83	82	81
15	323	147	386	91	218	15	39	166	76	64	70	54	86	81	87	76
Averages (rounded)	322	144	374	88	206	13	40	153	67	65	63	56	86	83	82	77

Note: All times are measured in seconds and rounded to the nearest second.

B.1 PROBLEMS

1. Consider an intersection of two one-way streets controlled by a traffic light. Assume that between 5 and 15 cars (varying probabilistically) arrive at the intersection every 10 sec in direction 1, and that between 6 and 24 cars arrive every 10 sec going in direction 2. Suppose that 36 cars per 10 sec can cross the intersection in direction 1 and that 20 cars per 10 sec can cross the intersection in direction 2 if the traffic light is green. No turning is allowed. Initially, assume that the traffic light is green for 30 sec and red for 70 sec in direction 1. Write a simulation algorithm to answer the following questions for a 60-min time period:

   a. How many cars pass through the intersection in direction 1 during the hour?

   b. What is the average waiting time of a car stopped when the traffic signal is red in direction 1? The maximum waiting time?

   c. What is the average length of the queue of cars stopped for a red light in direction 1? The maximum length?

   d. What is the average number of cars passing through the intersection in direction 1 during the time when the traffic light is green? What is the maximum number?

e. Answer Problems a–d for direction 2.

How would you use your simulation to determine the switching period for which the total waiting time in both directions is as small as possible? (You will have to modify it to account for the waiting times in direction 2.)

2. Modify Steps 20–32 in the elevator simulation algorithm so that loading of the first available elevator commences immediately upon its return. Thus, if TIME > return$_j$ so that elevator j is available for loading, then loading commences at time return$_j$ rather than TIME. Consider how you will now process customer i in Step 20 who has not yet quite arrived on the scene.

B.1 PROJECTS

1. Find a building in your local area that has from 4 to 12 floors that are serviced by 1 to 4 elevators. Collect data for the interarrival times (and, possibly, floor destinations) of the customers during a busy hour (like the morning rush hour), and build the interarrival and destination submodels based on your data (by constructing the cumulative histograms, as discussed in Section 7.3). Write a computer program incorporating your submodels into the elevator system algorithm to obtain results like those given in Table B.1.

Appendix C

THE REVISED SIMPLEX METHOD

For those of you familiar with matrix algebra, we demonstrate how to accomplish a pivot using matrix techniques. Any desired extreme point can be determined by first inverting a submatrix of the original tableau, followed by premultiplying the original tableau by the inverted submatrix. The method is called the **Revised Simplex Method** and has advantages of both speed and accuracy. In particular, round-off errors can be minimized in subsequent pivots because the Revised Simplex Method can use the original data to perform any desired pivot. We begin by illustrating pivoting by matrix inversion, using the carpenter's problem as presented in Section 9.3 for illustration.

PIVOTING BY MATRIX INVERSION AND MULTIPLICATION

We illustrate how to use matrix inversion and multiplication to move from Tableau 0 to Tableau 1 (see Section 9.3, Example 1, The Carpenter's Problem). For Tableau 1, we want the set of dependent variables to be (in order)

$$\{x_2, y_2, z\}$$

The corresponding columns from the *original tableau*, in the same order, are selected to form the matrix P:

$$P = \begin{bmatrix} 30 & 0 & 0 \\ 4 & 1 & 0 \\ -30 & 0 & 1 \end{bmatrix}$$

First, compute the inverse of P:

$$P^{-1} = \begin{bmatrix} \frac{1}{30} & 0 & 0 \\ -\frac{2}{15} & 1 & 0 \\ 1 & 0 & 1 \end{bmatrix}$$

Then obtain Tableau 1 by premultiplying the matrix corresponding to Tableau 0 by P^{-1}:

$$T^1 = P^{-1}T^0 = \begin{bmatrix} \frac{1}{30} & 0 & 0 \\ -\frac{2}{15} & 1 & 0 \\ 1 & 0 & 1 \end{bmatrix} \begin{bmatrix} 20 & 30 & 1 & 0 & 0 & 690 \\ 5 & 4 & 0 & 1 & 0 & 120 \\ -25 & -30 & 0 & 0 & 1 & 0 \end{bmatrix}$$

$$= \begin{bmatrix} \frac{2}{3} & 1 & \frac{1}{30} & 0 & 0 & 23 \\ \frac{7}{3} & 0 & -\frac{2}{15} & 1 & 0 & 28 \\ -5 & 0 & 1 & 0 & 1 & 690 \end{bmatrix}$$

THE REVISED SIMPLEX METHOD

Note that the columns for the matrix P are selected from the original tableau. Also note that the matrix P^{-1} premultiplies the original data. Any intersection point can be enumerated in this fashion. This method is known as the Revised Simplex Method and is beneficial when solving large linear programs. By returning to the original data to compute P^{-1}, the round-off error (which accumulates in successive pivots using other methods) is reduced. Let's illustrate the idea by computing Tableau 2 by premultiplying the original tableau by an appropriate pivot matrix.

Examining T^1 above, the optimality test determines x_1 as the entering variable. The feasibility test as demonstrated in Section 9.3 determines y_2 as the exiting variable. Thus the new dependent variables are

$$\{x_2, x_1, z\}$$

The columns from the *original tableau* corresponding to the dependent variables, in order, are selected to form the matrix P:

$$P = \begin{bmatrix} 30 & 20 & 0 \\ 4 & 5 & 0 \\ -30 & -25 & 1 \end{bmatrix}$$

We compute the inverse of P:

$$P^{-1} = \begin{bmatrix} \frac{1}{14} & -\frac{2}{7} & 0 \\ -\frac{2}{35} & \frac{3}{7} & 0 \\ \frac{5}{7} & \frac{15}{7} & 1 \end{bmatrix}$$

We now obtain Tableau 2 by premultiplying the matrix corresponding to Tableau 0 by the inverse of the pivot matrix corresponding to the set of dependent variables $\{x_2, x_1, z\}$:

$$T^2 = P^{-1}T^0 = \begin{bmatrix} \frac{1}{14} & -\frac{2}{7} & 0 \\ -\frac{2}{35} & \frac{3}{7} & 0 \\ \frac{5}{7} & \frac{15}{7} & 1 \end{bmatrix} \begin{bmatrix} 20 & 30 & 1 & 0 & 0 & 690 \\ 5 & 4 & 0 & 1 & 0 & 120 \\ -25 & -30 & 0 & 0 & 1 & 0 \end{bmatrix}$$

$$= \begin{bmatrix} 0 & 1 & \frac{1}{14} & -\frac{2}{7} & 0 & 15 \\ 1 & 0 & -\frac{2}{35} & \frac{3}{7} & 0 & 12 \\ 0 & 0 & \frac{5}{7} & \frac{15}{7} & 1 & 750 \end{bmatrix}$$

which is the optimal tableau (see Section 9.3).

The Revised Simplex Method is used (with various enhancements) to solve large problems where speed and accuracy are important.

Index

Photo Credits

This page constitutes an extension of the copyright page.